Practical Statistics for Biologists Workbooks

AN INTRODUCTION TO MAKING GRAPHS AND MAPS FOR BIOLOGISTS USING R

About The Authors: Dr Colin D. MacLeod graduated from the University of Glasgow with an honours degree in Zoology in 1994. He then spent a number of years outside of the official academic environment, working as, amongst other things, a professional juggler and magician to fund a research project conducting the first ever study of habitat preferences in a member of the genus *Mesoplodon*, a group of whales about which almost nothing was known at the time. He obtained a masters degree in marine and fisheries science from the University of Aberdeen in 1998 and completed a Ph.D. on the ecology of North Atlantic beaked whales in 2005, using techniques ranging from habitat modelling to stable isotope analysis. Since then he has spent time working as either a teaching or research fellow at the University of Aberdeen and has taught Geographic Information Systems (GIS) at the University of Aberdeen, the University of Bangor (as a guest lecturer) and elsewhere. He has been at the forefront of the use of habitat and species distribution modelling as a tool for studying and conserving cetaceans and other marine organisms, and he has co-authored over forty scientific papers on subjects as diverse as beaked whales, skuas, bats, lynx, climate change and testes mass allometry, many of which required the use of complex statistical analysis. In 2011, he created *Pictish Beast Publications* to publish a series of books, such as this one, introducing life scientists to key practical skills, and *GIS In Ecology* to provide training and advice on the use of GIS, data visualisation and spatial statistics in marine biology and ecology.

Dr Ross MacLeod graduated from the University of Glasgow with an honours degree in Zoology in 1999 and went on to study risk trade-off behaviours in birds for his doctoral degree at the Edward Grey Institute of Field Ornithology, University of Oxford, graduating in 2004. Since then, he has focused on behavioural ecology and biodiversity conservation research around the world, and has led projects in the UK, Bolivia and Peru. After working as a NERC-funded post-doctoral researcher at the University of St Andrews, he moved to the Institute of Biodiversity & Animal Health (IBAHCM), University of Glasgow as a Royal Society of Edinburgh Research Fellow, investigating how the impacts of environmental change on biodiversity can be predicted from knowledge of animals behavioural decisions and examining how biodiversity conservation can be delivered through sustainable development and rainforest regeneration. In 2018, he was appointed as a Lecturer in Behavioural Ecology at Liverpool John Moores University, with a research focus on forecasting future population and ecosystem impacts of environmental change. Throughout his career he has been involved in developing new ecological survey techniques and skills-based teaching approaches in the fields of biodiversity measurement, environmental monitoring, statistics and, more recently, GIS.

Cover Image: A long-beaked common dolphin (*Delphinus* sp) breaching. Information on relative prey sizes consumed by common dolphin is used to study the dietary niche occupied by this species, such as the frequency of different sizes of prey in the diet (left insert graph), comparing the relative prey sizes consumed by common dolphin with other sympatric cetacean species (central insert graph) and a comparison of the relative occurrence of prey in different prey size categories between common dolphin and harbour porpoise (right insert graph). Common dolphin photograph © Sergey Uryadnikov/Shutterstock.com

PSLS

Practical Statistics for Biologists Workbooks

AN INTRODUCTION
TO MAKING GRAPHS AND MAPS
FOR BIOLOGISTS USING R

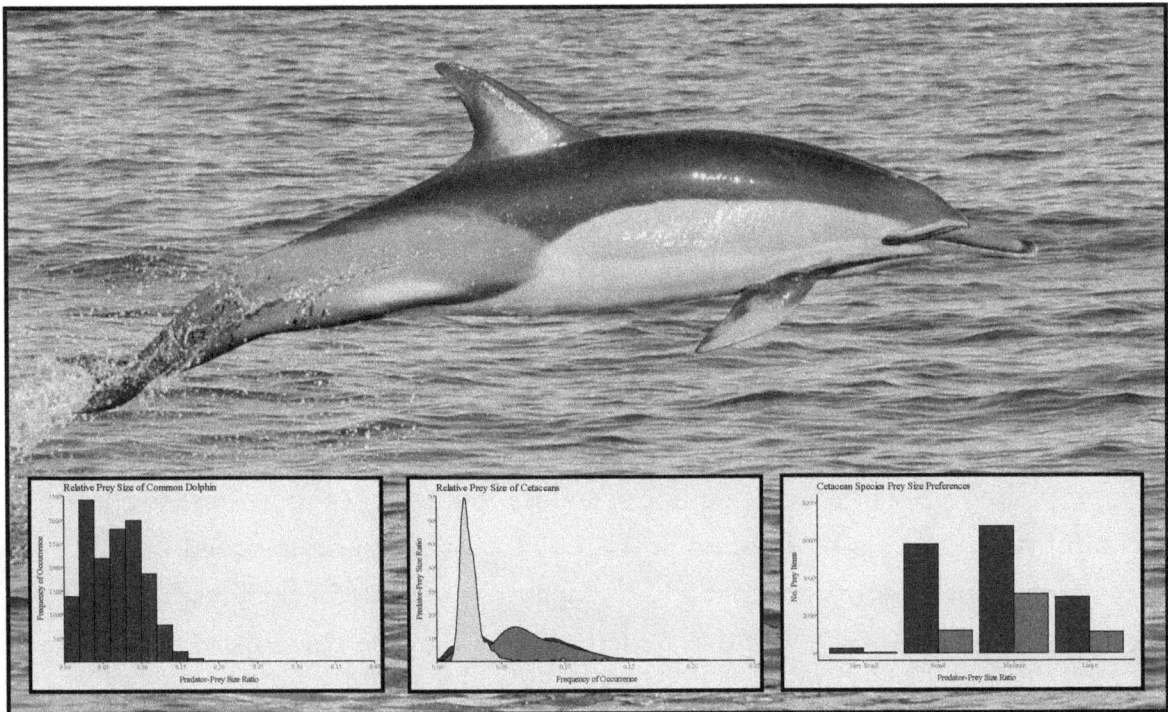

Colin D. MacLeod and Ross MacLeod

Pictish Beast
Publications

ISBN – 978-1-909832-08-4
Published by Pictish Beast Publications, Glasgow, UK.
Printed in the United Kingdom.
First Edition: 2022.

Trademarks

Warning And Disclaimer

'Don't Panic!'

From the cover of *The Hitchhiker's Guide To The Galaxy*

*This book is dedicated to those biologists
who wish to learn how to use R to create informative graphs and maps from
biological data without panicking too much.*

Table of Contents

Preface

This is the second book in the *Practical Statistics for Biologists Workbooks* series. It has been written to fill the gap between learning about the principles behind making informative data visualisations and learning how to actually create effective, high quality graphs and maps from biological data using R. Thus, this workbook focuses on developing the practical skills and knowledge needed to create such data visualisations in R, rather than providing information about the principles behind doing this.

Traditionally, the creation of graphs and maps have been taught to biologists using a knowledge-based approach. In this approach, individual components of data processing and visualisation, including the underlying principles, are taught as more or less separate entities. Each person is then expected to work out for themselves how to integrate these components so this knowledge can be applied to their own work. In contrast, this workbook uses a task-oriented learning (TOL) approach. Rather than focussing solely on knowledge acquisition, this approach focuses on teaching complex tasks, such creating graphs and maps, through their practical implementation. This means that instead of simply being shown how to do the individual components of data visualisation, you are provided with all the steps you need to do to complete a specific data visualisation task from start to finish. For example, when learning how to create a basic graph (see Exercise 1.1), you are not just provided with instructions for creating the graph itself, but also with details of how to import your data set into R, and how to check that it has been imported properly before you create your graph from it. This TOL approach allows you to start applying the workflows for the specific tasks provided in this workbook to your own data almost immediately without first having to learn a large amount of background information, as well as every possible command you can use to make graphs and maps in R. Of course, this doesn't mean it replaces the need to learn about the principles behind creating data visualisations or to gain a more detailed understanding of what can be achieved by using R. Instead, it provides the impetus to persist with such learning. Thus, task-oriented learning provides a quick-start approach on which you can build a more detailed understanding of how to produce high quality data visualisations from biological data.

--- *Chapter One* ---

Introduction

The aim of this workbook is to introduce biologists to the practical elements of creating graphs and maps using R statistical software. This means it is primarily aimed at undergraduate and postgraduate students who either wish to teach themselves how to create high quality and informative data visualisations in R or who are taking their first courses in how to use R. However, it will be just as useful for more experienced biologists who currently use other data visualisation software packages, but wish to learn how to use R to make effective and informative graphs and maps.

This workbook uses the same task-oriented learning (TOL) approach found in other books in the *PSLS* series, such as *An Introduction to Basic Statistics for Biologists using R*. The TOL Approach helps you learn how to carry out the types of tasks biologists need to be able to do on a regular basis in a practical and meaningful way without getting too tangled up in learning about the underlying theoretical basis for them. This workbook, therefore, does not aim to provide you with information about the principles behind data processing and visualisation. Instead, it focuses on building up practical experience and providing advice about how to create the types of graphs and maps that biologists need to be able to make on a regular basis.

This practical experience and advice comes in the form of a series of exercises which you can work through to learn how to complete specific data visualisation tasks. This might be something simple, such as creating a basic frequency distribution histogram, or something more complicated, like creating a graph of time series data or an animated map from raster data sets. No matter what, for each task, you are provided with all the steps you need to undertake to complete it, starting with getting your data into R and finishing with how to present your data visualisation to others. These exercises use a workflow approach based around flow diagrams to help you understand exactly what you need to do at each step in the process, and where you are using the same basic steps to complete different tasks. This allows you to see how more complicated tasks can be carried out by connecting together simpler individual steps.

The exercises in this workbook are divided into five groups. These are: 1. Creating your first graphs from biological data in R using the GGPlot package (Chapter Three); 2. Creating graphs displaying groups of data with GGPlot (Chapter Four); 3. Creating graphs displaying individual data points with GGPlot (Chapter Five); 4. Creating other types of graphs not covered in Chapters Three to Five (Chapter Six); and 5. Creating maps from biological data in R (Chapter Seven). Together, these represent most of the key tasks that biologists need to be able to do to start creating high quality data visualisations using R in a practical and meaningful way.

R was selected as the basis for the instructions provided in this book because it is free to download and because it is widely used by biologists around the world. While it is a command-driven package, which some people initially find off-putting, it is relatively simple to use, and with the right type of instructions, it is relatively easy for anyone to successfully start using it for data processing and visualisation in a short space of time. However, while the exact instructions will vary for other graphing and data visualisation software, the same basic steps outlined in the boxes on the left hand side of the flow diagrams are required to complete the tasks outlined in this workbook in almost all of them. Thus, it should be relatively easy to adapt the instructions provided here for use with other similar software.

The exercises start with learning how to make a basic graph, such as a frequency distribution histogram, in R using a package called GGPlot (Exercise 1.1), before moving on to customising how your graph looks (Exercise 1.2) and creating multi-series and multi-panel graphs (Exercise 1.4). The basic processes required to make graphs in R using this package learned in these initial exercises are then applied to creating specific types of graphs and other data visualisations. These include as bar graphs (Exercises 2.1 to 2.3), line graphs (Exercises 2.3), scatter plots (Exercise 3.1 to 3.3), graphs of time series data (Exercises 3.4), pie charts (Exercise 4.1), bubble graphs (Exercises 4.2) and maps of biological data (Exercises 5.1 to 5.3). This allows you to build up your data visualisation skills in a logical order as you work through this book one chapter at a time. However, the instructions provided in each chapter are sufficiently complete that you can work through the exercises in each one on their own without having to refer back to the contents of previous chapters. This means if your main aim from reading this book is to work out how to do a specific task, such as making a bar graph or making a map of biological data, you can go straight to the relevant chapter for that task and find out how to complete it. If you do this, however, it is worth taking the time at a later date to work through the earlier chapters too, as these will help widen your skills base and give you a better understanding of what you are doing in each individual step when creating more advanced data visualisations.

When you start using R you may find it rather frustrating, particularly if you are not already familiar with using command-driven software packages. This is because, rather than picking actions from a menu as you would do with a graphic user interface, you need to enter blocks of code before running them. To make things more complicated, the code in these blocks needs to be entered in a very precise manner, including using exactly the same uppercase and lowercase letters provided in the instructions for each exercise. This means that any typos you make will cause your code not to work, and R will not necessarily provide you with suggestions or indications as to what has gone wrong. Since the aim of this workbook is to help you get up and running with making biological data visualisations using R as quickly as possible, rather than testing your typing skills, you will find a text document called R_CODE_DATA_VISUALISATION_WORKBOOK.DOC in the compressed folder containing all the data required to complete these exercises (instructions for downloading this folder can be found at the start of Chapter Three). This document provides a copy of all the R code used in the flow diagrams that you will find at the start of each exercise. You can, therefore, choose to simply copy and paste the required code into R from this document without having to worry about making mistakes while typing it. Once you are familiar with how to complete a specific task, you can then work out how to modify this code to allow you to do other, related tasks, and you will be given the opportunity to do this as part of each exercise. To help with this, the blocks of code provided in the above document have been colour-coded. This not only makes it easier for you to work out what each different part does, it also makes it easier to work out which bits need be modified to make them do something different. If you wish to learn more about how to create R code to do specific tasks for yourself, we recommend reading *Getting Started with R: An Introduction for Biologists* by Andrew P. Beckerman, Dylan Z. Childs and Owen L. Petchey.

How The Exercises in This Workbook Are Structured:

The exercises in this workbook all follow a standard structure that has been developed to help you to understand what you need to do to complete a specific task in R, to gain experience in doing it, and to help you work out how to apply it to your own data. First, you are provided with a brief introduction to the task itself and what you will learn by completing it. Next, you will find a flow diagram with all the information you need to work through an example of the task using a specific data set. Once you have worked through this initial example, you will find details of how you can modify the code you used to produce different results, as well as examples of such modifications that you can work through. This will help you gain a deeper understanding of how you can adapt a specific workflow to your own data. This approach of providing detailed step-by-step workflows, along with examples of increasing complexity for you to work through by modifying the commands it contains, means that you can use these exercises to rapidly and efficiently increase your graphing and data visualisation skills. In addition, it provides a resource you can refer back to any time you wish to refresh your knowledge of how to do a specific task.

Why Are Some Instructions And Steps Repeated In Different Exercises?

As you work through the chapters in this workbook, you will quickly notice that there are some instructions and steps that are repeated in many different exercises. If you are not already familiar with the task-oriented learning (TOL) approach used in this book, you may think this repetition is unnecessary. It is not, and it does, in fact, perform a number of important functions which will help you master the use of R for making graphs and maps. First and foremost, it reminds you that there are certain key steps which you need to do each and every time you wish to do a task in R. These include steps like setting your working directory, importing a data set and checking that it has been loaded into R correctly. By repeating them in each individual exercise, it not only helps you to become familiar with these basic, but important, steps, it also serves to reinforce the importance of including them in every workflow that you carry out. Secondly, by including the instructions for the same steps in multiple exercises, it enables you work through a specific task, such as making a bar graph or a map, from start to finish. This means you can concentrate on learning all the steps you need to do to complete that task without becoming distracted by having to refer back to other sections of the book. Finally, by including the same steps in the flow diagrams for different exercises, it helps you see how the same processes are used to create very different types of graphs and maps This makes it much easier to understand how you can create your own custom workflows for types of data visualisations not included in this workbook by using a similar series of steps.

NOTE: As with many things in life, there may be more than one way to do the processes required to complete the exercises outlined in this workbook. The instructions presented here will work for the associated data sets, and this means they should also work in most other circumstances. However, if you find an alternative way to do them which works for you, or if you have someone who can show you how to do them in another way, feel free to do them differently.

Chapter Two

What You Need To Know To Get Started With R

What Is R?

R is an open source, and so freely available, analytical software package that is widely used by biologists for carrying out data analysis and visualisation. In fact, it has become so widely used that knowing how to use R can now be considered a critical skill for almost all biologists. However, unlike most other widely used software packages, R does not have a true graphic user interface (GUI) that allows you to simply click on menus and/or buttons to carry out specific tasks. Instead, it is primarily command-driven. This means that you need to enter lines of code in order to get it to do what you want it to do. Many biologists find this rather daunting, and at first it can be, but you should not allow this to put you off using it. This is because R is an amazingly powerful tool that, if you learn to use it properly, can make your life as a biologist so much easier. In addition, once you start using it, you will find that R is not as difficult, or as complicated, to use as it may at first seem, and the effort you put into learning it will be repaid many times over by the benefits of being able to use it to process, analyse and visualise your data more quickly and more efficiently than you could ever hope to do without it.

Where Can I Get R From?

As an open source software package, R can be freely downloaded from the R Project website at *www.r-project.org*. This workbook assumes that you will be using version 4.1.0 or later, but the instructions provided in it should work for almost any recent and future versions of R. However, if you are currently using an earlier version of R, you may have problems with some of the additional packages required to complete the exercises provided in this book if you do not already have them installed. As a result, it is strongly recommended that you update R to the latest version before you start working through these exercises.

How Can I Use R?

There are two common ways that biologists use R. These are using the native R user interface, which will be referred to in this workbook as RGUI, and by accessing it using a third-party user interface. The RGUI option is the one that automatically appears when you open R, and it looks like this:

This window (which, when you first open it, contains text detailing of the version of R you are using and various other bits of information) is known as the R CONSOLE window. To use the RGUI, you simply type the R code you wish to run on the command line of this window (the line that starts with a > symbol) and press ENTER on your keyboard. To save your work, or to reload a previous work session, you can select the appropriate option from the menus along the top of the user interface (see below for more details). As and when you call them up, other windows, such as DATA VIEWER windows (which allow you to view data tables that you have within R), R GRAPHCS windows (which will display graphs and other graphics generated by the R commands you enter into the R CONSOLE window) and an EDITOR window, will appear in the main RGUI window alongside the R CONSOLE window.

Of the third-party user interfaces available for R, the one that is the most widely used by biologists is called RStudio. The main reason that many biologists prefer to use RStudio rather than the native RGUI is because it makes it easier to follow exactly what you are doing, to edit the blocks of code you have used in the past to adapt them for use with new data sets and to archive them in a meaningful way so you have a record of exactly what you have done (see Appendix II of *An Introduction to Basic Statistics for Biologists using R*). As a result, we recommend that you use RStudio to access R, rather than RGUI, both for completing the exercises in this workbook and for your own analyses. However, regardless of how you access R, you will use the same commands to complete the same tasks. Any points of difference between using these two interfaces for the exercises in this workbook are highlighted within the instructions provided. This means that, as far as this workbook is concerned, it does not matter which interface you choose to use.

Unlike the native RGUI, RStudio needs to be downloaded and installed separately. However, it is also freely available and you can download it from *www.rstudio.com*. Once you have installed RStudio, it will automatically connect to R (as long as you have already installed R on your computer) and it will be ready to use. When you first open RStudio, its user interface should look like this:

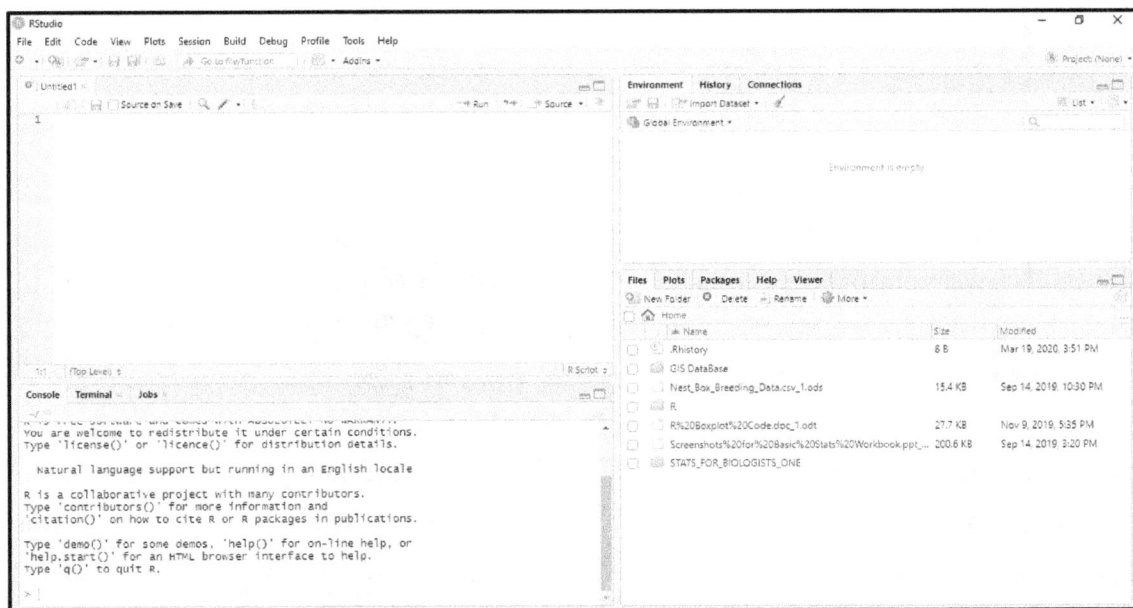

If your version of RStudio does not look like this, click on VIEW on the main menu bar, and select PANES> SHOW ALL PANES. This will make sure that all the required windows are set to display. If the upper left hand window is still not visible after you have done this, click on FILE on the main menu bar and select NEW FILE> R SCRIPT. This should make the missing window appear.

The RStudio user interface is divided into four windows or panes. These are: 1. The SCRIPT EDITOR window (the upper left hand window); 2. The R CONSOLE window (the lower left hand window); 3. The ENVIRONMENT/HISTORY window (the upper right hand window); 4. A window in the lower right hand corner that includes tabs for FILES (which allows you to browse files on your computer – this means the list of files and folders that appears in this tab will be unique to your computer and will not match those in the image on page 8), PLOTS (which displays any graphs or other graphics created by commands you run in the R CONSOLE window), PACKAGES (which allows you to access additional packages – see below for information on what an R package is) and HELP (where you can get help if you get stuck with anything while using R).

As with the native RGUI, you can enter the R code you need to run a specific command directly in the R CONSOLE window in RStudio. However, the main advantage of using RStudio is that you can also use the SCRIPT EDITOR window to enter, edit, prepare, annotate and store any R code you might wish to use before you run them in the R CONSOLE window. This makes it a much more flexible way to enter commands into R. In order to run a block of code from the SCRIPT EDITOR window, you need to transfer it to the R CONSOLE window. To do this, you simply select the appropriate code block in the SCRIPT EDITOR window and click on the RUN button at the top of it.

What Special Terms Are Used When Describing How To Make Graphs And Maps In R?

For the most part, the terms that are used when creating graphs and maps in R are the same as those you would use in any other data visualisation package. However, there are eight terms which are more specific to R that are useful to understand in order to be able to use it more easily. These are:

1. Code: In R, Code refers to the string of text and characters that you need to enter to make it do something, such as importing a data set, creating a graph or running a specific statistical test. A piece of Code is made up of a number of different components, including Commands, Elements, Arguments and Objects (see below). Longer pieces of code that contains multiple Commands will be referred to as a Block of Code or a Code Block. In this workbook, R code will always be highlighted using the `Courier New` font (this is the default font for the R CONSOLE window in both RGUI and RStudio).

2. **Commands:** In R, a Command is the name given to a piece of Code that allows you to do a specific function. For example, the `ggplot` command allows you to create a blank graph using the GGPlot package to which you can then add information using additional commands from it (see below), while the `geom_histogram` command allows you to create a histogram from a specific data set (see Exercise 1.1). Within this workbook, two sub-types of Commands will be referred to, where appropriate. These are Graphing Commands, which refer to Commands which determine what type of graph will be created from a specific data set, and Style Commands, which refer to Commands that are used to modify or customise the appearance of a graph or map. The Block of Code for a complex task may consist of the number of different Commands, these Commands can either start on a new line, or be attached to a previous Command using a + symbol (as will be the case when creating graphs in the exercises in Chapters Three to Six). When the + symbol is used to join two or more commands together, they can be considered linked together. For example, to make a graph using GGPlot, you need to use the `ggplot` command to create a blank GGPlot graph and then uses a + symbol to link a graphing command to it to define the type of graph that will be added to it (such as the `geom_histogram` command - see Exercise 1.1). An alternative way to apply multiple commands to a single graph, which allows you to shorten the individual lines of code that you need to enter, is provided on pages 43 to 45 of Exercise 1.2, and is used in Exercises 5.1 to 5.3 in Chapter Seven.

3. **Elements:** In R, an Element can be thought of as a sub-Command within a specific piece of code. Elements are used within a Command to allow you to specify exactly what elements of your data, or of an R object, the Command will be applied to. For example, when making a graph using GGPlot, the `aes` element can be used either in the `ggplot` command or the linked graphing command to tell R what data the graph created by it will be based on (see Exercise 1.1). In this workbook, one sub-type of Element will be referred to, where appropriate. This is a Style Element, which refers to an Element that is used to modify or customise the appearance of a graph or map.

4. **Arguments:** In R, an Argument is a term you include in a Command, or in an Element within a Command, in order to get it to work properly. For example, within the `ggplot` command or a graphing command, such as `geom_histogram`, you need to include a `data` argument to define exactly which data set the resulting graph will be made from. If you don't include all the required Arguments in the Code for a Command (and any included Elements), it will not work properly. This means that Arguments can be thought of as the settings you need to specify to make a Command, or an Element within a Command, do what you would like it to do. In this workbook, the single term Argument will refer to both required and optional arguments (referred to as Additional Arguments in the first workbook in this series) that can be included in a Command. One sub-type of Argument will be referred to, where appropriate. This is a Style Argument, which refers to an argument that is used to modify or customise the appearance of a graph.

5. **Objects:** An Object is something created by a Command in R that stores a specific piece of information. In this workbook, the Objects created by commands will usually be tables of data, graphs or maps, but they can also be the outputs of statistical tests. As such, you can think of R Objects as being equivalent to the tables used by other software packages, and the graphs and maps created from them. The only difference is that Objects are not saved separately from your R project and can only be accessed through R. If you set a Command to create a new Object, it will be saved in your analysis project. If you don't, while the results will be displayed on your screen, they will not be saved. Within R, the name of the Object that will be created by a specific Command (or set of commands) is linked to it by either an equals sign (=) or by an arrow created by using a less than sign (<) followed immediately by a hyphen (-) so it looks like this <-. This latter option will be used in this workbook to link each Object to the Command (or Commands) used to create it.

6. **Scripts:** A Script is a collection of R Commands that allows you to do a more complex task from start to finish. They are often stored in a specific file type called an R Script file (which has the extension .R) that can be imported into R, although you can also save them as other file formats, such as text files (.txt). In other analytical software, such as SPSS, Script files are known by other names, such as Syntax files.

7. Libraries: A Library is a block of pre-existing Code that tells R how to process a data set when certain Commands are entered into it. This Code runs in the background each time you use one of these Commands and all you will see is the results it returns. This means that in order to be able to use a specific Command from a specific Library, the Library itself has to first be loaded into your current analysis project. For most basic Commands, the Libraries required for them to work will be loaded into R by default. However, for other Commands, you need to manually load their associated Library each time you wish to use them. This can be done using the `library` Command. If you cannot get a specific Command to work, you should check that its associated Library has been loaded into your current R project (see pages 17 and 18 for details of how to do this).

8. Packages: A Package is a collection of Libraries that allows you to do a specific set of tasks in R. Before you can load a Library from a specific Package into an analysis project and use the associated Commands, you need to install the Package itself. This can be done using the `install.packages` Command. If you cannot load a specific Library that you wish to use into R you should check that the Package which contains it has been installed in your current version of R (see pages 17 and 18 for details of how to do this).

How Do I Get Started With Using R?

The very first thing you should do before you start any data analysis project in R is to create a new and dedicated folder for it on your computer. This folder will then be used to store all the data you are going to use as part of that analysis project, as well as all the outputs that are generated during it and all the code you used to run commands within it. This folder is what is known as the WORKING DIRECTORY for your project. It is good practice to create a new WORKING DIRECTORY folder for each project you wish to carry out in R (and you will do this as part of the practical exercises in this workbook – see Chapter Three). This is because it prevents data, commands and results of different projects getting mixed up. After you have created a WORKING DIRECTORY folder, you need to copy or transfer all the data you are going to use for that specific project into it.

Once you have created a WORKING DIRECTORY folder for a specific analysis project, you need to copy the address for this folder to the clipboard of your operating system. To do this on a computer running a Windows operating system, open Windows Explorer and navigate to your newly created WORKING DIRECTORY folder. After you have opened it, click on the folder icon at the left hand end of the ADDRESS BAR at the top of the Windows Explorer window to reveal the full address of this folder. You can then copy this address by selecting it and pressing the CTRL and C keys on your keyboard at the same time. **NOTE:** You will need to modify folder addresses before you can use them in R. Specifically, Windows uses backslashes (which look like this \) as the separators between the parts of the addresses. However, R uses slashes (which look like this /). This means that you will need to replace all the backslashes (\) in the address of your WORKING DIRECTORY folder with slashes (/) before R will recognise it. For example, a Windows address that reads C:\USERS\DOCUMENTS\ANALYSIS would need to be modified for use in R so that it reads C:/USERS/DOCUMENTS/ANALYSIS.

To copy the address of your newly created WORKING DIRECTORY folder on a computer running a Mac operating system, open Finder and navigate to the location of your WORKING DIRECTORY folder. Open the folder and then press the CMD and I keys on your keyboard at the same time. This will open the GET INFO window where you will find the address for this folder alongside WHERE. You can then select it and copy it by pressing the CMD and C keys on your keyboard at the same time. **NOTE:** You may find that you have to modify Mac OS folder addresses before you can use them in R. Specifically, you may need to drop the part that says MAC HD at the start and replace the folder separators between the parts of the addresses with slashes (which look like this /). For example, a Mac OS address that reads MAC HD▶USERS▶ADMIN▶DOCUMENTS▶ANALYSIS would need to be modified for use in R so that it reads /USERS/ADMIN/DOCUMENTS/ANALYSIS.

Once you have created your WORKING DIRECTORY folder and copied its address, you can open your preferred user interface (either the native RGUI or RStudio). As soon as it opens, you should save a copy of the file where you are going to enter your R commands in this folder. This file is known as your WORKSPACE file. It will have the extension .RDATA and it will contain all the R objects you will create during your project, such as the tables of data you import and graphs you make from them. In RGUI, you can do this by clicking on the FILE menu at the top of the main RGUI window and selecting SAVE AS. This will allow you to save your project as an R WORKSPACE file under a specific name in the WORKING DIRECTORY folder you have created for your specific project. In

RStudio, you can save your WORKSPACE file in your WORKING DIRECTORY folder by clicking on the SESSION menu at the top of the main RStudio window and selecting SAVE WORKSPACE AS.

If you are using RStudio, you can also save the contents of your SCRIPT EDITOR window as an R SCRIPT file (which has the extension .R). This file will contain all the code that you have used in your project meaning that if you wish to re-run it, or run a modified version of it, you do not need to type it in again. To save the contents of your SCRIPT EDITOR window as an R SCRIPT file, click on the FILE menu at the top of the main RStudio window and select SAVE AS. This will allow you to save it using a specific name in your WORKING DIRECTORY folder.

After you have saved the contents of your WORKSPACE and the contents of your SCRIPT EDITOR window (if you are using RStudio), the next thing, you need to do is to clear any data that are currently held in the temporary memory of R itself. This prevents you from accidentally using the wrong data set when you start a new analysis project. You can do this by using the following command:

```
rm(list=ls())
```

If you are using the native RGUI, you can simply type this code after the command prompt at the bottom of the R CONSOLE window (it looks like this: >). If you are using RStudio, you can type this command into the SCRIPT EDITOR window (the upper left hand window) and run it in the R CONSOLE window (the lower left hand window) by selecting it and then clicking on the RUN button at the top of this window.

Next, you need to tell R where the WORKING DIRECTORY you are going to use for your analysis is located on your computer. This is done using the `setwd` command. This command should be followed by the address of the WORKING DIRECTORY folder you have created for your project (and into which you have already saved or transferred all the data you are going to use for your analysis - see above). For example, if your WORKING DIRECTORY folder has the address C:\STATS_FOR_BIOLOGISTS_TWO, this command would read:

```
setwd("C:/STATS_FOR_BIOLOGISTS_TWO")
```

If you are using RGUI, you can set your WORKING DIRECTORY by typing `setwd("` after the command prompt in the R CONSOLE window. Next, paste or type the address of your WORKING DIRECTORY folder after this command (remembering to modify the folder separators in this address as outlined on pages 12 and 13 so R can recognise it) and then type a second quotation mark followed by closing bracket, like this `")`. Finally, press the ENTER key on your keyboard to run this command. Alternatively, you can click on the FILE menu at the top of the main RGUI window and select CHANGE DIR.

If you are using RStudio, you can set your WORKING DIRECTORY by typing setwd(" into the SCRIPT EDITOR window. Next, paste or type the address of your WORKING DIRECTORY folder after this command (remembering to modify the folder separators in this address as outlined on pages 12 and 13 so R can recognise it) and then type a second quotation mark followed by a closing bracket, like this `")`. Finally, select the whole command then click on the RUN button at the top of the SCRIPT EDITOR window to run it in the R CONSOLE window. Alternatively, you can click on the SESSION menu at the top of the main RStudio window and select SET WORKING DIRECTORY> CHOOSE DIRECTORY.

To check that you have set R to use the correct WORKING DIRECTORY, you can use the `getwd()` command. When you run this command, it should return the same address as the folder you wish to use as your WORKING DIRECTORY.

Once you have set your WORKING DIRECTORY, you are ready to start entering the commands you need to process and/or visualise your data. This may include creating an initial graph from your data (Exercises 1.1), creating a high quality graph for inclusion in a manuscript or presentation (Exercise 1.2), creating multi-series and multi-panel graphs (Exercise 1.4), creating graphs displaying groups of data (Exercise 2.1 to 2.5), creating graphs displaying individual data points (Exercises 3.1 to 3.6), and creating other types of graphs and maps (Exercises 4.1 to 5.3). In all cases, when entering your code, remember that if you are using the native RGUI user interface you can simply paste or type the required R code after the command prompt at the bottom of the R CONSOLE window and then press the ENTER key on your keyboard. If you are using RStudio, you can paste or type the required R code into the SCRIPT EDITOR window (the upper left hand window) before selecting it and then clicking on the RUN button at the top of it to run this code in the R CONSOLE window.

If you wish to load a project you have been working on into RGUI, you can click on the FILE menu and select LOAD to load the contents of a specific WORKSPACE file, including all the objects previously created in it, into R (as long as you remembered to save it the last time you stopped working on it). If you wish to load the contents of a WORKSPACE file into RStudio, click on the SESSION menu at the top of the main window and select LOAD WORKSPACE. If you are using RStudio and you also wish to load an R SCRIPT file containing R code you previously saved from the SCRIPT EDITOR window, click on the FILE menu and select OPEN FILE. **NOTE:** If you re-load an R SCRIPT file into RStudio without also loading the associated WORKSPACE file, you will need to re-run all the R code you have previously run (such as the commands to import a specific data set) to re-create the required R objects before you can apply any new blocks of code based on them.

What Do I Do If I Get Stuck When Using R?

There will be times when using R that you will find yourself getting stuck. When this happens, the most important thing to remember is 'Don't Panic!'. It is completely normal to get stuck when using R and it happens to everyone at some point or other. In fact, when trying to do something new in R, you should always assume that it will take you an average of three attempts to do it successfully (the first time to do it wrong, the second time to work out why it went wrong the first time, and third time to do it right). As a result, any time you manage to do something in less than three attempts should be considered a bonus. Of course, one of the main advantages of R is that once you get a specific set of commands working, you can save them and use them repeatedly without having to trouble-shoot them all over again. In addition, you can share the R code you have created with others to allow them to repeat the same analysis without having to build the same code for themselves from scratch. However, this does require that you keep a good record of exactly what code you have used and what it does (see Appendix II of *An Introduction to Basic Statistics for Biologists using R*). When you get stuck trying to use a particular command or piece of R code, the situation can generally be resolved using one of the following five solutions:

1. **Check For Typos (Including The Mis-Use Of Capital Letters):** Typos in file names, R object names, variable names, commands and the arguments you can include in them are the number one cause of problems you will encounter with R. As a result, this is the first thing you should check for when anything goes wrong or doesn't work properly.

2. **Check You Are Using The Correct Files, Data Sets, R Objects And Variables:** If you are copying individual commands or blocks of code from someone else or from your own R scripts, it is very easy to forget to change one or more of the settings, meaning that a particular command within the code block ends up trying to use the wrong information. In addition when using multiple commands within a single block of code, you need to check that the appropriate data sets, objects and variables are referred to in the correct command within it to give the desired results. Therefore, this is the second thing you should check when a block of code doesn't give you the results you were expecting.

3. **Check The Help Files And/Or Documents For Commands You Are Using To Ensure You Are Using Them Correctly:** Sometimes, you may find that a particular command simply will not be able to do what you wish it to do or that you need to change the settings you are using for it. As a result, if you find that a particular command does not work, and you don't find any typos in it and you are certain that it is being applied to the correct data, you should carefully read through the help files and/or the documentation for the command to ensure you are using it correctly. You can access the help files for a particular command in R by entering the word 'help' followed by the name of the command in brackets. For example, to get help with a command called `read.table`, you would enter the code `help(read.table)`. Alternatively, you can enter a question mark (?) followed by the command you are looking for help with, like this: `?read.table`. Both of these options will bring up the help file for that command which will provide detailed advice about how you can use it. You can find similar information in the documentation for each command, which can be found at *www.rdocumentation.org*. This information can be located using an internet search engine by entering the name of the command followed by the words *R Documentation*.

4. **Check That All The Required Packages And Libraries Have Been Installed In Your Version Of R:** If you find that you cannot access a specific command (or the help file for it), this may be because you don't have all the required packages and libraries installed.

You can check which packages are installed in your copy of R using the `library()` command (**<u>NOTE</u>:** Even though you are checking for installed packages, you do this with a command with the term `library` in it). This will open a new window called R PACKAGES AVAILABLE containing a list of all the installed packages. If you are missing a package that you require to use a specific command, you will need to download and install it using the `install.packages` command. If you wish to check which libraries have been loaded into a specific project, you can use the `(.packages())` command (**<u>NOTE</u>:** Even though you are checking for installed libraries, you do this with a command with the term `packages` in it). This will return a list of all the libraries that have been loaded into your current R project. If the library containing the command you wish to use has not been loaded into R, you can use the `library` command to load it into your current R project.

5. **Look For A Solution And/Or Advice Online:** There will be occasions when you simply cannot work out the root cause of a problem or you just don't understand how to resolve an error message you are getting when you try to do something in R. In these situations, the best solution is usually to either use an internet search engine or post a question on one of the many online forums available for asking questions about using R and about how to create various kinds of data visualisations with it, such as Stack Overflow (*www.stackoverflow.com*).

--- Chapter Three ---

How To Create Your First Graphs In R Using GGPlot

As a biologist, one of the first types of graph which you are likely to want to make from your data is a frequency distribution histogram. This is a graph which shows how records in a data set are distributed across a range of values for a continuous variable. As a result, in this chapter, you will use this type of graph to learn the basics of how to make graphs and other types of data visualisations in R. Simple frequency distribution histograms can be created in R with the `hist` command (see Exercise 2.1 in *An Introduction to Basic Statistics for Biologists using R*). However, such histograms are very basic and they are limited in terms of the ways that data can be divided into bins (or bars) to be plotted on them, the ways that they can be customised, and the different ways that the distributional information can be displayed. As an alternative, many biologists use a more advance graphing package called GGPlot to create not only frequency distribution histograms, but also most other types of graphs and data visualisations they wish to make from their data in R.

The GGPlot package contains an incredibly powerful set of tools for making many different types of graphs and for modifying almost every part of them so that they look exactly the way you wish them to look. This ability to customise almost every possible aspect of the contents and appearance of a graph comes at a cost. This cost is that while the code used to create basic graphs may be relatively simple, and usually consisting of just two commands (see Exercise 1.1 of this workbook), the additions you need to make to this code to fully customise the resulting graph can substantially increase the complexity of the code required to make it (see Exercise 1.2). As a result, using GGPlot can feel quite daunting for those setting out to use it for the first time. However, while using GGPlot can require a very large block of code to produce a graph with a specific appearance (such as that detailed in the instructions to authors from a scientific journal), the underlying structure of the code required to do this is actually quite simple, and with a bit of time and experience, it becomes relatively easy to construct the code required to fully customise any type of graph. The key to understanding how to structure the code required to do this is to understand that each individual element of the graph that you wish to alter can be assigned either its own command, or its own argument(s) within a command. This means that the long, complex block of code needed to create a final graph with a specific appearance can be built up step by step by adding the code required to modify each individual element that you wish to modify one bit at a time, until you have all the code required to produce a graph with the exact appearance you are looking to achieve.

In order to learn how you can do this, in this chapter, you will work through a series of exercises based around producing frequency distribution graphs using GGPlot. In these exercises, you will learn how to customise many different elements of GGPlot graphs, including how to specify exactly how your data are displayed (as will be done in Exercise 1.1), how the various elements of the graph, such as the axes, the labels and the background, are drawn to produce a publication quality graph (see Exercise 1.2), how the frequency distribution data are plotted (see Exercise 1.3), and how multiple data series can be displayed either on the same graph or on separate graphs that can then be combined into a multi-panel figure (see Exercise 1.4). While you will learn how to do all of these tasks with GGPlot using frequency distribution graphs as examples, the same skills can be applied to almost all the other types of graph and data visualisation that can be made with this package. Thus, by the end of this chapter, you will not only know how to make frequency distribution graphs in a variety of different ways, but you will also have learned how you can customise the various elements of a graph using the tools in the GGPlot package to create high quality and informative data visualisations.

Before you start the exercises in this chapter, you first need to create a WORKING DIRECTORY folder on your computer and load the necessary data into it. To do this on a computer with a Windows operating system, open Windows Explorer and navigate to the location where you would like to create the folder (such as your C:\ drive or your DOCUMENTS folder). Next, right click anywhere in this location and select NEW> FOLDER. Now call this folder STATS_FOR_BIOLOGISTS_TWO by typing this into the folder name section to replace what it is currently called (which will most likely be NEW FOLDER). To create a WORKING DIRECTORY folder on a computer running a Mac operating system, open Finder and navigate to the location where you would like to create the folder (such as your DOCUMENTS folder or your DESKTOP). Next, click on FILE>

NEW FOLDER, and then type the name STATS_FOR_BIOLOGISTS_TWO before pressing the ENTER key on your keyboard.

Once you have created your WORKING DIRECTORY folder, you are ready to download the data sets you will use for the exercises in this workbook from *www.gisinecology.com/stats-for-biologists-2*. After you have downloaded the compressed folder containing the required data by following the instructions provided on that page, you need to extract all the data files from it and copy them into the folder called STATS_FOR_BIOLOGISTS_TWO that you have just created.

Next, you need to check that the required data have been extracted to the correct folder. If you are using a computer with a Windows operating system, you can use Windows Explorer to open your newly created WORKING DIRECTORY folder and examine its contents. If all the files from the compressed folder are present in it (there should be a total of 90 of them), you can click on the folder icon at the left hand end of the ADDRESS BAR at the top of the WINDOWS EXPLORER window to reveal its full address. Note this address down somewhere as you will need it to set this folder as your WORKING DIRECTORY during the exercises provided in this workbook (see pages 12 and 13 for details of how to modify folder addresses so they will be recognised by R).

If you are using a computer with a Mac operating system, you can use Finder to open your newly created WORKING DIRECTORY folder and examine its contents. If all the required data files are present in it (there should be a total of 90 of them), press the CMD and I keys on your keyboard at the same time. This will open the GET INFO window where you will find its address (which is also called the pathway). Note this address down somewhere as you will need it to set this folder as your WORKING DIRECTORY during the exercises provided in this workbook (see pages 12 and 13 for details of how to modify folder addresses so they will be recognised by R).

After you have loaded the required data into your WORKING DIRECTORY folder, you can open RGUI or RStudio, depending on which option you wish to use (see Chapter 2 for more details). Once you have opened your preferred R user interface, you need to create a file called CHAPTER_THREE_EXERCISES where you will save the results of your analyses from your R CONSOLE window as you work through this chapter. To do this using RGUI, click on the FILE menu and select SAVE WORKSPACE. To do this in RStudio, click on SESSION and select SAVE WORKSPACE AS. In both cases, save it as a WORKSPACE file with the name CHAPTER_THREE_EXERCISES.RDATA in your WORKING DIRECTORY folder (this will be the one called STATS_FOR_ BIOLOGISTS_TWO that you have just created). If you are using RStudio, you will also want to save the contents of your SCRIPT EDITOR window (where you will enter and edit the R code you will use to carry out specific commands). If this window is not already visible, click on the FILE on the main menu bar of RStudio and select NEW FILE> R SCRIPT. To save the contents of your SCRIPT EDITOR window, click on the FILE menu and select SAVE AS. Save your file as an R SCRIPT file with the name CHAPTER_ THREE_EXERCISES.R in your WORKING DIRECTORY folder. As you work through the exercises in this chapter, remember to regularly save the contents of your R CONSOLE window (which will contain the R objects you have created up to that point) to your WORKSPACE file and, if you are using RStudio, the contents of your SCRIPT EDITOR window to your R SCRIPT file.

Finally, you need to remove any data that are currently held in R's temporary memory. To do this, enter the following command into R:

```
rm(list=ls())
```

If you are using RGUI, you can simply type this code after the command prompt at the bottom of the R CONSOLE window (it looks like this: >) and then press the ENTER key on your keyboard to run it. If you are using RStudio, you can type this command into the SCRIPT EDITOR window (the upper left hand window). To run this command, select it and then click on the RUN button at the top of this window. This will run it in the R CONSOLE window (the lower left hand one in the main RStudio user interface). You are now ready to start the exercises in this chapter.

EXERCISE 1.1: HOW TO CREATE A BASIC GRAPH IN R USING GGPLOT:

The first step in making a graph with the exact appearance that you wish it to have using GGPlot is to create a basic version of it using a short block of code. This code will consists of two commands, one to create a blank GGPlot graph (called `ggplot`) and a second to set the type of graph that will be added to it, such as a histogram, a bar graph, a line graph or a scatter plot, which will be referred to here as the graphing command. These two commands will contain a number of elements and arguments that will not only determine what variable(s) will be used to make the graph, but also exactly how they will be displayed, and these should be modified before you move on to customising any other elements of your graph, such as the colours used for it, the scales used for the axes and the fonts used for the labels (which you will learn how to do in Exercise 1.2). In this exercise, you will learn how make a basic graph by creating a frequency distribution histogram from a specific data set using a graphing command called `geom_histogram`. Within this command, you will set how the data are divided into bins (or bars) in a number of different ways by modifying the arguments and settings that are used in it. By doing this, you can obtain frequency distribution histograms that look quite different, and as a result, the decisions you make when deciding how to do this will be an important for determining exactly how your final graph will look. To create your basic frequency distribution histogram and set how the data will be divided into bins in order to be displayed on it, work through the flow diagram that starts at the top of the next page.

The data that you will use for this exercise, and the others in this chapter, come from a study of the dietary preferences of a range of whale and dolphin species (collectively known as cetaceans). Rather than looking at the species of prey consumed, this study looked at the size distribution of prey and how this varied between different cetacean species. In order to do this, it calculated a predator-prey size ratio (or PPSR for short) for all prey items recovered from the stomachs of each species. This is calculated by dividing the size of the prey found in an individual's stomach contents by its body length. A PPSR of 0.1 means that a specific prey item was 10% of the body length of the individual cetacean that it was recovered from, while a PPSR of 0.01 would mean that it was only 1% of the predator's body length. By comparing the frequency distribution histograms of the PPSR of different cetacean species, their prey size preferences can be identified and compared. This, in turn, allows cetacean species to be divided into different ecological guilds based on the relative sizes of prey that they consume which are linked to other aspects of cetacean ecology, such as how they capture their prey (see MacLeod *et al.* 2006. *MEPS.* 236: 295-307 for more details).

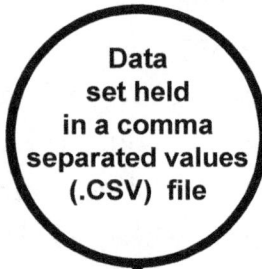

Data set held in a comma separated values (.CSV) file

For this example, the data set you will use is stored in a file called `cetacean_prey_sizes.csv` that is located in the WORKING DIRECTORY folder you created during the introduction to this chapter.

1. Set the WORKING DIRECTORY for your analysis project

Before you start any analysis in R, you first need to set the WORKING DIRECTORY. To do this, enter the text `setwd("` and then type the address of your WORKING DIRECTORY, using slashes (/) as the folder separators, before entering a second quotation mark followed by a closing bracket, like this `")`. For example, if your WORKING DIRECTORY has the address C:\STATS_FOR_BIOLOGISTS_TWO, your `setwd` command should look like this:

```
setwd("C:/STATS_FOR_BIOLOGISTS_TWO")
```

If you are using RGUI, enter your `setwd` command in the R CONSOLE window (remembering to use the address of your own WORKING DIRECTORY folder in it) and then press the ENTER key on your keyboard. If you are using RStudio, enter your `setwd` command into the SCRIPT EDITOR window. To run it, select it and then click on the RUN button at the top of this window. You will enter all the remaining commands for this exercise in a similar manner, depending on the user interface you are using.

To check that your WORKING DIRECTORY has been set properly, enter the command `getwd()` and carefully check that the address it returns is the same as the one for the STATS_FOR_BIOLOGISTS_TWO folder you created at the start of this chapter.

Before you move on to step 2, make sure that all the data you wish to use in your analysis project are located in this WORKING DIRECTORY folder. In this case, this is a file called `cetacean_prey_sizes.csv`. **NOTE**: If the data you are going to import into R in step 2 are not located in the WORKING DIRECTORY you set in this step, the import code provided in the next step will not work.

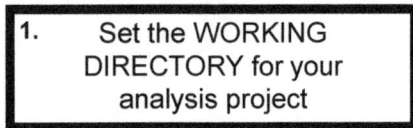

The `read.table` command provides the easiest way to load data held in a .CSV file (and stored in the WORKING DIRECTORY you set in step 1) into R so you can analyse it. To do this for the data set being used in this example, enter the following command into R:

```
cetacean_prey_sizes <-
read.table(file="cetacean_prey_sizes.csv",
    sep=",",as.is=FALSE, header=TRUE)
```

This code has to be entered exactly as it is written here or it will not work. If you wish to use the copy-and-paste approach for entering this command, copy the text directly below CODE BLOCK 1 in the document R_CODE_DATA_ VISUALISATION_WORKBOOK.DOC and paste it into R.

This command will create a new object in R called `cetacean_prey_sizes` which will contain the data from the specified .CSV file. To import a different .CSV file into R, all you need to do is change the file name in the `file` argument to the name of the one you wish to import. You can also use whatever name you wish for the R object which will be created by this command. To do this, simply replace `cetacean_prey_sizes` at the start of the first line of the above code with the name you wish to use for it. **NOTE:** If your .CSV data set uses a semicolon as the column separator, you would need to replace the `sep=","` argument with `sep=";"`.

Whenever you import any data into R you need to check that they have loaded correctly. First, you need to check that all the required columns are present in the R object you just created. To do this, enter the following command into R:

```
names(cetacean_prey_sizes)
```

This is CODE BLOCK 2 in the document R_CODE_DATA_ VISUALISATION_WORKBOOK.DOC. This command will return the names used for each column in the R object you just created. For this example, the names should be: `rissos_dolphin, pilot_whale, common_dolphin, white_beaked_dolphin, striped_dolphin, bottlenose_dolphin, atlantic_white_sided_ dolphin, harbour_porpoise, northern_ bottlenose_whale, pygmy_sperm_whale, sowerbys_beaked_whale, sperm_whale` **and** `cuviers_beaked_whale`.

Next, you should view the contents of the whole table using the `View` command. This is done by entering following code into R:

```
View(cetacean_prey_sizes)
```

This is CODE BLOCK 3 in the document R_CODE_DATA_ VISUALISATION_WORKBOOK.DOC. This command will open a DATA VIEWER window where you can examine your data set and check that the correct data have been loaded into R.

2. Load your data into R using the `read.table` command

3. Check the data have loaded into R correctly by checking the names of the columns and by viewing it

25

4. If required, download and install the `ggplot2` package into your version of R

Once your data have been successfully imported into R, you are ready to create your first graph. However, you will only be able do this if you have the `ggplot2` package installed in your copy of R. To check whether you already have this package installed, enter the following command into R:

```
library()
```

This will open the R PACKAGES AVAILABLE window which will have an alphabetical list of all the packages already installed in your version of R. Scroll down and see if the `ggplot2` package is on this list. If it is, you can go straight to step 5 (where you will load its command library into your analysis project). If is isn't, you can download and install it by entering the following command into R:

```
install.packages("ggplot2")
```

This is CODE BLOCK 4 in the document R_CODE_DATA_ VISUALISATION_WORKBOOK.DOC. If, when you run it, you are provided with any additional instructions, follow them as outlined on your screen.

5. Install the `ggplot2` command library into your analysis project

After you have ensured that you have the `ggplot2` package installed in your version of R, you are ready to install the command library it contains into your analysis project. To do this, enter the following command into R:

```
library(ggplot2)
```

This is CODE BLOCK 5 in the document R_CODE_DATA_ VISUALISATION_WORKBOOK.DOC. This `library` command will load the `ggplot2` command library into your analysis project. **NOTE:** In order to be able to use the commands contained in this library, you will need to run this installation command in each new R project you create. If, when you try to run the R code provided in step 6, you get an error message saying that R cannot find the `ggplot` function, come back to this step and re-run this command.

After you have successfully loaded the `ggplot2` command library into your analysis project, you are ready to use it to create a basic graph from the data set you imported into R in step 2. In this example, this will be a frequency distribution histogram. To do this, enter the following block of code into R:

```
ggplot(data=cetacean_prey_sizes,
aes(x=common_dolphin)) + geom_histogram()
```

This is CODE BLOCK 6 in the document R_CODE_DATA_ VISUALISATION_WORKBOOK.DOC, and it contains two commands separated by a + symbol. These are the `ggplot` command and the `geom_histogram` graphing command. In the `ggplot` command, the `data` argument is used to set the R object containing the data that will be plotted on the graph. In this case, it will the one called `cetacean_prey_sizes` created in step 2 of this exercise. The column of data which will be plotted on the X axis of the resulting graph is set using the `x` argument of the `aes` element of this command. In this case, it is the column called `common_dolphin` in the `cetacean_prey_sizes` data set.

The second command in this code block sets the type of graph that will be created from the data specified in the `ggplot` command. In this case, this graphing command will be `geom_histogram`. This means the block of code will create a histogram from the data set and X variable specified in the `ggplot` command. Initially, this `geom_histogram` command will have no elements or arguments in it.

When you run this block of code, you may get a warning message that states: `'stat_bin()' using 'bins = 30'. Pick better value with 'binwidth'`. This means that the default number of bins (or bars) for the `geom_histogram` command (which is 30) is not appropriate for your data. For this example, you need to add a `bins` argument to the `geom_histogram` command and then re-run it. To do this, edit the above code so it looks like this (the newly added `bins` argument is highlighted in **bold**):

```
ggplot(data=cetacean_prey_sizes,
aes(x=common_dolphin)) +
geom_histogram(bins=10)
```

This is CODE BLOCK 7 in the document R_CODE_DATA_ VISUALISATION_WORKBOOK.DOC, and it adds the argument `bins=10` to the `geom_histogram` command. This means it will create a histogram with approximately 10 bins (or bars) on it. **NOTE:** This argument only provides a target number of bins for your histogram and the actual number will be the closest one that produces an even distribution of bars along the X axis of your graph. Once you have finished editing this code block, you can run it again to create the final version of your first histogram.

6. Create a basic graph based on the data in your data set using the commands from the `ggplot2` package

Basic graph created from a biological data set

At the end of the first part of this exercise, the frequency distribution histogram of the predator-prey size ratios (PPSRs) for common dolphin produced by working through the above flow diagram should look like this (**NOTE:** Even though you set the `bins` argument to `10`, it only has nine bins or bars on it – three of which are very short and are barely visible – as this argument provides a target number of bins rather than setting the absolute number):

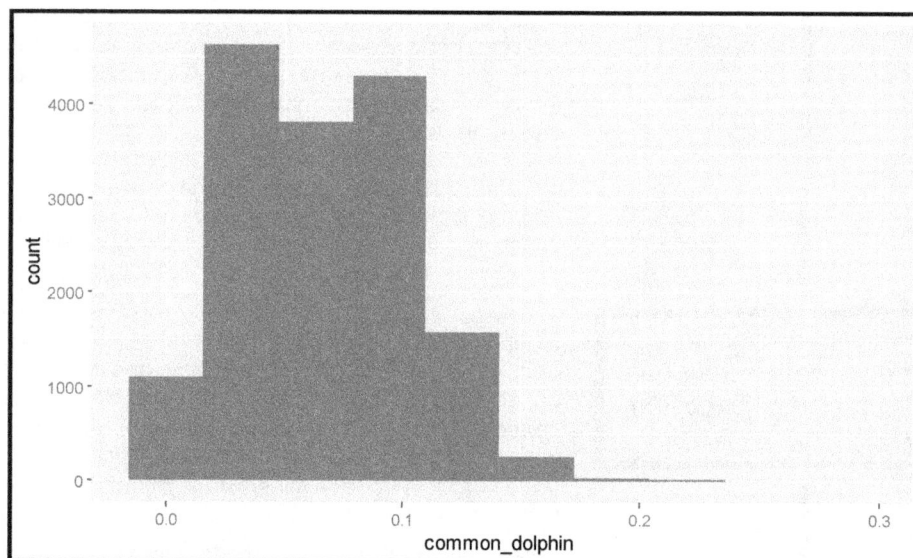

Once you have created your basic graph, you may decide that you need to change how it looks in order for it to have the exact appearance that you would like it to have. In Exercises 1.2 to 1.4 you will learn how to customise elements of the graph such as the scales used for the axes, the colours used on it and its background, how to change the type of graph used to display the data, and how to display data from multiple data series on it. However, in the next part of this exercise, you will learn how to change the way the data are displayed on the specific type of basic graph you have created. This is done by changing arguments within the commands used to make it. For example, you can change the arguments in the `geom_histogram` command to alter the way that the data on your frequency distribution histogram are divided into bins. This allows you to create histograms with different numbers and/or widths of bins (or bars) displayed on it. Information on three commonly used ways you can do this can be found in the table below, while more information on how to do this, as well as other arguments that can be used with the `geom_histogram` command, can be found at *ggplot2.tidyverse.org/reference/geom_ histogram.html*.

Argument	How To Use It
bins	This argument allows you to specify a target number of bins (or bars) for your frequency distribution histogram. In all cases, the number included in this argument needs to be a whole number. For example, including the argument `bins=10` in the `geom_histogram` command will result in a frequency distribution histogram with approximately ten bars on it, while including the argument `bins=5` will result in a frequency distribution histogram with approximately five bars on it.
binwidth	This argument allows you to specify the width (or range of values) that will be included in each bin or bar of your frequency distribution histogram. The numbers included in this argument will be in the same units as the continuous variable being plotted on the X axis of your histogram. For example, including the argument `binwidth=0.02` in the `geom_histogram` command will result in a frequency distribution histogram where each bin contains values with a range of 0.02, while including the argument `binwidth=0.1` will result in a frequency distribution histogram where each bin contains values with a range of 0.1.
breaks	This argument allows you to specify the exact values where breaks will occur between different bins (or bars) on your frequency distribution histogram. The numbers included in this argument will be in the same units as the continuous variable being plotted on the X axis of your histogram. There are a number of ways that these break values can be set. This includes providing a minimum, maximum and bin width value, and by providing a complete list of break values. For example, including the argument `breaks(seq(0,0.3, by=0.02))` in the `geom_histogram` command will result in a frequency distribution histogram that begins at a value 0, and ends at a value of 0.3. Between these two limits there will be a series of bins, each with a width of 0.02. Alternatively, including the argument `breaks=c(0.1,0.2,0.3)` in the command `geom_histogram` will result in a frequency distribution histogram with breaks between the bins at 0.1, 0.2 and 0.3. When specifying the break values in this way, they do not need to be equally distributed, and you can use this option to create histograms with unequal bin widths. For example, including the argument `breaks=c(0.05,0.1,0.2,0.4)` will result in a frequency distribution histogram with bins that start at the lowest value provided in the `breaks` argument (in this case, 0.05) and finish at the highest specified value (in this case, 0.4). Between these minimum and maximum values, there will be bins with widths, from left to right, of 0.05, 0.1 and 0.2 respectively.

To explore how you can use these arguments to set the number and/or width of the bins displayed on a frequency distribution histogram, you will modify the arguments in the `geom_histogram` command from the above flow diagram in a number of different ways. Firstly, rather than creating a frequency distribution histogram with a target number of bins, you will produce one where each bin has a specific width. This is done by replacing the `bins` argument in this command with the `binwidth` argument. To do this, you will need edit the final code from step 6 (this is CODE BLOCK 7 in the document R_CODE_DATA_VISUALISATION_WORKBOOK.DOC) so that it looks like the code at the top of the next page (the required modifications are highlighted in **bold**).

```
ggplot(data=cetacean_prey_sizes,aes(x=common_dolphin)) +
            geom_histogram(binwidth=0.02)
```

You can modify this code either by editing it in the R CONSOLE window of RGUI or through the SCRIPT EDITOR window of RStudio (depending on which interface you are using). If you are entering commands directly into the R CONSOLE window, you can use the UP arrow on your keyboard to bring commands and code blocks you have previously run during the same session back on to the command line of this window, and then use the LEFT and RIGHT arrows to scroll through and edit them. In this case, use the UP arrow to bring the previous version of the code block used to create your basic frequency distribution histogram back onto the command line and edit it so that it looks like the one above. Once you have finished modifying this code block, you can run it by pressing the ENTER key on your keyboard. If you are using RStudio, you can copy and paste the original code block in the SCRIPT EDITOR window before editing the new version to include the required modifications. After you have done this, select the modified version of the code block and click on the RUN button to run it in the R CONSOLE window.

Once you have run this new version of the R code for creating a frequency distribution histogram, you should have a graph that looks like this:

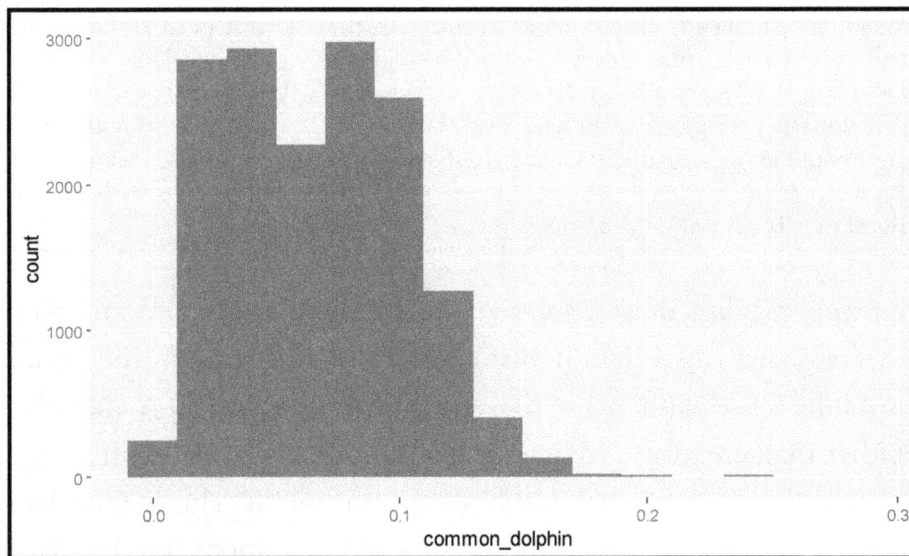

On this graph, each bin as a width of 0.02, resulting in a total of 13 bins, rather than nine, as there were on the graph created using the bins=10 argument in the geom_

histogram command. If you wish to produce a frequency distribution histogram with a different bin width, you would simply need to change the number provided in the binwidth argument. However, this argument does not allow you to specify exactly where each bin on the graph starts and ends. In order to be able to do this, you would need to specify the minimum and maximum values to be used for your bins along with the desired bin width. This is done using the breaks argument in the geom_histogram command rather than the binwidth argument. For example, to create a frequency distribution histogram where the bins have a width of 0.02, starting at a value of 0 and ending at a value of 0.3, you would need to edit the above R code so that it looks like this (the required modifications are highlighted in **bold**):

```
ggplot(data=cetacean_prey_sizes,aes(x=common_dolphin)) +
        geom_histogram(breaks=seq(0,0.3,by=0.02))
```

Once you have run this new version of the R code for creating a frequency distribution histogram, you should have a graph that looks like this:

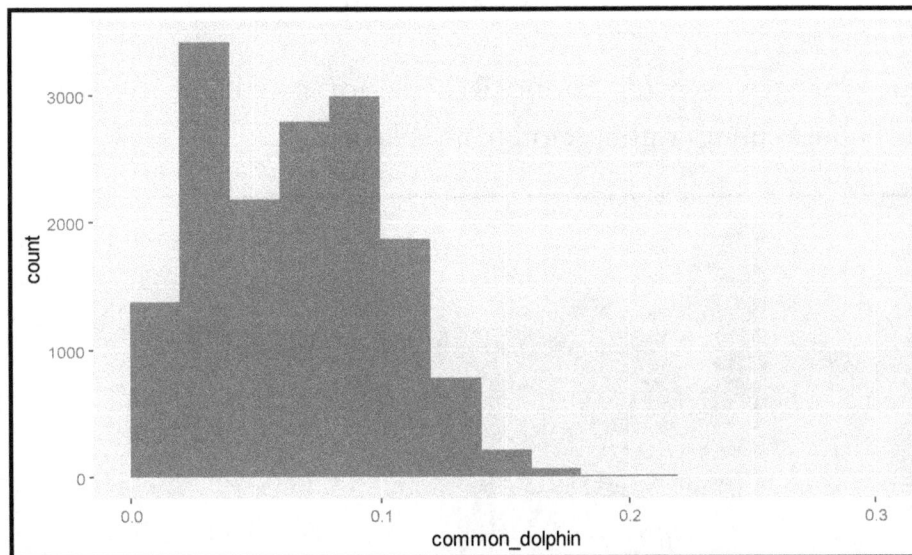

On this graph, each bin still has a width of 0.02 (set using the by term in the breaks argument). However, unlike the previous graph, these bins start at a minimum value of 0 (set by the first value in the breaks argument) and finish at a maximum value of 0.3 (set by the second value). By varying the three parameters in the breaks argument, you can create a frequency distribution histogram with any range and bin width you wish. For example, on this histogram, the bins above a value of 0.26 are empty as they contain no data. This means they could safely be removed from the graph by decreasing the maximum value from 0.3 to 0.26.

The one limitation of setting the bins for a histogram using the `breaks` argument in this way is that it requires all the bins to have the same width. If you wish to create a frequency distribution histogram where different bins have different widths, you can do this by specifying the exact breaks you wish to use between each bin. This is done by providing a list of the break values you wish to use in the `breaks` argument of the `geom_histogram` command. For example, to create a frequency distribution histogram with breaks between the bins at PPSR values of 0.005, 0.01, 0.02, 0.04, 0.06, 0.08, 0.10, 0.15, 0.20 and 0.3 you would need to edit the above R code so that it looks this (the required modifications are highlighted in **bold**):

```
ggplot(data=cetacean_prey_sizes,aes(x=common_dolphin)) +
geom_histogram(breaks=c(0.005,0.01,0.02,0.04,0.06,0.08,0.1,
                        0.15,0.20,0.30))
```

When you run this new version of the R code for creating a frequency distribution histogram, you should end up with a graph that looks like this:

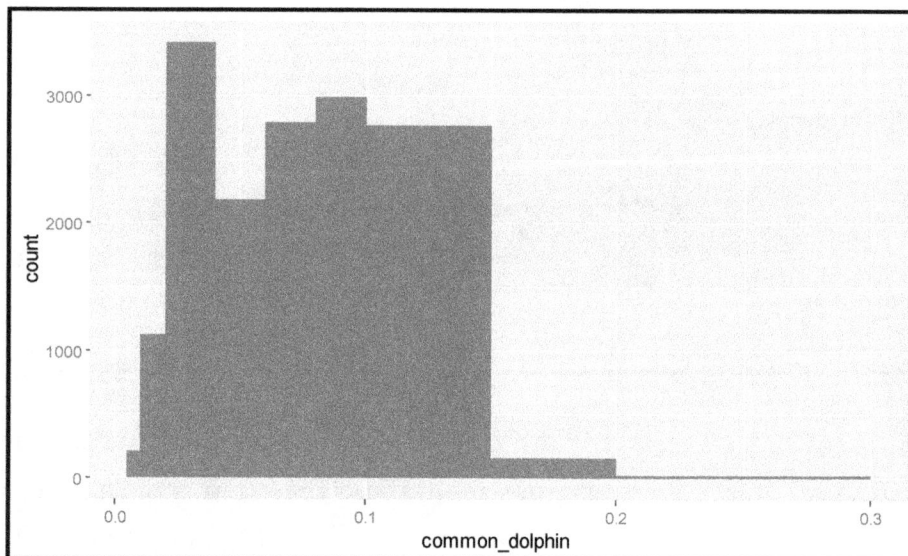

EXERCISE 1.2: HOW TO CREATE A PUBLICATION QUALITY GRAPH USING GGPLOT:

Once you have created a basic graph, during which you have determined which data you wish to plot on it, what type of graph you wish to make and how you wish these data to be displayed on the selected type of graph, you will almost certainly want to modify other elements of the graph. This can include changing the colours used to display the data, the limits, scale and labels used for the axes, the fonts used on it and the formatting of the background of the plot area. By customising these elements of a graph, you can turn it from a basic graph into one that is suitable for inclusion in a presentation or for publication. This is where using the commands from the GGPlot package, as opposed to the basic graphing functions that come with R, really come into its own, and within the GGPlot commands, you can customise almost every possible element of a graph.

The different elements of your graph can be modified by adding a series of new commands, elements and arguments to the R code used to create your basic graph (see Exercise 1.1), and the same general process and modifications can be applied to almost any type of graph. While these can be added all at once, it is much easier to concentrate on modifying one element of your graph at a time (e.g. the colours use to display your data or the scale and limits of the axes), and only once you are happy with it should you move onto the modifying another one. This allows you to gradually build up the complete code block that you will need to use to create your final publication quality graph and it will make it easier to trouble-shoot any problems you encounter while creating it.

Some of the modifications to the code used to create a basic graph will involve adding elements and arguments to the existing graphing command (these will primarily modify exactly how the information on your graph is displayed and will do things like change the colours used for the symbols on it). However, others will involve adding completely new commands to your code block. These new commands will alter other elements of your graph, such as the colour of the axes lines, the fonts used, the scales used for the axes and their labels. In this workbook, additions to the code which change how a graph appears will be referred to as style elements, style arguments and style commands.

To demonstrate how this can be done, in this exercise you will modify a variety of elements of one of the frequency distribution histograms created in Exercise 1.1 to turn it into a publication quality graph. As a result, before you can start this exercise, you will need to have completed the previous exercise in this chapter. The graph that you will modify is the one created by the following block of code from page 31:

```
ggplot(data=cetacean_prey_sizes,aes(x=common_dolphin)) +
        geom_histogram(breaks=seq(0,0.3,by=0.02))
```

To undertake this modification process for the graph created by this code block, work through the flow diagram that starts at the top of the next page.

NOTE: The block of code for creating a high quality graph that you will build up by working through the steps in this flow diagram will quickly become very long and unwieldy as you add more arguments and commands to it with each successive step. While you can re-type the entire block of code that you need to complete each step (including the commands and arguments from the previous ones), this is time-consuming, repetitive, and has a high risk of including typos which will stop the code from functioning properly. As a result, we recommend you use a 'copy-and-paste' short cut to avoid you having to type the entire block of code from scratch each time. The blocks of code required to complete each step in the flow diagram can be found in a document called R_CODE_DATA_ VISUALISATION_WORKBOOK.DOC. This document is in the compressed file containing all the data for the exercises in this book that you downloaded at the start of this chapter. If you open this document, you can copy the entire block of code needed to complete a specific step in the flow diagram and paste it into RStudio or the RGUI (depending on which interface you are using to access R). Once you have pasted the required code into your preferred interface, you can then run it as usual. A code block number for the code for each step is provided in the flow diagram and you can use this to locate the appropriate block of code in the above word document.

If you do not wish to use this 'copy-and-paste' short cut, then rather than re-typing the entire block of code needed for each new step, you can edit the code you used for the previous step and add the required the new arguments and commands to it before running it again. To help you do this, the new code that you need to add to the code block from the previous step is highlighted in **bold**. If you are using RStudio, you can do this in the SCRIPT

EDITOR window. If you are using the RGUI, you can use the UP arrow on your keyboard to bring the last block of code you ran back on to the command line of your R CONSOLE window, and then use the LEFT and RIGHT arrows to scroll through and edit it.

Data set held in an existing R object in your analysis project

For this example, the data set you will use is stored in the R object called `cetacean_prey_sizes` created in Exercise 1.1 of this workbook.

Creating a publication quality graph using GGPlot involves customising the various individual elements of your graph one at a time. As a result, the first step in creating a publication quality graph is to create an initial one using the basic graphing commands on which you can then build up all the additional customised elements. In this example, you will start by creating a basic frequency distribution histogram from an existing R object with a specified set of bins along its X axis. To do this, enter the following block of code into R:

```
ggplot(data=cetacean_prey_sizes,aes(x=
common_dolphin)) + geom_histogram(breaks=
seq(0,0.3,by=0.02))
```

1. Create an initial graph from your data set using basic GGPlot graphing commands

This is CODE BLOCK 8 in the document R_CODE_DATA_ VISUALISATION_WORKBOOK.DOC, and it contains two commands separated by a + symbol. These are the `ggplot` command and the `geom_histogram` graphing command. The `ggplot` command sets the data set which will be used for the graph. This is done using the `data` argument and, in this case, it will the R object called `cetacean_prey_ sizes` created in Exercise 1.1. The column of data which will be plotted on the X axis of the resulting graph is set using the `x` argument of the `aes` element of this `ggplot` command. In this case, it is the column called `common_dolphin` in the `cetacean_prey_sizes` data set.

The second command in this code block, `geom_ histogram`, sets the type of graph which will be created from the data specified in the `ggplot` command. In this case, it will be a histogram. Within this command, you can set the way that your data will be divided into individual bins (or bars) on your histogram (see Exercise 1.1). For this example, you will use the `breaks` argument to set it to create a series of bins with a width of `0.02` between a value of `0` and `0.3` for the variable being plotted on the X axis (and stored in the column called `common_dolphin`). In this case, this is the predator-prey size ratios, or PPSRs for short, for common dolphin.

Once you have created an initial frequency distribution histogram, you can start modifying it to make it more suitable for inclusion in a publication. To do this, you will start by customising the colours that are used for the bars that are displayed on it. This is done by adding three additional style arguments to the `geom_histogram` graphing command from step 1. These are `col`, `fill` and `alpha`. To do this for the graph being created in this example, edit the code from step 1 so that it looks like this (the newly added style arguments are highlighted in **bold**):

```
ggplot(data=cetacean_prey_sizes,aes(x=
common_dolphin)) + geom_histogram(breaks=
seq(0,0.3,by=0.02),col="black",fill=
"blue",alpha=1)
```

---

2. Customise the colours used for the initial graph created in step 1

This is CODE BLOCK 9 in the document R_CODE_DATA_VISUALISATION_WORKBOOK.DOC. **NOTE:** We strongly recommend that you copy and paste the appropriate block of code for each step in this flow diagram from this document rather than trying to type it in from scratch.

The new `col` argument in the `geom_histogram` command in this block of code sets the colour of the outlines of the bars displayed on the graph. In this case, it sets the colour for the outlines to `black`. The `fill` argument sets the colour for the bars themselves. In this case, it sets the `fill` colour to `blue`. **NOTE:** A full list of the colour terms you can use with the `col` and `fill` arguments can be obtained by entering the command `colours()` into R. Finally, the `alpha` argument sets the level of transparency for the bars. These values can range from `0` (completely transparent) to `1` (completely opaque). In this case, the value will be set `1`, meaning that the bars will completely opaque. Once you have finished editing this code block, you can run it again to create an updated version of your histogram.

To add custom labels for the title and axes for your graph, you need to add a new command to the code block from step 2. This is the `labs` command, and it will be the first of a series of style commands you will add to your R code over the next five steps. To do this for the graph being created in this example, edit the above code so that it looks like this (the newly added style command is highlighted in **bold**):

```
ggplot(data=cetacean_prey_sizes,aes(x=
common_dolphin)) + geom_histogram(breaks=
seq(0,0.3,by=0.02), col="black",
fill="blue",alpha=1) + labs(title="Relative
Prey Size of Common Dolphin",x="Predator-
Prey Size Ratio",y="Frequency of
Occurrence")
```

---

3. Customise the labels for the title, the X axis and the Y axis of your graph

---

This is CODE BLOCK 10 in the document R_CODE_DATA_VISUALISATION_WORKBOOK.DOC. The new `labs` command contains three arguments. The `title` argument provides the text to be used for the title at the top of your graph, the `x` argument provides the label for the X axis, and the `y` argument provides the label for the Y axis. Once you have finished editing this code block, you can run it again to create an updated version of your histogram.

4. Format the limits, ticks, intervals and axes-crossing values for the X and Y axes of your graph

To set the limits of your X and Y axes, you need to add another pair of new style commands to the code block from step 3. These are the `scale_x_continuous` and the `scale_y_continuous` commands. To do this for the graph being created in this example, edit the above code so that it looks like this (the newly added style commands are highlighted in **bold**):

```
ggplot(data=cetacean_prey_sizes,aes(x=
common_dolphin)) + geom_histogram(breaks=
    seq(0,0.3,by=0.02), col="black",
fill="blue",alpha=1) + labs(title="Relative
Prey Size of Common Dolphin",x="Predator-
    Prey Size Ratio",y="Frequency of
Occurrence") + scale_x_continuous(limits=
c(0,0.4),breaks=seq(0,0.4,0.05),expand=
    c(0,0)) + scale_y_continuous(limits=
    c(0,3500),breaks=seq(500,3500,500),
            expand=c(0,0))
```

This is CODE BLOCK 11 in the document R_CODE_DATA_ VISUALISATION_WORKBOOK.DOC. In the new `scale_ x_continuous` command, the `limits` argument sets the minimum and the maximum values to be displayed on the X axis. In this case, these are `0.0` and `0.4` respectively. **NOTE:** It is this argument that sets the value at which the Y axis will cross the X axis. The `breaks` argument sets the intervals for the values that will be displayed on the X axis. This is done by setting the lowest value to be displayed (in this case `0`), the highest value to be displayed (in this case, `0.4`) and the intervals between these values for intermediate labels (in this case, `0.05`). Finally, the `expand` argument sets how far the axis extends beyond the lower limit (in this case, `0`) and the upper limit (in this case, also `0`) specified by the `limits` argument.

In the new `scale_y_continuous` command, the `limits` argument sets the minimum and the maximum values to be displayed on the Y axis. In this case, these are `0` and `3500` respectively. **NOTE:** It is this argument that sets the value at which the X axis will cross the Y axis. The `breaks` argument sets the intervals for the values that will be displayed on the Y axis. This is done by setting the lowest value to be displayed (in this case `500`), the highest value to be displayed (in this case, `3500`) and the intervals between these values for intermediate labels (in this case, `500`). Finally, the `expand` argument sets how far the axis extends beyond the lower limit (in this case, `0`) and the upper limit (in this case, also `0`) specified by the `limits` argument. Once you have finished editing this code block, you can run it again to create an updated version of your histogram.

There are number of different ways that the axes of your graph can be formatted. This includes setting the colour they are drawn in, setting the distance between the axes label and its numbers, and setting the margin around the plot so that the numbers on the upper limits of the axes are not cut off. This formatting is done by adding a series of `theme` style commands containing different elements and arguments to the code block from step 4. To do this for the graph being created in this example, edit the above code so that it looks like this (the newly added style commands are highlighted in **bold**):

```
ggplot(data=cetacean_prey_sizes,
aes(x=common_dolphin)) + geom_histogram(
breaks=seq(0,0.3,by=0.02),col="black",
fill="blue",alpha=1) + labs(title="Relative
Prey Size of Common Dolphin",x="Predator-
Prey Size Ratio",y="Frequency of
Occurrence") + scale_x_continuous(limits=
c(0,0.4),breaks=seq(0,0.4,0.05),expand=
c(0,0)) + scale_y_continuous(limits=
c(0,3500),breaks=seq(500,3500,500),expand=
c(0,0)) + theme(axis.line=(element_line(
colour="black"))) + theme(axis.text.x=
element_text(margin=margin(t=0,r=0,b=5,
l=0))) + theme(axis.text.y=element_
text(margin=margin(t=0,r=0,b=0,l=5))) +
theme(plot.margin=margin(t=10,r=10,b=10,
l=10))
```

5. Format the style of the X and Y axes of your graph

This is CODE BLOCK 12 in the document R_CODE_DATA_VISUALISATION_WORKBOOK.DOC. The first `theme` command uses the `colour` argument in the `axis.line` element to set the colour that the axes are drawn in. In this example, it is set to `black`. The second `theme` command uses the `margin` argument in the `axis.text.x` element to set the margin between the label for the X axis and the numbers on it. Within this argument, you can set the top (`t`), right (`r`), bottom (`b`) and left (`l`) margins. In this case, only the bottom margin is set to a non-zero value to increase the gap between the label and the numbers of the X axis.

The third `theme` command uses the `margin` argument in the `axis.text.y` element to set the margin between the label for the Y axis and the numbers on it. Within this argument, you can set the top (`t`), right (`r`), bottom (`b`) and left (`l`) margins. In this case, only the left margin is set to a non-zero value to increase the gap between the label and the numbers of the Y axis.

The final `theme` command uses the `margin` argument in the `plot.margin` element to set the margins around the plot area. This prevents the numbers on the axes from being partially or fully cut off. As before, within this argument, you can set the top (`t`), right (`r`), bottom (`b`) and left (`l`) margins. In this case, a value of `10` is used for all four margins in this argument. Once you have finished editing this code block, you can run it again to create an updated version of your histogram.

Before you can set the font which will be used for your graph, you first need to load the required fonts into R. To do this, first check if you already have the `extrafont` package installed in your version of R by entering the code `library()`. If this package is not listed in the R PACKAGES AVAILABLE window that opens, enter the follow code into R:

```
install.packages("extrafont")
library(extrafont)
```

This is CODE BLOCK 13 in the document R_CODE_DATA_ VISUALISATION_WORKBOOK.DOC. This block of code installs the `extrafont` package into R and then loads the associated command library into your analysis project. If the `extrafont` package is listed in the R PACKAGES AVAILABLE window, you do not need to re-install it by running the above `install.packages` command. Instead, you can simply run the `library(extrafont)` command to load its command library into your project.

To set the font for your graph, you can add another new `theme` style command to the end of your current code block with arguments that specify the `colour`, `size` and `font` to be used for the final graph. To do this for the graph being created in this example, edit the code from step 5 so that it looks like this (the newly added style command is highlighted in **bold**):

```
ggplot(data=cetacean_prey_sizes,aes(x=
common_dolphin)) + geom_histogram(breaks=
seq(0,0.3,by=0.02),col="black",fill="blue",
alpha=1) + labs(title="Relative Prey Size of
Common Dolphin",x="Predator-Prey Size
Ratio",y="Frequency of Occurrence") +
scale_x_ continuous(limits=c(0,0.4),breaks=
seq(0,0.4,0.05),expand=c(0,0)) + scale_
y_continuous(limits=c(0,3500),breaks=
seq(500,3500,500),expand=c(0,0)) +
theme(axis.line=(element_line(colour
="black"))) + theme(axis.text.x=element_
text(margin=margin(t=0,r=0,b=5,l=0)))+
theme(axis.text.y=element_text(margin=
margin(t=0,r=0,b=0,l=5))) + theme(plot.
margin=margin(t=10,r=10,b=10,l=10)) +
theme(text=element_text(colour="black",
size=12,hjust=0.5,family="serif"))
```

This is CODE BLOCK 14 in the document R_CODE_DATA_ VISUALISATION_WORKBOOK.DOC. This newly added `theme` command sets the colour of the font to `black`, the size to `12`, the position of the text to the middle (using the `hjust=0.5` argument) and the font to Times New Roman (using the `family="serif"` argument). This is all done within the `text` element of this command. Once you have finished editing this code block, you can run it again to create an updated version of your histogram.

6.	Format the font to be used for your graph

7. Format the background of your graph

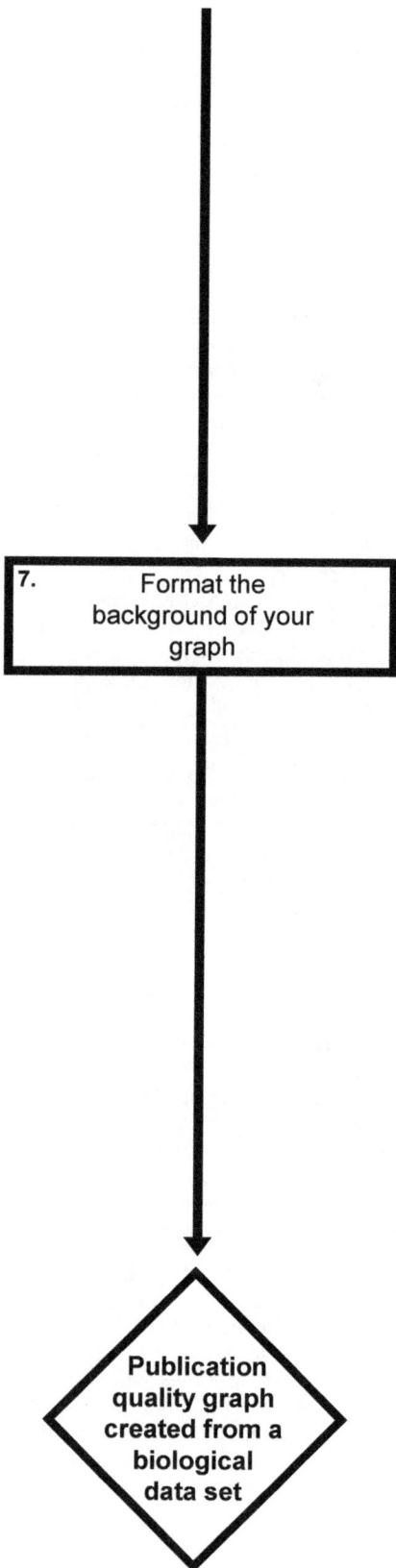

Finally, you need to format the background of your graph. This is done by adding one last `theme` style command containing the `panel.background` argument to the code block. To do this for the graph being created in this example, edit the code from step 6 so that it looks like this (the newly added style command is highlighted in **bold**):

```
ggplot(data=cetacean_prey_sizes,aes(x=
common_dolphin)) + geom_histogram(breaks=
seq(0,0.3,by=0.02),col="black",fill="blue",
alpha=1) + labs(title="Relative Prey Size of
   Common Dolphin",x="Predator-Prey Size
      Ratio",y="Frequency of Occurrence") +
scale_x_continuous(limits=c(0,0.4),breaks=
   seq(0,0.4,0.05),expand=c(0,0)) + scale_
   y_continuous(limits=c(0,3500),breaks=
     seq(500,3500,500),expand=c(0,0)) +
   theme(axis.line=(element_line(colour=
   "black"))) + theme(axis.text.x=element_
  text(margin=margin(t=0,r=0,b=5,l=0)))+
   theme(axis.text.y=element_text(margin=
  margin(t=0,r=0,b=0,l=5))) + theme(plot.
   margin=margin(t=10,r=10,b=10,l=10)) +
   theme(text=element_text(colour="black",
     size=12,hjust=0.5,family="serif")) +
   theme(panel.background=(element_rect(
                fill="white")))
```

This is CODE BLOCK 15 in the document R_CODE_DATA_ VISUALISATION_WORKBOOK.DOC. The new `theme` command added to this code block sets the background colour of the graph to white by using the term `element_ rect(fill="white"))` in the `panel.background` argument. Once you have finished editing this code block, you can run it again to create the final version of your histogram.

Publication quality graph created from a biological data set

At the end of the first part of this exercise, you should have a final, publication quality frequency distribution histogram of the predator-prey size ratios (PPSRs) for common dolphin that looks like this:

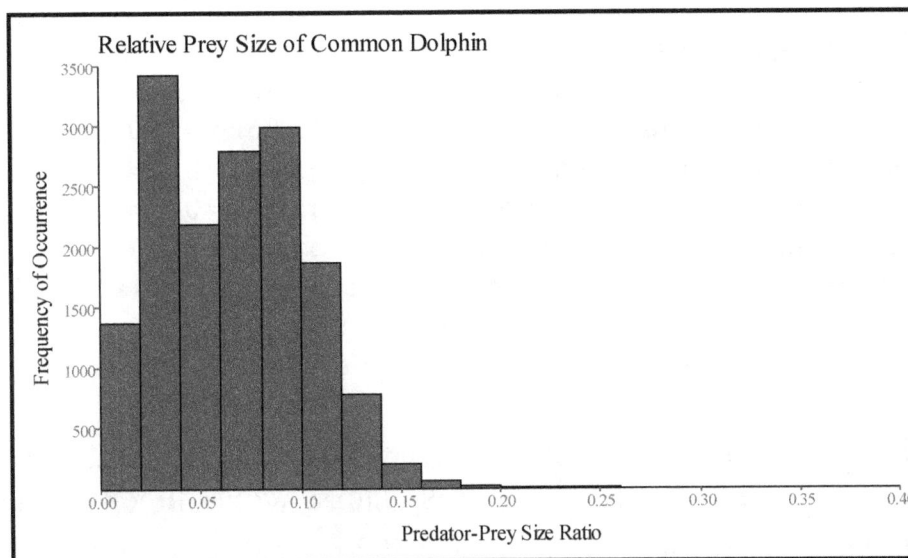

When building up the block of code required to make the above publication quality graph, you used separate `theme` commands to customise a range of different elements of the graph, such as the colour of the axes lines, the font and the colour of the background panel. This approach was used to help you understand how the code required to customise exactly how a final graph will look can be created by adding new pieces of code to modify each individual element. However, the result of this is a very long and unwieldy block of code. Luckily, there are a number of ways that this code can be streamlined to make it easier to manage.

Firstly, you can combine the elements and arguments from each individual `theme` command into a single combined `theme` command. While this makes it more difficult to track exactly which piece of code does what (and so to re-use the individual parts of the overall code block when making other graphs), it does shorten the length of the code that you need to input. For example, if you do this for all the `theme` commands in the code from step 7 of the above flow diagram, you end up with a block of code that looks like the one provided at the top of the next page (the new combined `theme` command containing all the style elements and arguments that were previously in individual `theme` commands is highlighted in **bold**).

```
ggplot(data=cetacean_prey_sizes,aes(x=common_dolphin)) +
  geom_histogram(breaks=seq(0,0.3,by=0.02),col="black",
fill="blue",alpha=1) + labs(title="Relative Prey Size of
Common Dolphin",x="Predator-Prey Size Ratio",y="Frequency
of Occurrence") + scale_x_continuous(expand=c(0,0),limits =
  c(0,0.4),breaks=seq(0,0.4,0.05)) + scale_y_continuous(
expand=c(0,0),limits=c(0,3500),breaks=seq(500,3500,500)) +
      theme(axis.line=(element_line(colour="black")),
  axis.text.x=element_text(margin=margin(t=0,r=0,b=5,l=0)),
  axis.text.y=element_text(margin=margin(t=0,r=0,b=0,l=5)),
      plot.margin=margin(t=10,r=10,b=10,l=10),text=
      element_text(colour="black",size=12,hjust=0.5,
          family="serif"),panel.background=
              (element_rect(fill="white")))
```

You can modify the code from step 7 of the above flow diagram (this is CODE BLOCK 15 in the document R_CODE_DATA_VISUALISATION_WORKBOOK.DOC) so that it looks like this either by editing it in the R CONSOLE window of RGUI or through the SCRIPT EDITOR window of RStudio (depending on which interface you are using). If you are entering commands directly into the R CONSOLE window, you can use the UP arrow on your keyboard to bring the last block of code you ran back on to the command line of this window, and then use the LEFT and RIGHT arrows to scroll through and edit it. After you have finished modifying the required code block, you can run it by pressing the ENTER key on your keyboard. If you are using RStudio, you can make a copy of the block of code from step 7 in the SCRIPT EDITOR window before editing it. Once you have done this, select the modified version of the code block and click on the RUN button to run it in the R CONSOLE window.

Despite only having one single combined theme command, the above block of code will produce exactly the same graph as the code with the individual theme commands provided in step 7. However, even with this streamlining, handling such a large block of code can be difficult. In particular, as the entire block of code is entered on a single line in both RStudio and the RGUI, you will find that you need to scroll a long way to the right in order to review or edit it. If you wish to avoid having to do this, you can split your block of code across a number of different lines. To do this, all you need to do is to put a return in the code immediately after a + symbol. This lets R know that the block of code is not yet

42

complete. For example, for the above block of code, you could divide it into four parts (one for making the histogram, a second for adding labels to it, a third for setting the scales for the axes, and a fourth for formatting the rest of the graph), each of which can be entered on a separate line, like this:

```
ggplot(data=cetacean_prey_sizes,aes(x=common_dolphin)) +
  geom_histogram(breaks=seq(0,0.3,by=0.02),col="black",
            fill="blue",alpha = 1) +

  labs(title="Relative Prey Size of Common Dolphin", x=
"Predator-Prey Size Ratio",y="Frequency of Occurrence") +

  scale_x_continuous(expand=c(0,0),limits=c(0,0.4),
   breaks=seq(0,0.4,0.05)) + scale_y_continuous(expand=
   c(0,0),limits=c(0,3500),breaks=seq(500,3500,500)) +

  theme(axis.line=(element_line(colour="black")),
 axis.text.x=element_text(margin=margin(t=0,r=0,b=5,l=0)),
 axis.text.y=element_text(margin=margin(t=0,r=0,b=0,l=5)),
    plot.margin=margin(t=10,r=10,b=10,l=10),text=
    element_text(colour="black",size=12,hjust=0.5,
        family="serif"),panel.background=
           (element_rect(fill="white")))
```

Secondly, you can make long blocks of code easier to enter, and understand, by separating out the individual commands, or groups of commands, and applying them to the graph individually. This is done by creating an R object to contain the initial graph (in this case, one called cetacean_histogram), and then applying each new part of the code to the contents of this object. For example, you can use this approach to divide the above block of code for creating a frequency distribution histogram from the cetacean prey size data set into four parts. The first part would make the initial frequency distribution histogram and save it in a specific R object, and would look like this (new additions to the code provided above have been highlighted in **bold**):

```
cetacean_histogram <- ggplot(data=cetacean_prey_sizes,
 aes(x=common_dolphin)) + geom_histogram(breaks=seq(0,0.3,
        by=0.02),col="black",fill="blue",alpha=1)
```

To check that this section of code has resulted in a graph with the intended appearance, once it has been run, you then need to call the R object created by it. This is done by entering the name of the object into R. In this case, enter the following object name into R:

```
cetacean_histogram
```

The second part adds the labels to the graph using the `labs` command, and would look like this (new additions to the code provided above have been highlighted in **bold**):

```
cetacean_histogram <- cetacean_histogram +
  labs(title="Relative Prey Size of Common Dolphin",
x="Predator-Prey Size Ratio",y="Frequency of Occurrence")
```

Again, once this command has been run, enter the name of the R object to display the graph and allow you to examine how it has changed, like this:

```
cetacean_histogram
```

The third part contains the `scale_x_continuous` and `scale_y_continuous` commands that determines the scales for the X and Y axes, and would look like this:

```
cetacean_histogram <- cetacean_histogram
+scale_x_continuous(expand=c(0,0),limits=c(0,0.4),
  breaks=seq(0,0.4,0.05)) + scale_y_continuous(expand=
  c(0,0),limits=c(0,3500),breaks=seq(500,3500,500))
```

Now enter the name of the R object containing the graph again (`cetacean_histogram`) to allow you to see the changes running this code has made to the graph.

The final part contains the `theme` command with all the arguments that format all the other aspects of the final graph, and would look like this:

```
cetacean_histogram <- cetacean_histogram + theme(axis.line=
  (element_line(colour="black")),axis.text.x=element_text(
  margin=margin(t=0,r=0,b=5,l=0)),axis.text.y=element_text(
  margin=margin(t=0,r=0,b=0,l=5)),plot.margin=margin(t=10,r=
  10,b=10,l=10),text=element_text(colour="black",size=12,
    hjust=0.5,family="serif"),panel.background=
        (element_rect(fill="white")))
```

Again, enter the name of the R object containing the graph (`cetacean_histogram`) so you can see what the final frequency distribution histogram created by these series of commands looks like. You will use this approach in Chapter Seven where you will learn how to make maps from your data in R using a variety of different packages, including GGPlot.

While creating the graph outlined in the above flow diagram, you have learned how to modify the most common elements of a graph that biologists need to customise to produce a publication quality graph. However, it is possible to modify many other elements in a similar manner. A full list of every possible modification to the elements of a graph that you can make using GGPlot can be found at *ggplot2.tidyverse.org/reference/*. As you will see if you examine this list, customising how a graph looks in this manner can not only take a lot of time, but also require the use of a very large and relatively complicated block of code. Luckily, this process can be made simpler by the existence of pre-existing themes which will set many of these elements using a single new command in the code block used to create a graph. This also has the advantage of substantially shortening the code required to make a graph that conforms to a specific style. The commands for the pre-existing themes most commonly used by biologists are provided in the table below.

Pre-existing Theme Command	Description
`theme_grey()`	This pre-existing theme has a grey background with white grid lines on it. There are no lines for the axes and no border around the plot area.
`theme_bw()`	This pre-existing theme has a white background with grey grid lines on it. There are black lines for the axes and a black border around the plot area.
`theme_light()`	This pre-existing theme has a white background with light grey grid lines on it. There are light grey lines for the axes and around the plot area.
`theme_dark()`	This pre-existing theme has a dark grey background with grey grid lines on it. There are no lines for the axes or around the plot area.
`theme_classic()`	This pre-existing theme has a white background with no grid lines on it. There are black lines for the axes, but none around the plot area.

To explore how you can use these pre-existing themes to customise the appearance of a graph, in the next part of this exercise you will add the appropriate command to the code block used in step 3 from the above flow diagram to apply the classic pre-existing theme to the graph created by it. To do this, edit the code from step 3 of this flow diagram (this is CODE BLOCK 10 in the document R_CODE_DATA_VISUALISATION_

WORKBOOK.DOC) so that it looks like this (the required modification is highlighted in **bold**):

```
ggplot(data=cetacean_prey_sizes,aes(x=common_dolphin)) +
  geom_histogram(breaks=seq(0,0.3,by=0.02),col="black",
fill="blue",alpha=1) + labs(title="Relative Prey Size of
Common Dolphin",x="Predator-Prey Size Ratio",y="Frequency
           of Occurrence") + theme_classic()
```

Once you have run this new version of the R code that includes the `theme_classic` command you should have a graph that looks like this:

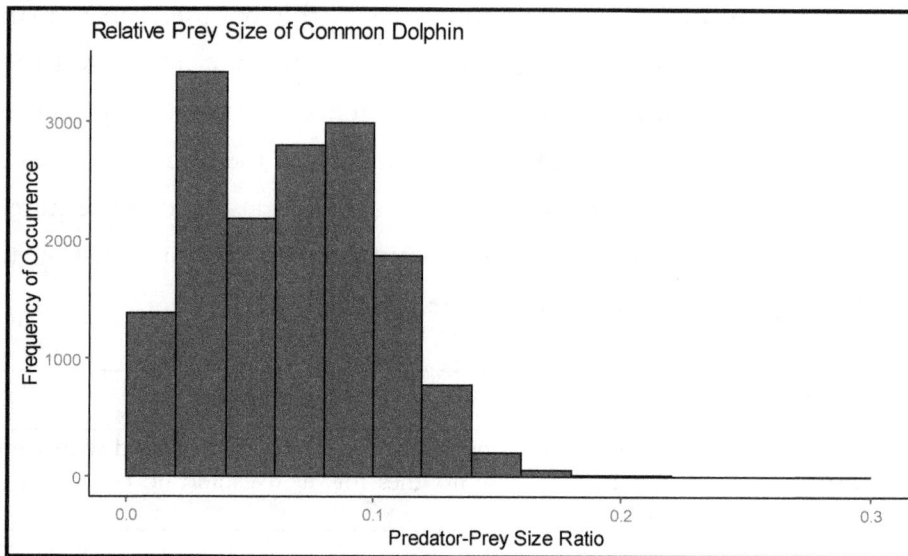

As you can see from this graph, by using the `theme_classic` command in place of the code you previously added in steps 4 to 7 of the above flow diagram, you have greatly shortened the code block required to make it while still managing to customise how it looks. However, if a pre-existing theme does not give your graph quite the appearance you are looking for, you can also use a combination of pre-existing theme settings and custom settings on the same graph to tweak exactly how it looks. To do this, you simply need to add the additional commands required to modify the specific setting you wish to customise from the pre-existing theme you are using. For example, to modify the gap between the axes and their labels, and change the font used for them, you would add the required commands (see steps 5 and 6 in the above flow diagram) after the `theme_classic` command in the code block provided on page 46 so that it looks like this (the required modifications are highlighted in **bold**).

```
ggplot(data=cetacean_prey_sizes,aes(x=common_dolphin)) +
   geom_histogram(breaks=seq(0,0.3,by=0.02),col="black",
fill="blue", alpha = 1) + labs(title="Relative Prey Size of
 Common Dolphin",x="Predator-Prey Size Ratio",y="Frequency
   of Occurrence") + theme_classic() + theme(axis.text.x=
       element_text(margin=margin(t=0,r=0,b=5,l=0))) +
 theme(axis.text.y=element_text(margin=margin(t=0,r=0,b=0,
   l=5))) + theme(axis.title=element_text(family="serif",
        face="bold",colour="black",size=12,hjust=0.5))
```

NOTE: Rather than trying to type this new section of code from scratch, you can copy and paste it from CODE BLOCKS 12 and 14 in the word document R_CODE_DATA_ VISUALISATION_WORKBOOK.DOC). However, if you are doing this, you will need to add the `face="bold"` argument to the final theme command to make the font for the axes labels bold.

Once you have run this version of the R code you should have a graph that looks like this:

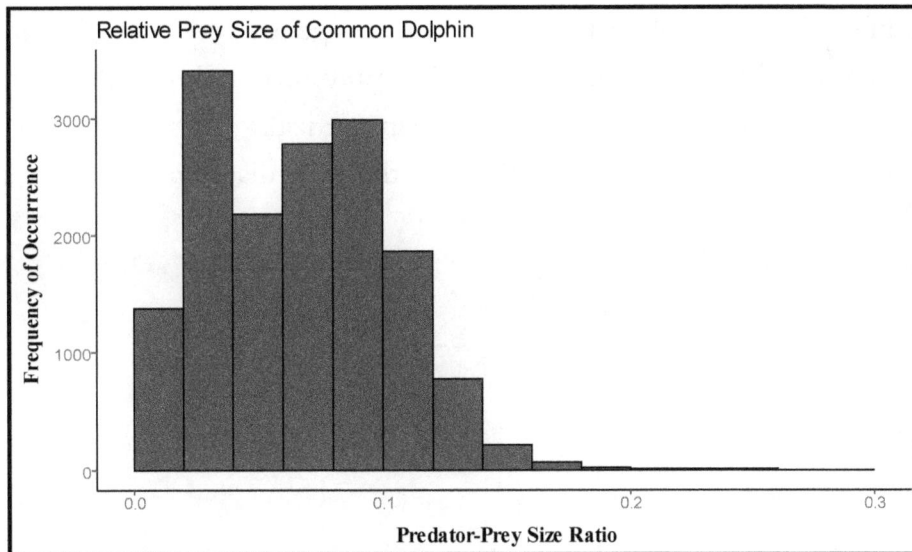

EXERCISE 1.3: HOW TO CREATE DIFFERENT TYPES OF GRAPHS FROM THE SAME DATA SET:

After you have created an initial graph from your data (see step 1 of Exercise 1.2), you may decide that rather than just modifying the appearance of a specific graph type (as you did in the rest of Exercise 1.2), you would rather display your data in a different way. This can be done either by changing the arguments in the command used to create your initial graph, or by changing the graphing command to a different one. For example, you may decide to create a histogram with the bars running horizontally rather than vertically, to display your frequency distribution as a line connecting the values for each bin rather than as a set of bars, or to plot the density of records in each bin rather than the count of records, as was the case in Exercises 1.1. and 1.2. In this exercise, you will learn how you can modify the type of graph created by a block of code by changing the arguments within the existing commands in it or by replacing the graphing command with a different one. To do this, you will modify the block of code used to create a histogram of the predator-prey size ratio (PPSR) data for common dolphin with a minimum value of 0.0, a maximum value of 0.3 and a bin width of 0.2. The original version of this code block was created towards the end of Exercise 1.1 and can be found on page 31 of this workbook.

The first modification you will make to it is to change the code so that it produces a frequency distribution histogram of the predator-prey size ratios (PPSRs) for common dolphin where the bars run horizontally rather than vertically (as is the case for a standard histogram). This can be done by working through the flow diagram that starts at the top of the next page.

Existing block of code that you wish to alter to produce a graph that displays your data in a different way

For this example, the existing block of code that you wish to alter can be found on page 31 of this workbook.

1. **Alter an existing block of code to display your data in a different way**

When you have an existing block of code, you can change the arguments and commands included in it to alter the way that your data are displayed on the graph it will create. In this example, you will change the histogram created by the code from page 31 so that the bars run horizontally rather than vertically. This is done by changing the argument used in the aes element of the ggplot command to specify the axis on which the data that will be plotted. To do this, edit this block of code so that it looks like this (the required modification to the original code from page 31 is highlighted in **bold**):

```
ggplot(data=cetacean_prey_sizes,
aes(y=common_dolphin)) + geom_histogram(
     breaks=seq(0,0.3,by=0.02))
```

This is CODE BLOCK 16 in the document R_CODE_DATA_ VISUALISATION_WORKBOOK.DOC. In this block of code, the x argument in the aes element of the ggplot command has been changed to y. This means that rather than creating a frequency distribution histogram with vertical bars on it, it will produce one where the distribution is plotted along the Y axis rather than the X axis, so that the bars run horizontally instead.

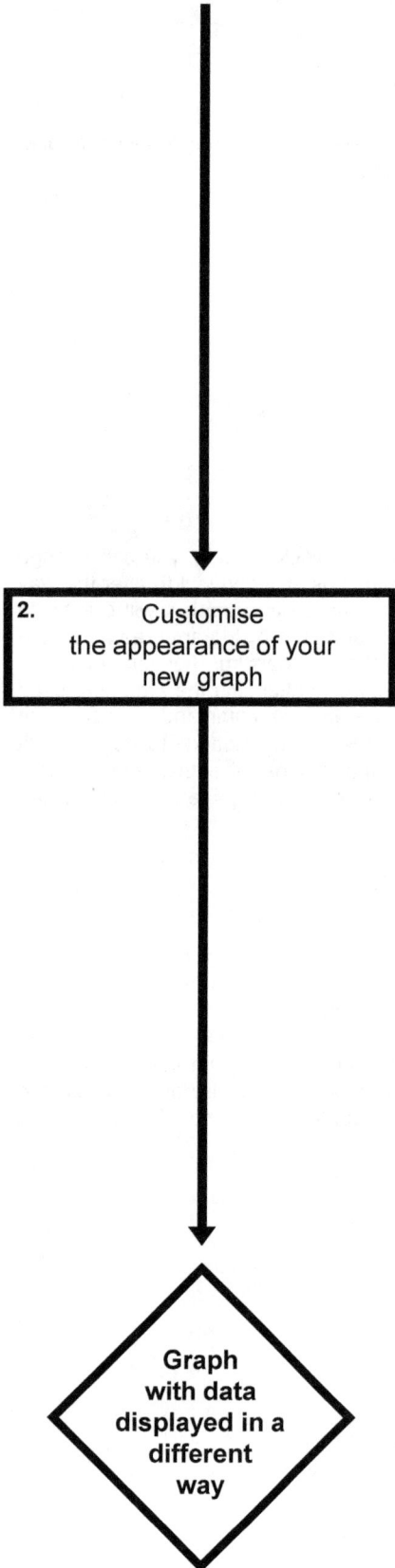

Once you have created a new graph which displays your data in a different way, you can finalise how you wish it to look. This is done by adding new arguments to the existing commands and/or new commands to the code used in step 1 in order to change its appearance. To do this for the graph being created in this example, edit the code from step 1 so that it looks like this (the newly added arguments and commands are highlighted in **bold**):

```
ggplot(data=cetacean_prey_sizes,
      aes(y=common_dolphin)) +
geom_histogram(breaks=seq(0,0.3,by=0.02),
   col="black",fill="blue",alpha=1) +
labs(title="Relative Prey Size of Common
   Dolphin",y="Predator-Prey Size Ratio",
      x="Frequency of Occurrence") +
   ylim(c(0.0,0.4)) + xlim(c(0,3500)) +
              theme_classic()
```

2. Customise the appearance of your new graph

This is CODE BLOCK 17 in the document R_CODE_DATA_ VISUALISATION_WORKBOOK.DOC. In this block of code, several new arguments and commands have been added to the code used in step 1. Firstly, arguments have been added to the `geom_histogram` to change how the bars on your histogram are displayed. These arguments specify that the outline of the bars used to represent the bins will be black (set by the `col="black"` argument), the fill colour for the bars will be blue (set by the `fill="blue"` argument) and the bars will be completely opaque (set by the `alpha=1` argument).

Secondly, a `labs` command has been added to the code block to specify what labels will be added for the main title (set by the `title="Relative Prey Size of Common Dolphin"` argument), the Y axis (set by the `y="Predator-Prey Size Ratio"` argument) and the X axis (set by the `x="Frequency of Occurrence"` argument).

Thirdly, `xlim` and `ylim` commands have been added to specify the minimum and maximum limits of the X and Y axes respectively. **NOTE:** This is an alternative way to set the limits of your axes to that used in step 4 of Exercsie 1.2.

Finally, the `theme_classic` command has been added so that the appearance of all other elements on the final graph will be determined by the settings used by the pre-existing classic theme. Once you have finished editing this code block, you can run it again to create the final version of your new histogram.

Graph with data displayed in a different way

At the end of the first part of this exercise, you should have a new frequency distribution histogram of the predator-prey size ratios (PPSRs) for common dolphin that looks like this:

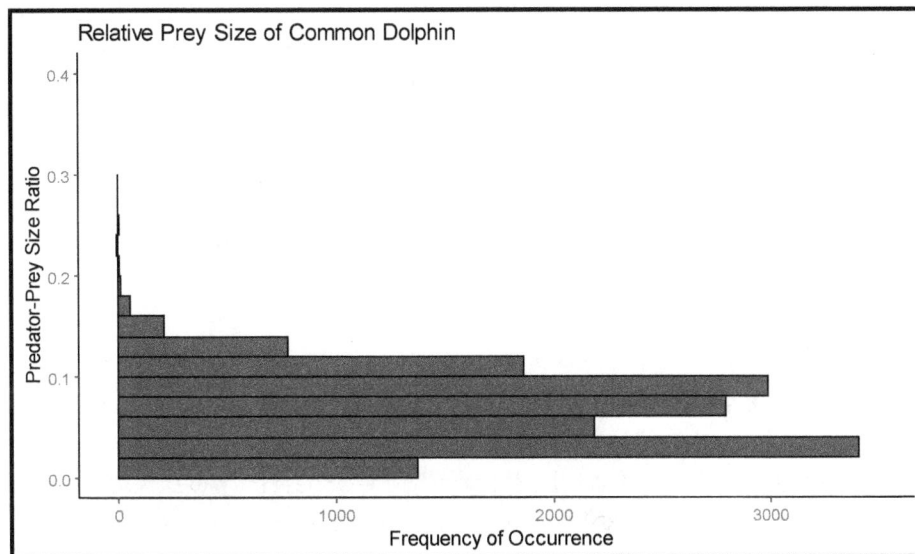

While creating this graph, you have learned how to modify the commands used to make a standard frequency distribution histogram to create a customised graph which displays the frequency distribution of the data in a different way. This was done by modifying the argument that determines which axis the data were displayed on to give a histogram with horizontal bars on it rather than the standard vertical bars. However, as well as modifying the axis that the data are plotted on, it is also possible to alter how the frequency distribution data themselves are displayed on the graph. This is done by replacing the `geom_histogram` command with another graphing command. Graphing commands from the GGPlot package that can be used to create the types of graphs commonly used by biologists include `geom_histogram` (used to create frequency distribution histograms – see above flow diagram), `geom_freqpoly` (used to create line graphs of frequency distributions – see below), `geom_density` (used to create filled density graphs to represent frequency distributions – see below), `geom_bar` (used to create bar graphs – see Exercises 2.1 to 2.3), `geom_point` (used to create graphs with points on them – see Exercises 2.4 to 3.3), `geom_line` (used to create line graphs - see Exercises 3.4 and 4.4) and `geom_boxplot` (used to create box plots – see Exercise 2.6). A full list of possible graphing commands available in this package can be found at *ggplot2.tidyverse.org/ reference/*.

To explore how you can use these alternative commands to create different types of graphs from a specific data set, in the next part of this exercise you will modify the code block used in step 2 of the above flow diagram to create graphs which display frequency distribution data in two different ways. Firstly, you will create a line graph of the frequency distribution of the common dolphin PPSR data rather a histogram. This is done by replacing the `geom_histogram` command with the `geom_freqpoly` command. To do this, edit the code from step 2 of the above flow diagram (this is CODE BLOCK 17 in the document R_CODE_DATA_VISUALISATION_WORKBOOK.DOC) so that it looks like this (the required modifications are highlighted in **bold**):

```
ggplot(data=cetacean_prey_sizes,aes(x=common_dolphin)) +
geom_freqpoly(breaks=seq(0,0.3,by=0.02),col="black",alpha=
   1) + labs(title="Relative Prey Size of Common Dolphin",
x="Predator-Prey Size Ratio",y="Frequency of Occurrence") +
   xlim(c(0.0,0.4)) + ylim(c(0,3500)) + theme_classic()
```

You can modify the code from step 2 of the above flow diagram either by editing it in the R CONSOLE window of RGUI or through the SCRIPT EDITOR window of RStudio (depending on which interface you are using). If you are entering commands directly into the R CONSOLE window, you can use the UP arrow on your keyboard to bring the last block of code you ran back on to the command line of this window, and then use the LEFT and RIGHT arrows to scroll through and edit it. After you have finished modifying the required code block, you can run it by pressing the ENTER key on your keyboard. If you are using RStudio, you can make a copy of the block of code from step 2 in the SCRIPT EDITOR window before editing it. Once you have done this, select the modified version of the code block and click on the RUN button to run it in the R CONSOLE window. After you have run this new version of the R code to create a frequency distribution line graph, you should have a graph that looks like the image at the top of the next page.

Note: When you run this code, you will get an error message saying *Removed 1 row(s) containing missing values (geom_path).* This is okay and is the result of the code being asked to start the line at a value of zero when there are no zero values in the data set.

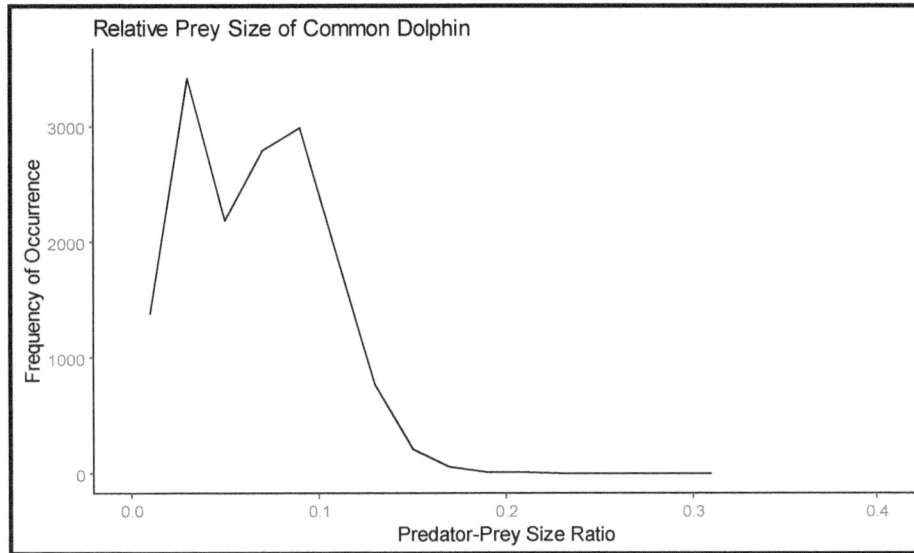

Relative Prey Size of Common Dolphin

Similarly, to create a graph showing the density of records rather than the count of records in different bins, edit the above code block to replace the `geom_freqpoly` command with the with `geom_density` command so that it looks like this (the required modifications are highlighted in **bold** – you also need to edit the maximum value in the `ylim` argument):

```
ggplot(data=cetacean_prey_sizes,aes(x=common_dolphin)) +
    geom_density(col="black",fill="blue",alpha=1) +
    labs(title="Relative Prey Size of Common Dolphin",x=
    "Predator-Prey Size Ratio",y="Density of Occurrence") +
    xlim(c(0.0,0.4)) + ylim(c(0,13)) + theme_classic()
```

NOTE: When using the `geom_density` command, you do not need to specify a term to set the number of bins. Instead, the distribution of the data is determined by the density function used by this command. The default setting for command, which is used when you do not specify any other function (as was the case in this example), is a kernel density based on a Gaussian distribution (see *www.rdocumentation.org/packages/ggplot2versions/3.3.2/ topics/geom_density* for more details). Once you have run this modified R code, you should have a density distribution graph which looks like the image at the top of the next page.

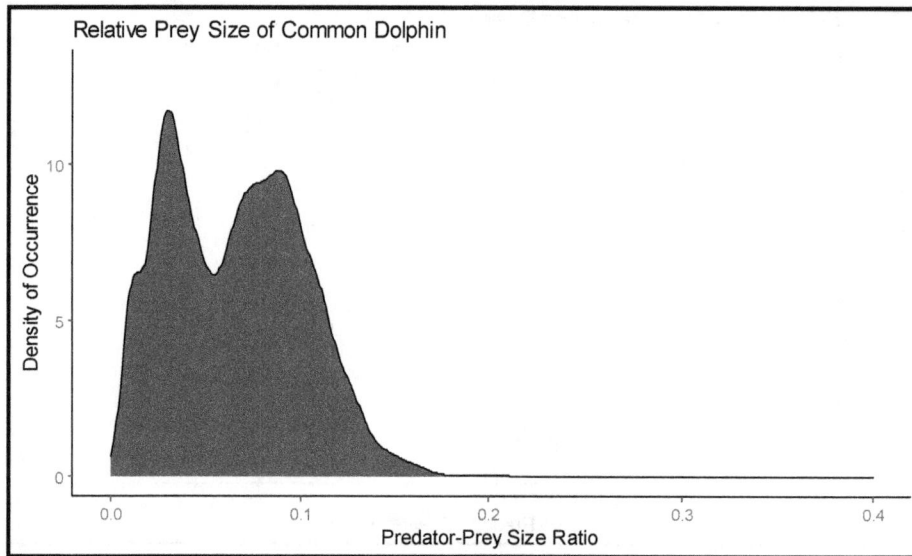

Relative Prey Size of Common Dolphin

EXERCISE 1.4: HOW TO CREATE GRAPHS THAT SHOW DATA FROM MORE THAN ONE DATA SERIES:

So far for the exercises in this chapter, you have been working with a single series of data. However, there will be many occasions where you wish to display more than one series of data at the same time, either on the same graph, or on separate graphs in a multi-panel figure. This is done in the same way regardless of the type of graph you are creating. However, the approach you need to use will vary depending on whether the data for the different data series are contained in separate columns in a single data set, in the same column in a single data set with information identifying the group the data from each row belongs to in a second column, or in separate data sets held in different R objects. You will learn how to create multi-series graphs from the first two types of data sets in this exercise.

In order to learn how to create figures which display data from multiple series of data, you will make density distribution graphs using data from four different species contained in the cetacean prey sizes data set which was used in Exercises 1.1 to 1.3. These species are the common dolphin, the harbour porpoise, the pilot whale and the sperm whale. These species were selected as they have markedly different density distributions of their predator-prey size ratios. You will start by creating a density distribution graph that shows the distributions of the PPSR data for both the common dolphin and harbour porpoises. In this case, the data for these two species are held in different columns within the same data set. To create this graph, work through the following flow diagram:

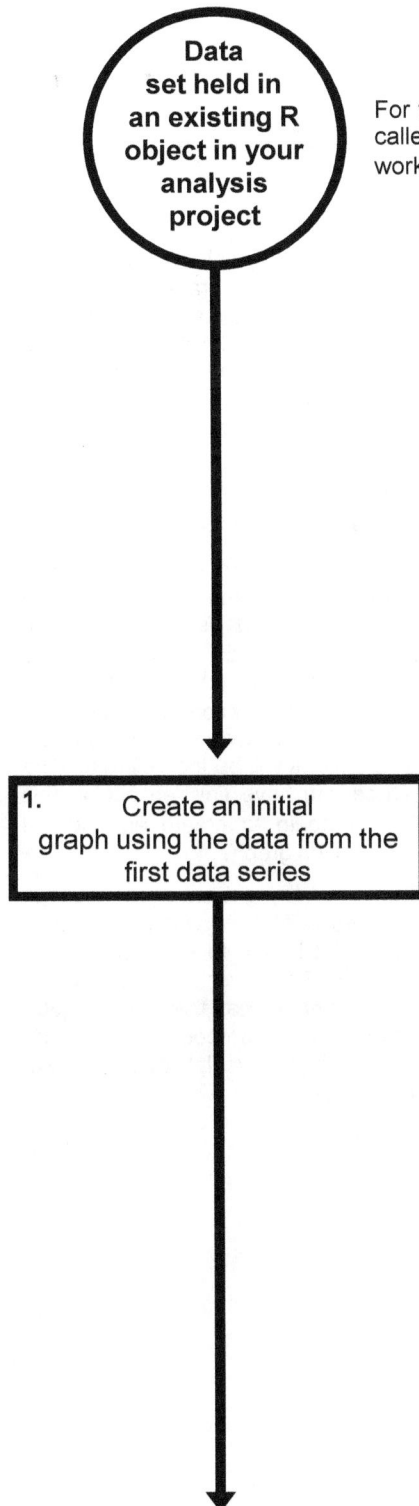

Data set held in an existing R object in your analysis project

For this example, the data set you will use is stored in the R object called `cetacean_prey_sizes` created in Exercise 1.1 of this workbook.

When displaying data from multiple data series on the same figure, the first step is to create an initial graph for a single data series. Additional data series can then be added to this initial graph to create one which displays multiple data series, or it can be used as a template for creating separate graphs for each series that can be brought together in a multi-panel figure. In this example, you will start by creating an initial density distribution graph which shows the density of records for the PPSRs of prey from common dolphin. To do this, enter the following block of code into R:

```
ggplot(data=cetacean_prey_sizes) +
geom_density(aes(x=common_dolphin))
```

This is CODE BLOCK 18 in the document R_CODE_DATA_VISUALISATION_WORKBOOK.DOC, and it contains two commands separated by a + symbol. These are the `ggplot` command and the `geom_density` command. The `ggplot` command sets the name of the R object which contains the data that will be used for the graph. This is done using the `data` argument and, in this case, it will be the R object called `cetacean_prey_sizes` created in step 2 of Exercise 1.1. **NOTE:** Unlike previous exercises in this chapter, the column of data to be plotted on the graph is not set in the `ggplot` command. Instead, it is set in the `geom_density` command (see below). This is required to allow multiple data series from different columns in the same R object to be plotted on the same graph.

1. Create an initial graph using the data from the first data series

The second command in this code block, `geom_density`, sets the type of graph that will be created from the data specified in the `ggplot` command. In this case, it will be a density distribution graph. Within this command, you will need to specify the column which contains the data for the first series you wish to plot on your graph. This is done using the `x` argument in the `aes` element of the `geom_density` command. In this case, it will be the column called `common_dolphin`.

NOTE: To make multi-series graphs from data held in different R objects rather than different columns in the same data set, you would need to move the `data` argument, as well as the `x` argument, from the `ggplot` command to the `aes` element of the individual graphing commands and specify the separate data sets to be used for each data series.

55

2. Add a second
data series to your initial
graph

Once you have created your initial graph, you can add a second data series to it by adding a new graphing command to the code block from step 1. In this example, the second data series you will add will be the PPSR data for harbour porpoise. To add this new data series to your graph, edit the block of code from step 1 so that it looks like this (the newly added graphing command is highlighted in **bold**):

```
ggplot(data=cetacean_prey_sizes) +
geom_density(aes(x=common_dolphin)) +
geom_density(aes(x=harbour_porpoise))
```

This is CODE BLOCK 19 in the document R_CODE_DATA_ VISUALISATION_WORKBOOK.DOC, and it adds a second `geom_density` command to the code used to create the initial graph in step 1. This is separated from the original code using a + symbol. Within this second `geom_density` command, the column containing the second series of data you wish to add to the graph is specified in the `x` argument of the `aes` element. In this case, it will be the column called `harbour_porpoise`. Once you have finished editing this code block, you can run it again to create an updated version of your density distribution graph.

NOTE: When you do this, you may get an error message saying *Removed 10303 rows containing non-finite values (stat_density)*. This is okay and is due to the fact that the harbour porpoise data have 10,303 fewer records than the common dolphin data. This means that there are 10,303 blank cells at the end of the harbour porpoise data column have been removed when doing the calculations required to make a density distribution graph.

When you add the second series of data to your graph, you will notice that it uses the same default colours as the first series. This makes it difficult to tell the two series apart. As a result, the next step you need to do is to customise the colours used for each series on your graph. This is done by adding two new arguments to each of the two graphing commands. These are `fill` and `col`. In addition, a new command, `scale_fill_manual`, is added which sets the colours used for the fill for each species. To do this for the graph being created in this example, edit the code from step 2 so that it looks like this (the newly added arguments and command are highlighted in **bold**):

```
ggplot(data=cetacean_prey_sizes) +
geom_density(aes(x=common_dolphin,fill=
   "Common dolphin"),col="black") +
geom_density(aes(x=harbour_porpoise,fill=
   "Harbour porpoise"),col="black") +
scale_fill_manual(values=c("blue","red"))
```

This is CODE BLOCK 20 in the document R_CODE_DATA_ VISUALATION_WORKBOOK.DOC. The new `col` arguments in the `geom_density` commands set the colours for the outlines of the each data series. In this case, the outline colour for both data series are set to `black`. The `fill` arguments are added to the `aes` element of the `geom_density` commands and they set the category labels for each data series that will be used on the legend of the graph. In this case, it sets the `fill` category label for the `common_dolphin` data to `Common dolphin`, and for the `harbour_porposie` data to `Harbour porpoise`.

Finally, the `scale_fill_manual` command is used to set the actual fill colours that will be used for the two data series. In this case, they are `blue` for the data defined in the first `geom_density` command (which is `common_ dolphin`), and `red` for data defined in the second `geom_density` command (which is `harbour_porpoise`). Once you have finished editing this code block, you can run it again to create an updated version of your density distribution graph.

3. Customise the colours used for each series on your graph

After you have set the data series so that they are displayed using different colours, you will notice that while you can now tell the two series apart, you may not be able to see the first data series because it is covered up by parts of the second one. In order to change this, you need to set the level of transparency that will be used for each series. This is done by adding an `alpha` argument to each of the two graphing commands. To do this for the graph being created in this example, edit the code from step 3 so that it looks like this (the newly added arguments are highlighted in **bold**):

```
ggplot(data=cetacean_prey_sizes) +
geom_density(aes(x=common_dolphin,fill=
"Common dolphin"),col="black",alpha=1) +
geom_density(aes(x=harbour_porpoise,fill=
"Harbour porpoise"),col="black",alpha=
0.75) + scale_fill_manual(values=
c("blue","red"))
```

4. Set the level of transparency to be used for each series on your graph

This is CODE BLOCK 21 in the document R_CODE_DATA_ VISUALISATION_WORKBOOK.DOC. The newly added `alpha` arguments can have a value between 0 (completely transparent) and 1 (completely opaque). For the first `geom_density` command (the one for the common dolphin data series), a value of 1 is used for the `alpha` argument. This will make it opaque. This is okay as it is the series that is displayed at the back of the graph. For the second `geom_density` command (the one for the harbour porpoise series), a value of 0.75 is used. This will make it semi-transparent, allowing you to see the distribution of the first data series behind it. Once you have finished editing this code block, you can run it again to create an updated version of your density distribution graph.

5. **Modify the legend that tells the reader which series is which**

After you have set the colours and levels of transparency for your data series, you need to modify the legend that tells the reader which data series is which. This is done by adding two new commands to the code block. These are a `theme` command and the `labs` command. To do this for the graph being created in this example, edit the code from step 4 so that it looks like this (the newly added commands are highlighted in **bold**):

```
ggplot(data=cetacean_prey_sizes) +
geom_density(aes(x=common_dolphin,fill=
"Common dolphin"),col="black",alpha=1) +
  geom_density(aes(x=harbour_porpoise,
  fill="Harbour porpoise"),col="black",
  alpha=0.75) + scale_fill_manual(values=
c("blue","red")) + theme(legend.position=
      c(0.85,0.85)) + labs(fill="Species")
```

This is CODE BLOCK 22 in the document R_CODE_DATA_ VISUALISATION_WORKBOOK.DOC. The newly added `theme` command sets the position of the legend on your plot area using the `legend.position` argument, while the `labs` command provides a title which will be used for the legend created by the `fill` arguments in each `geom_density` commands. Once you have finished editing this code block, you can run it again to create an updated version of your density distribution graph.

6. Customise how the other elements of your final graph will look

Once you have formatted how your data series will look and the legend that tells the reader which data series is which, you can modify the other elements of your graph as outlined in Exercise 1.2. To do this, edit the code from step 5 so that it looks like this (the newly added style arguments and commands are highlighted in **bold**):

```
ggplot(data=cetacean_prey_sizes) +
geom_density(aes(x=common_dolphin,fill=
"Common dolphin"),col="black",alpha=1) +
geom_density(aes(x=harbour_porpoise,fill=
"Harbour porpoise"),col="black",alpha=
0.75) + scale_fill_manual(values=
c("blue","red")) + theme(legend.position=
c(0.85,0.85)) + labs(fill="Species",
title="Relative Prey Size of Cetaceans",
x="Predator-Prey Size Ratio",y="Density of
Occurrence") + xlim(c(0.0,0.4)) +
theme(axis.line=(element_line(colour=
"black"))) + theme(panel.background=
(element_rect(fill="white")))
```

This is CODE BLOCK 23 in the document R_CODE_DATA_ VISUALISATION_WORKBOOK.DOC. This new block of code adds three new arguments to the existing `labs` command. The `title` argument provides the text to be used for the title at the top of your graph, the `x` argument provides the label for the X axis, and the `y` argument provides the label for the Y axis. If you wish any of these labels not to be present on your graph, use the term `""` in its associated argument in the `labs` command. In addition, it adds the `xlim` command which contain the arguments that set the lower and upper limits for the X axis. Finally, it also adds two new `theme` commands. The first of these includes the `axis.line` element that sets the colour for the axes to `black`, while the second contains the `panel.background` element that sets the fill colour for the plot area of the graph to `white`. Once you have finished editing this code block, you can run it again to create the final version of your multi-series density distribution graph.

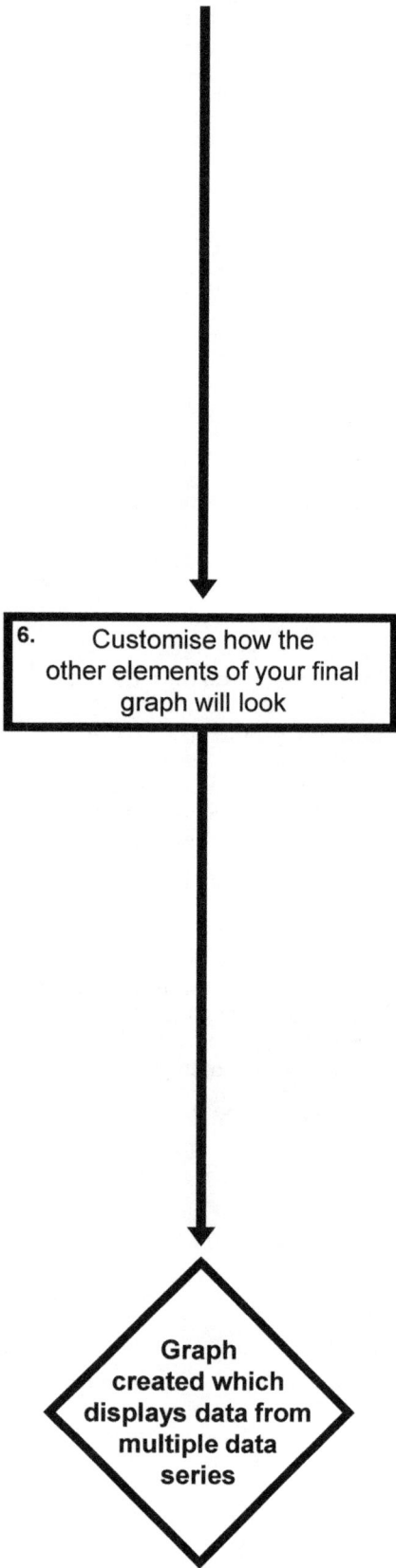

Graph created which displays data from multiple data series

At the end of the first part of this exercise, you should have a multi-series density distribution graph of cetacean predator-prey size ratios (PPSRs) that looks like this:

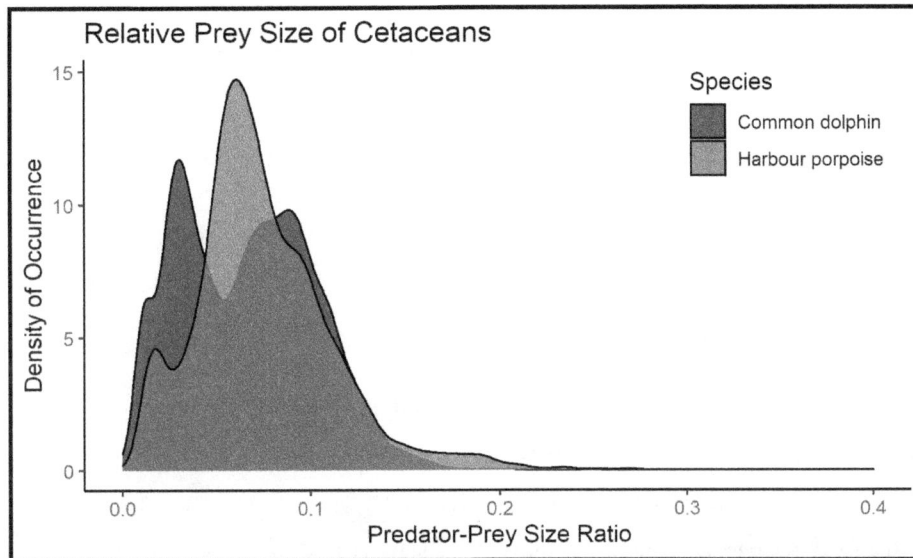

If you need to add more data series to the same graph, this can be done by adding additional graphing commands to your code block, and adding the colours you wish to use to display each additional data set to the `scale_fill_manual` command. To explore how to do this, you will add data for two new species to the above graph. These are the pilot whale and the sperm whale. To add the data for your first new species (the pilot whale) to the graph created by the block of code from step 6 of the above flow diagram, you will need to add a third `geom_density` command to it that specifies the information required to plot a data series for this species and add a new colour to the `scale_fill_manual` command to determine the colour that will be used to display it (in this case, green). To do this, edit the code block from step 6 of the above flow diagram (this is CODE BLOCK 23 in the document R_CODE_DATA_VISUALISATION_WORKBOOK.DOC) so that it looks like the code provided at the top of the next page and enter it into R (the newly added graphing command and other required modifications are highlighted in **bold**).

```
ggplot(data=cetacean_prey_sizes) + geom_density(aes(x=
  common_dolphin,fill="Common dolphin"),col="black",
   alpha=1) + geom_density(aes(x=harbour_porpoise,
  fill="Harbour porpoise"),col="black",alpha=0.75) +
  geom_density(aes(x=pilot_whale,fill="Pilot whale"),
  col="black",alpha=0.75) + scale_fill_manual(values=
   c("blue","red","green")) + theme(legend.position=
c(0.85,0.85)) + labs(fill="Species",title="Relative Prey
   Size of Cetaceans",x="Predator-Prey Size Ratio",
   y="Density of Occurrence") + xlim(c(0.0,0.4)) +
  theme(axis.line=(element_line(colour="black"))) +
  theme(panel.background=(element_rect(fill="white")))
```

When you run this modified block of code in R, it should produce a graph that looks like this (**NOTE:** This time you will get two warning messages about rows being removed, the first related to number of the blank cells in the `harbour_porpoise` data column and the second is related to the number of blank cells in the `pilot_whale` data column – as before, for this data set this is okay):

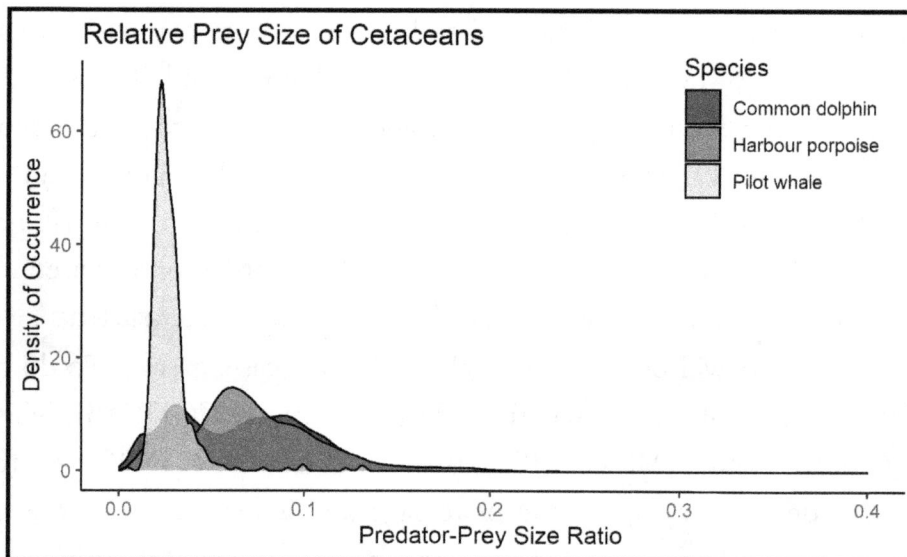

From this graph, you can see that the pilot whale has a very different density distribution of predator-prey size ratios than the other two species. Specifically, pilot whales consume a much narrower range of prey sizes relative to their body size than either common dolphin or harbour porpoises. While this results in a much higher peak density of predator-prey size ratios, this peak is not sufficiently strong that it would be inappropriate to display the density distributions of all three species on the same graph.

You can now add a fourth `geom_density` command to the code block for your multi-series graph to add a density distribution for a fourth, and final, species, the sperm whale. To do this, edit the above block of code so that it looks like this (the newly added graphing command and other required modifications are highlighted in **bold**).

```
ggplot(data=cetacean_prey_sizes) + geom_density(aes(x=
    common_dolphin,fill="Common dolphin"),col="black",
  alpha=1) + geom_density(aes(x=harbour_porpoise,fill=
    "Harbour porpoise"),col="black",alpha=0.75) +
geom_density(aes(x=pilot_whale,fill="Pilot whale"),col=
"black",alpha=0.75) + geom_density(aes(x=sperm_whale,fill=
    "Sperm whale"),col="black",alpha=0.75) +
scale_fill_manual(values=c("blue","red","green","yellow"))+
theme(legend.position=c(0.85,0.85)) + labs(fill="Species",
  title="Relative Prey Size of Cetaceans",x="Predator-Prey
Size Ratio",y="Density of Occurrence") + xlim(c(0.0,0.4)) +
    theme(axis.line=(element_line(colour="black"))) +
    theme(panel.background=(element_rect(fill="white")))
```

When you run this modified block of code in R, it should produce a graph that looks like this (**NOTE:** This time you will get three warning about rows being removed – this is okay):

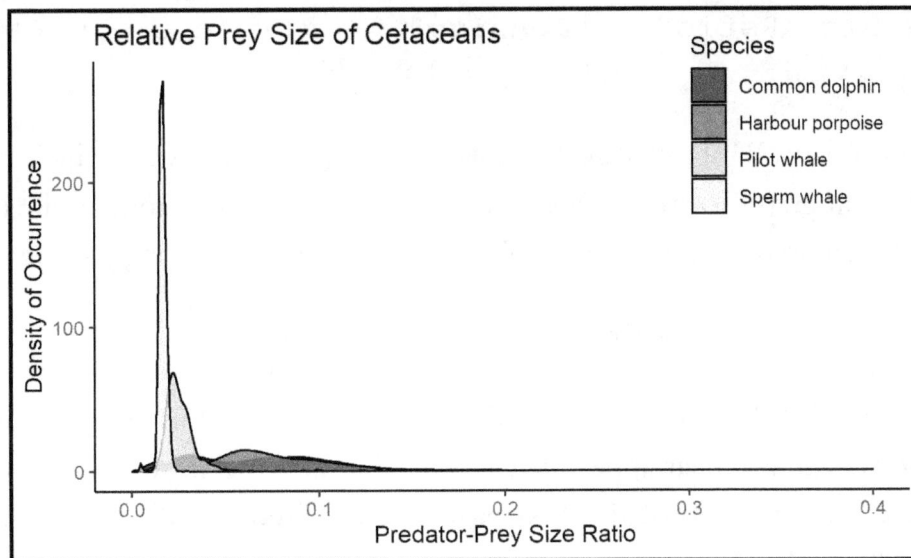

From this graph, you can see that the density distribution of predator-prey size ratios for the sperm whale is even narrower than that of the pilot whale. In fact, its density distribution is so narrow, and its peak density so high, that when it is plotted on the same graph as the other species, you can barely make out any information about the density distributions for common dolphin and harbour porpoise. As a result, it is probably inappropriate to try to show the density distributions for all four species on the same graph. Instead, if you wish to show the all four of these density distributions in the same figure, it would be better to create a multi-panel figure that includes separate graphs for each of the four species. In order to do this, you first need to create individual density distribution graphs for each species and store each one in its own R object. Once you have done this, you can use the ggarrange command from a package called ggpubr to bring them together into a single, multi-panel figure.

To explore how to use the ggarrange command to create a multi-panel figure featuring a number of different graphs, you can start by creating a density distribution graph for the common dolphin using a variation on the code used to create the density distribution graph for this species from the end of Exercise 1.3 (see page 53). To do this, enter the following block of code into R (the modifications required to the code from page 53 are highlighted in **bold – NOTE:** You will also need to remove the ylim command from this code block):

```
common_dolphin_graph <- ggplot(data=cetacean_prey_sizes,
aes(x=common_dolphin)) + geom_density(col="black",fill=
"blue",alpha=1)+ labs(title="Common Dolphin",x="Predator-
          Prey Size Ratio",y="Density") + xlim(c(0.0,0.4)) +
                        theme_classic()
```

This will create a new R object called common_dolphin_graph which contains the density distribution graph for this species. As a result, it will not automatically be displayed in R. Instead, if you wish to view it, you will need to enter the following code into R:

```
common_dolphin_graph
```

The above block of code can then be modified to create a density distribution graph for harbour porpoise with a similar appearance. To do this, edit the above code block so that it looks like the code at the top of the next page (the required modifications from the code used to create the common dolphin graph are highlighted in **bold**) and then enter it into R.

```
harbour_porpoise_graph <- ggplot(data=cetacean_prey_sizes,
    aes(x=harbour_porpoise)) + geom_density(col="black",
fill="red",alpha=1)+ labs(title="Harbour Porpoise",
       x="Predator-Prey Size Ratio",y="Density") +
           xlim(c(0.0,0.4)) + theme_classic()
```

The graph created by this block of code can be viewed by entering the following code into R:

```
harbour_porpoise_graph
```

This process can then be repeated again for the next species, which in this case is the pilot whale. To do this, edit the above block of code so that it looks like this (the required modifications are highlighted in **bold** – in this case, the code for viewing the resulting graph is included on a new line at the end of the code block to allow the resulting graph to be viewed automatically):

```
pilot_whale_graph <- ggplot(data=cetacean_prey_sizes,
    aes(x=pilot_whale)) + geom_density(col="black",
fill="green",alpha=1)+ labs(title="Pilot Whale",
       x="Predator-Prey Size Ratio",y="Density") +
           xlim(c(0.0,0.4)) + theme_classic()
                  pilot_whale_graph
```

This code block is then modified one last time to create the individual density distribution graph for the sperm whale. To do this, edit the above block of code so that it looks like this (the required modifications are highlighted in **bold**).

```
sperm_whale_graph <- ggplot(data=cetacean_prey_sizes,
    aes(x=sperm_whale)) + geom_density(col="black",
fill="yellow",alpha=1)+ labs(title="Sperm Whale",
       x="Predator-Prey Size Ratio",y="Density") +
           xlim(c(0.0,0.4)) + theme_classic()
                  sperm_whale_graph
```

After you have created the four individual graphs, you are ready to bring them together into a single multi-panel figure. To do this, first check if you already have the ggpubr package installed in your version of R by entering the code `library()`. If this package is not listed in the R PACKAGES AVAILABLE window that opens, enter the follow code into R:

```
install.packages("ggpubr")
library(ggpubr)
```

This block of code installs the `ggpubr` package into R and then loads the associated command library into your analysis project. If the `ggpubr` package is listed in the R PACKAGES AVAILABLE window, you do not need to re-install it by running the above `install.packages` command. Instead, you can simply run the `library(ggpubr)` command to load its command library into your project.

At this point, you are ready to use the `ggarrange` command from this package to bring the four graphs together into a single, multi-panel figure. To do this, enter the following command into R:

```
ggarrange(common_dolphin_graph,
harbour_porpoise_graph,pilot_whale_graph,sperm_whale_graph,
labels=c("A","B","C","D"),ncol=2,nrow=2)
```

In this command, the names of the R objects containing the four graphs to be included in the multi-panel figure (`common_dolphin_graph`, `harbour_porpoise_graph`, `pilot_whale_graph` and `sperm_whale_graph`) are listed first. Next, the `labels` argument is used to provide labels that will be used for each graph. In this case, these are A, B, C and D. Finally the `ncol` and `nrow` arguments are used to define how the four graphs will be arranged. In this case, they will be arranged in a matrix that is two columns wide (`ncol=2`) and two rows long (`nrow=2`). The multi-panel figure created by this command should look like the image at the top of the next page.

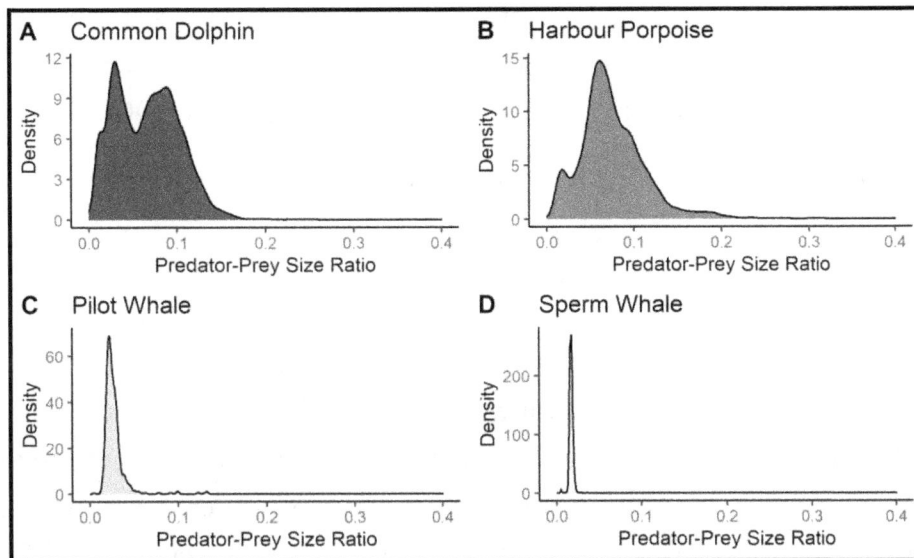

The above examples assume that the data for the different data series are held in different columns in the same data set. However, in many cases the data for each series will be held in a single column, with a second column providing information about which group each individual data point belongs to. In order to create a multi-series graph from such data, the commands used in the above flow diagram need to be modified to tell R that this is the case. To explore how you can do this, for the final part of this exercise, you will modify the code blocks used in the above flow diagram in order to make a multi-series graph from a data set that uses this format. Before you can do this, you will need to import a new data set using the following code block containing three commands, each of which needs to be entered on a separate line (this ensures that they are run one after another rather than all at the same time):

```
two_species_diets <- read.table(file=
"two_species_diets.csv",sep=",",as.is=TRUE,header=TRUE)
              names(two_species_diets)
              View(two_species_diets)
```

This imports a data set called `two_species_diets.csv` into R as an object called `two_species_diets`. This data set contains two columns, one called `ppsr` which contains the predator-prey size ratio for each individual prey item recorded from the stomach contents of two species of cetacean, and one called `species` which tells you whether each individual prey item was recorded in a common dolphin or a harbour porpoise. Next, you will need to modify the code used to create the basic density distribution graph from step 1 of the above flow diagram (this is CODE BLOCK 18 in the document R_CODE_ DATA_VISUALISATION_WORKBOOK.DOC) so that it looks like this (the required modifications are highlighted in **bold**):

```
ggplot(data=two_species_diets) + geom_density(aes(x=ppsr,
                      fill=species))
```

As before, in this block of code, the `ggplot` command is used to specify which data set contains the information that will be plotted on the resulting frequency distribution graph (using the `data` argument), and the `geom_density` command is used to set which data will be displayed on the X axis (using the `x` argument). However, a second argument also needs to be added to the `geom_density` command at this stage to tell R which data belong to which data series. This is the `fill` argument, and in this case, it tells R that the column called `species` contains the group identifier for each row of data in the `ppsr` column. This will create a density distribution graph with both series of data displayed on it. Once you have obtained this graph, you can then customise how your final multi-series graph will look by modifying the above code in a similar way to what you did in steps 3 to 6 of the above flow diagram. For this example, you will use the same settings that were used for the final graph produced by the working through the flow diagram for this exercise. To do this, edit the above code so that it looks like this (newly added style arguments and commands are highlighted in **bold** – these additions can be copied and pasted from the end of CODE BLOCK 23 from the document R_CODE_DATA_VISUALISATION_ WORKBOOK.DOC used in Step 6 of the above flow diagram):

```
ggplot(data=two_species_diets) + geom_density(aes(x=ppsr,
   fill=species),alpha=0.75) + scale_fill_manual(values=
 c("blue","red")) + theme(legend.position=c(0.85,0.85)) +
     labs(fill="Species",title="Relative Prey Size of
   Cetaceans",x="Predator-Prey Size Ratio",y="Density of
   Occurrence") + xlim(c(0.0,0.4)) + theme(axis.line=
 (element_line(colour="black"))) + theme(panel.background=
             (element_rect(fill="white")))
```

Once you have run this new version of the R code for creating a multi-series density distribution graph from data held in a single column, you should have a graph that looks like this:

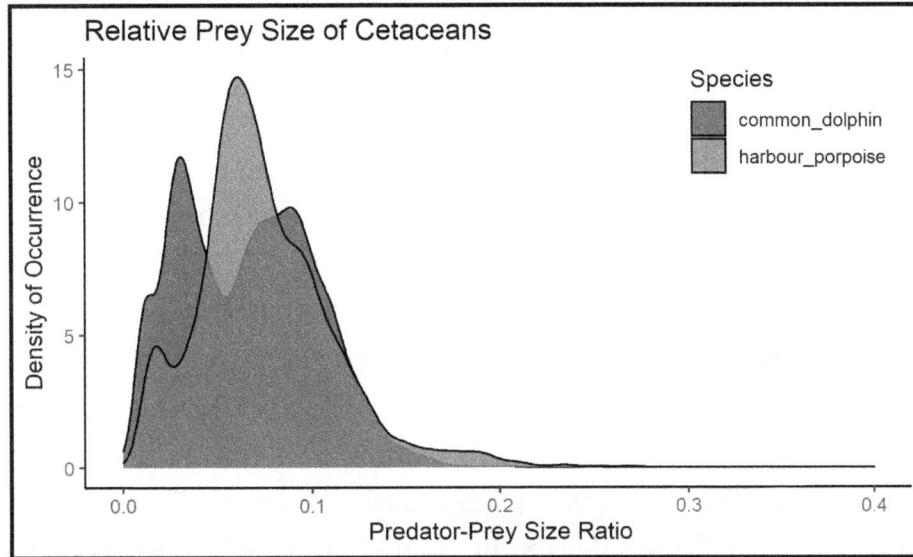

If you examine this graph, you will see that while it is very similar to the one generated from working through the above flow diagram, it differs in one key way. This is the names given to the two data series in the legend. In the flow diagram, these names were manually set using the `fill` argument in the `aes` element of the `geom_density` command for each species. However, when creating a graph where the data for multiple data series are in the same column, you cannot do this. This is because you need to use this argument to specify the column which contains the groupings that identify which data series each row of data belongs to. This means that the names provided on the legend for each data series are taken directly from the group names in the column set in this `fill` argument. If you wish to manually set the data series names for a graph created from a data set with this structure, you need to add a `labels` argument to the `scale_fill_manual` command. For the above graph, you could add the argument `labels=c("Common dolphin", "Harbour porpoise")` to this command to modify the labels used for the two data series on the legend. This would then create a graph that is identical to the one you created when working through the flow diagram for this exercise.

Chapter Four

How To Create Graphs Displaying Groups Of Data With GGPlot

Once you have learned the basics of using the various commands in the GGPlot package by making and customising a range of different types of frequency distribution graphs, you can then explore how to use this knowledge to make other types of graphs. In this chapter, you will learn how to make graphs which compare data from different groups, rather than looking at the distribution of a single continuous variable. As part of these exercises, you will learn how to use R to divide a continuous variable into categories which can then be used to create graphs such as bar graphs. In addition, you will learn how to generate tables of summary statistics, such as means and standard deviations, for the data from these categories that can be used as the values to be plotted on these graphs. The instructions for the exercises in this chapter will contain all the steps required to import and process a data set before creating a graph from it, as well as how to customise the look of your final graph. While this approach of providing complete instructions for each type of graph does result in some repetition of the initial steps between exercises (such as those for setting the WORKING DIRECTORY, importing data and checking they have been loaded into R correctly), it is designed to make it easier to apply the same workflows to your own data. When working through the exercises in this chapter, you can avoid having to repeat these initial steps, if you are confident that you have already completed them for a previous exercise, by skipping ahead to the later ones.

If you have not already done so, before you start the exercises in this chapter, you first need to create a WORKING DIRECTORY folder on your computer and load the necessary data into it (**NOTE:** If you have already created this folder and downloaded data for a previous chapter in this workbook, you do not need to do this again). To do this on a computer with a Windows operating system, open Windows Explorer and navigate to the location where you would like to create the folder (such as your C:\ drive or your DOCUMENTS folder). Next, right click anywhere in this location and select NEW> FOLDER. Now call this folder STATS_FOR_BIOLOGISTS_TWO by typing this into the folder name section to replace what it is currently called (which will most likely be NEW FOLDER). To create a WORKING DIRECTORY folder on a computer running a Mac operating system, open Finder and navigate to the location where you would like to create the folder (such as your DOCUMENTS folder or your DESKTOP). Next, click on FILE> NEW FOLDER, and then type the name STATS_FOR_BIOLOGISTS_TWO before pressing the ENTER key on your keyboard.

Once you have created your WORKING DIRECTORY folder, you are ready to download the data sets you will use for the exercises in this workbook from *www.gisinecology.com/ stats-for-biologists-2*. After you have downloaded the compressed folder containing the required data by following the instructions provided on that page, you need to extract all the data files from it and copy them into the folder called STATS_FOR_BIOLOGISTS_TWO that you have just created.

Next, you need to check that the required data have been extracted to the correct folder. If you are using a computer with a Windows operating system, you can use Windows Explorer to open your newly created WORKING DIRECTORY folder and examine its contents. If all the files from the compressed folder are present in it (there should be a total of 90 of them), you can click on the folder icon at the left hand end of the ADDRESS BAR at the top of the WINDOWS EXPLORER window to reveal its full address. Write this address down as you will need it to set this folder as your WORKING DIRECTORY during the exercises provided in this workbook (see pages 12 and 13 for details of how to modify folder addresses so they will be recognised by R).

If you are using a computer with a Mac operating system, you can use Finder to open your newly created WORKING DIRECTORY folder and examine its contents. If all the required data files are present in it (there should be a total of 90 of them), press the CMD and I keys on your keyboard at the same time. This will open the GET INFO window where you will find its address (which is also called the pathway). Write this address down somewhere as you will need it to set this folder as your WORKING DIRECTORY during the exercises provided in this workbook (see pages 12 and 13 for details of how to modify folder addresses so they will be recognised by R).

After you have loaded the required data into your WORKING DIRECTORY folder, you can open RGUI or RStudio, depending on which option you wish to use (see Chapter 2 for more details). Once you have opened your preferred R user interface, you need to create a file called CHAPTER_FOUR_EXERCISES where you will save the results of your analyses from your R CONSOLE window as you work through this chapter. To do this using RGUI, click on the FILE menu and select SAVE WORKSPACE. To do this in RStudio, click on SESSION and select SAVE WORKSPACE AS. In both cases, save it as a WORKSPACE file with the name CHAPTER_FOUR_EXERCISES.RDATA in your WORKING DIRECTORY folder (this will be the one called STATS_FOR_BIOLOGISTS_TWO that you have just created). If you are using RStudio, you will also want to save the contents of your SCRIPT EDITOR window (where you will enter and edit the R code you will use to carry out specific commands). To do this, click on the FILE menu and select SAVE AS. Save your file as an R SCRIPT file with the name CHAPTER_FOUR_EXERCISES.R in your WORKING DIRECTORY folder. As you work through the exercises in this chapter, remember to regularly save the contents of your R CONSOLE window (which will contain the R objects you have created up to that point) to your WORKSPACE file and, if you are using RStudio, the contents of your SCRIPT EDITOR window to your R SCRIPT file.

Finally, you need to remove any data that are currently held in R's temporary memory. To do this, enter the following command into R:

```
rm(list=ls())
```

If you are using RGUI, you can simply type this code after the command prompt at the bottom of the R CONSOLE window (it looks like this: >) and then press the ENTER key on your keyboard to run it. If you are using RStudio, you can type this command into the SCRIPT EDITOR window (the upper left hand window). To run this command, select it and then click on the RUN button at the top of this window. This will run it in the R CONSOLE window (the lower left hand one in the main RStudio user interface). You are now ready to start the exercises in this chapter.

EXERCISE 2.1: HOW TO CLASSIFY DATA INTO GROUPS AND CREATE A BAR GRAPH BASED ON THEM:

Bar graphs differ from histograms in one key way. This is that they display information about groups of data rather than for a continuous variable. This means that the data may need to be divided into groups before they can be used to make a bar graph. In this exercise, you will learn how to use R to divide a continuous variable into categories and then use these categories to create a bar graph showing the number of records within a data set which fall into each one. In order to be able to do this, you need to have your data arranged in a spreadsheet or table where each row contains data from a single record in your data set. In this table, there also needs to be a column which contains the values for the continuous variable you wish to divide into categories and then create a bar graph from.

For this exercise, you will start by dividing data on the prey size preferences of a cetacean species, the common dolphin, into two groups before creating a bar graph based on the counts of data in each one. This will be done using a new data set called `cetacean_prey_sizes_2.csv`. While this contains the same data that were used for the exercises in chapter 3, it differs in its structure. Specifically, all the prey size data are contained in a single column (called `ppsr`), while a second column (called `species`) provides information about which species each individual prey item was recorded from. As a result, before you can start processing the data for common dolphin, you will first need to create a new data set that contains only these data. To process the common dolphin data and create a bar graph based on prey size categories from it, work through the flow diagram that starts at the top of the next page.

NOTE: If you have not already done so for an earlier exercise, you will need to download the `ggplot2` package and install it in your version of R before you start working through the instructions in this flow diagram. If you do not do this, the `library` command used to load the command library from this package into your analysis project in step 6 of this flow diagram will not work. You can check if you already have the `ggplot2` package installed in your version of R using the command `library()`. If you find you need to install the this package, this can be done by entering the following command into R:

```
install.packages("ggplot2")
```

Data set held in a comma separated values (.CSV) file

For this example, the data set you will use is stored in a file called `cetacean_prey_sizes_2.csv` that is located in the WORKING DIRECTORY folder you created during the introduction to this chapter.

Before you start any analysis in R, you first need to set the WORKING DIRECTORY. To do this, enter the text `setwd("` and then type the address of your WORKING DIRECTORY, using slashes (/) as the folder separators, before entering a second quotation mark followed by a closing bracket, like this `")`. For example, if your WORKING DIRECTORY has the address C:\STATS_FOR_BIOLOGISTS_TWO, your `setwd` command should look like this:

```
setwd("C:/STATS_FOR_BIOLOGISTS_TWO")
```

1. Set the WORKING DIRECTORY for your analysis project

If you are using RGUI, enter your `setwd` command in the R CONSOLE window (remembering to use the address of your own WORKING DIRECTORY folder in it) and then press the ENTER key on your keyboard. If you are using RStudio, enter your `setwd` command into the SCRIPT EDITOR window. To run it, select it and then click on the RUN button at the top of this window. You will enter all the remaining commands for this exercise in a similar manner, depending on the user interface you are using.

To check that your WORKING DIRECTORY has been set properly, enter the command `getwd()` and carefully check that the address it returns is the same as the one for the STATS_FOR_BIOLOGISTS_TWO folder you created at the start of this chapter.

Before you move on to step 2, make sure that all the data you wish to use in your analysis project are located in this WORKING DIRECTORY folder. In this case, this is a file called `cetacean_prey_sizes_2.csv`. **NOTE:** If the data you are going to import into R in step 2 are not located in the WORKING DIRECTORY you set in this step, the import code provided in the next step will not work.

74

```
2.    Load your data into
R using the read.table
        command
```

The `read.table` command provides the easiest way to load data held in a .CSV file (and stored in the WORKING DIRECTORY you set in step 1) into R so you can analyse it. To do this for the data set being used in this example, enter the following command into R:

```
cetacean_prey_sizes_2 <-
read.table(file="cetacean_prey_sizes_2.csv",
    sep=",",as.is=FALSE,header=TRUE)
```

This code has to be entered exactly as it is written here or it will not work. If you wish to use the copy-and-paste approach for entering this command, copy the text directly below CODE BLOCK 24 in the document R_CODE_DATA_VISUALISATION_WORKBOOK.DOC and paste it into R.

This command will create a new object in R called `cetacean_prey_sizes_2` which will contain the data from the specified .CSV file. To import a different .CSV file into R, all you need to do is change the file name in the `file` argument to the name of the one you wish to import. You can also use whatever name you wish for the R object which will be created by this command. To do this, simply replace `cetacean_prey_sizes_2` at the start of the first line of the above code with the name you wish to use for it. **NOTE:** If your .CSV data set uses a semicolon as the column separator, you would need to replace the `sep=","` argument with `sep=";"`.

```
3.    Check the data have
loaded into R correctly by
checking the names of the
columns and by viewing it
```

Whenever you import any data into R you need to check that they have loaded correctly. First, you need to check that all the required columns are present in the R object you just created. To do this, enter the following command into R:

```
names(cetacean_prey_sizes_2)
```

This is CODE BLOCK 25 in the document R_CODE_DATA_VISUALISATION_WORKBOOK.DOC. This command will return the names used for each column in the R object you just created. For this example, the names should be: `record_no`, `ppsr` and `species`.

Next, you should view the contents of the whole table using the `View` command. This is done by entering following code into R:

```
View(cetacean_prey_sizes_2)
```

This is CODE BLOCK 26 in the document R_CODE_DATA_VISUALISATION_WORKBOOK.DOC. This command will open a DATA VIEWER window where you can examine your data set and check that the correct data have been loaded into R.

4. Create a new R object containing just the subset of data you wish to create a bar graph from

In the data set being used in this example, there are data from a total of four cetacean species (the common dolphin, the harbour porpoise, the pilot whale and the sperm whale). However, for this example, you only wish to create a bar graph from the common dolphin data. This means that before you can do anything else, you will need to create a new R object which contains only the data from common dolphin. In R, this is done using the `subset` command. To do this for the data set being used in this example, enter the following code into R:

```
common_dolphin_ppsr_data <-
subset(cetacean_prey_sizes_2,species==
          "common_dolphin")
```

This is CODE BLOCK 27 in the document R_CODE_DATA_ VISUALISATION_WORKBOOK.DOC. This command will create a new R object called `common_dolphin_ppsr_ data` which will only contain the rows from the `cetacean_ prey_sizes_2` data set that have the text `common_ dolphin` in the `species` column.

Next, you should view the contents of your new table using the `View` command. This is done by entering following code into R:

```
View(common_dolphin_ppsr_data)
```

This is CODE BLOCK 28 in the document R_CODE_DATA_ VISUALISATION_WORKBOOK.DOC. This command will open a DATA VIEWER window where you can examine your data set and check that it contains the required subset of data. In this example, it should only contain data from common dolphin and not from any of the other species contained in the original data set.

5. Create a new column which contains the categories that will be used for the X axis of your bar graph

In this example, you will use the `ifelse` command to create your new column containing the categories which will be used for the X axis of your bar graph. This command is found in the `dplyr` package. This means that you will first need to check if you already have this package installed in your version of R by entering the code `library()`. If this package is not listed in the R PACKAGES AVAILABLE window that opens, enter the follow code into R:

```
install.packages("dplyr")
```

This is CODE BLOCK 29 in the document R_CODE_DATA_ VISUALISATION_WORKBOOK.DOC. When you run this command, follow any additional instructions that appear on your screen in order to download and install the specified package. If the `dplyr` package is listed in the R PACKAGES AVAILABLE window, you do not need to re-install it by running the above command.

Once you have ensured that the `dplyr` package is installed in your version of R, you then need to load its command library into your analysis project. This is done by entering following code into R:

```
library(dplyr)
```

This is CODE BLOCK 30 in the document R_CODE_DATA_ VISUALISATION_WORKBOOK.DOC.

Now that the required `dplyr` command library has been loaded into your analysis project, you can use the `ifelse` command to create a new column with two categories in it which will then be used as the groups for the X axis of your bar graph. In this command, you provide a logical statement based on a specific column in your data set. You also need to provide two values, one to be entered in the new column for rows where this statement is true, and a second to be used for rows where it is not. To do this for the data set being used in this example, enter the following command into R:

```
common_dolphin_ppsr_data$ppsr_cat=
ifelse(common_dolphin_ppsr_data$ppsr<0.05,
'1','2')
```

This is CODE BLOCK 31 in the document R_CODE_DATA_ VISUALISATION_WORKBOOK.DOC. This command creates a new column called `ppsr_cat` in the data set called `common_dolphin_ppsr_data`. This new column is then filled with values based on the logical statement in the `ifelse` command. In this case, this logical statement is `common_dolphin_ppsr_data$ppsr<0.05`. If this statement is true for any given row in the data set called `common_dolphin_ppsr_data` (i.e. if the value in the `ppsr` column is `<0.05`), it will put a value of `1` (the first value after the `ifelse` logical statement) in the new `ppsr_cat` column. If it is not true (i.e. if the value is equal to or greater than 0.05), the new column for that row will be filled with a value of `2` (the second value after the `ifelse` statement).

6. Load the command library for `ggplot2` package into your analysis project

Before you can make a bar graph based on the categories you have just created, you need to load the `ggplot2` command library into your analysis project. To do this, enter the following command into R:

```
library(ggplot2)
```

This is CODE BLOCK 32 in the document R_CODE_DATA_VISUALISATION_WORKBOOK.DOC.

After you have successfully loaded the `ggplot2` command library into your analysis project, you are ready to use it to create your bar graph based on the count of records in each category created in step 5. To do this, enter the following block of code into R:

```
ggplot(data=common_dolphin_ppsr_data,
aes(x=ppsr_cat)) + geom_bar(stat="count") +
scale_x_discrete(labels=c("Small","Large"))
```

This is CODE BLOCK 33 in the document R_CODE_DATA_VISUALISATION_WORKBOOK.DOC, and it contains three commands separated by + symbols. These are the `ggplot` command, the `geom_bar` command, and the `scale_x_discrete` command. The `ggplot` command sets the data set which will be used for the graph. This is done using the `data` argument and, in this case, it will be the R object called `common_dolphin_ppsr_data` created in step 4 of this exercise. The column of data that will be plotted on the X axis of the resulting graph is set using the `x` argument of the `aes` element of this `ggplot` command. In this case, it is the column called `ppsr_cat` in the `common_dolphin_ppsr_data` data set created in step 5.

7. Create your initial bar graph based on the count of records in each category created in step 5

The second command in this code block, `geom_bar`, sets the type of graph that will be created from the data specified in the `ggplot` command. In this case, it will be a bar graph. Within this command, the values to be plotted on the Y axis are set by the `stat` argument. In this case, the argument used is `stat="count"`, meaning that the height of the bar for each category of the X axis will be the count of the number of rows in that category in the data set defined in the `data` argument of the `ggplot` command (in this case, `common_dolphin_ppsr_data`).

The third command is `scale_x_discrete`. This command sets the labels that will be used for each group on the X axis of the resulting bar graph. Without this command, each group would be labelled with the values from the `ppsr_cat` column, in this case, `1` and `2`. However, It is much better to include this command to provide more descriptive labels. In this case, the groups will be labelled as `small` (for the group with a value of `1` in the `ppsr_cat` column) and `large` (for the group with a value of `2` in this column).

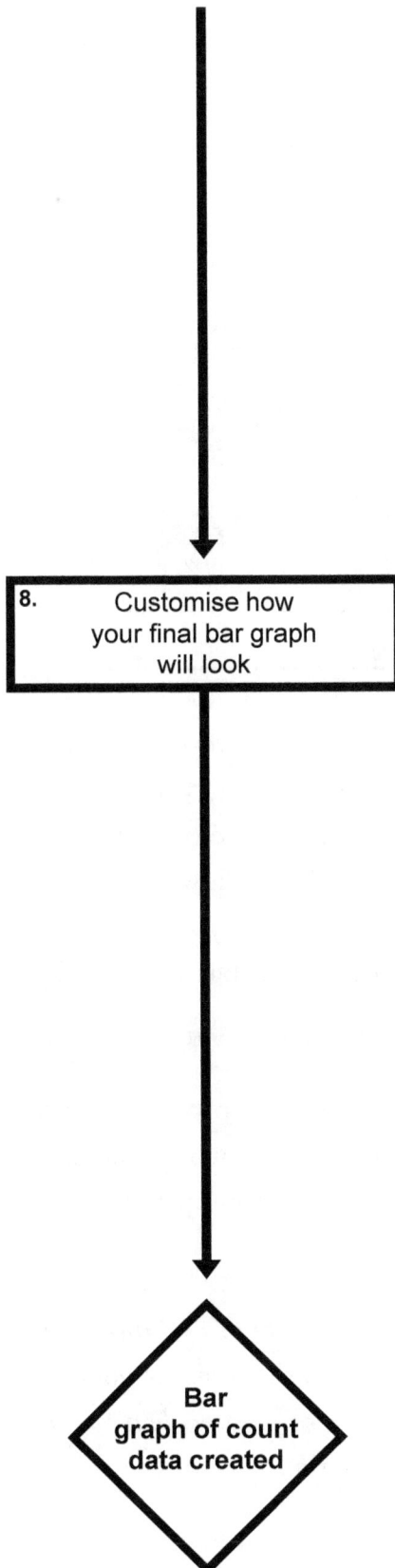

Once you have created your initial bar graph, and you are happy with how the categories are displayed on it, you can customise how your final graph will look. This can be done by adding a number of new style commands to the block of code used to produce it. To do this, edit the code from step 7 so that it looks like this (the newly added style commands are highlighted in **bold**):

```
ggplot(data=common_dolphin_ppsr_data,
aes(x=ppsr_cat)) + geom_bar(stat="count") +
    scale_x_discrete(labels=c("Small",
"Large")) + labs(title="Relative Prey Size
  of Common Dolphin",x="Predator-Prey Size
  Ratio Category",y="Number of Prey Items") +
    ylim(c(0,10000)) + theme_classic()
```

This is CODE BLOCK 34 in the document R_CODE_DATA_ VISUALISATION_WORKBOOK.DOC, and it adds three new style commands separated by + symbols to the code block used to create your initial bar graph in step 7. These are the `labs` command, the `ylim` command and the `theme_ classic` command. The `labs` command sets the labels that will be used tor the title of the graph (using the `title` argument), the X axis (using the `x` argument) and the Y axis (using the `y` argument). The `ylim` command sets the minimum and maximum values that will be displayed on the Y axis of the graph. In this case, these will be `0` for the minimum value and `10,000` for the maximum value. Finally, the `theme_classic` command sets the remaining style elements of the final graph to those of the pre-existing classic theme. Once you have finished editing this code block, you can run it again to create the final version of your bar graph.

8. Customise how your final bar graph will look

Bar graph of count data created

At the end of the first part of this exercise, you should have a bar graph of count data that looks like this:

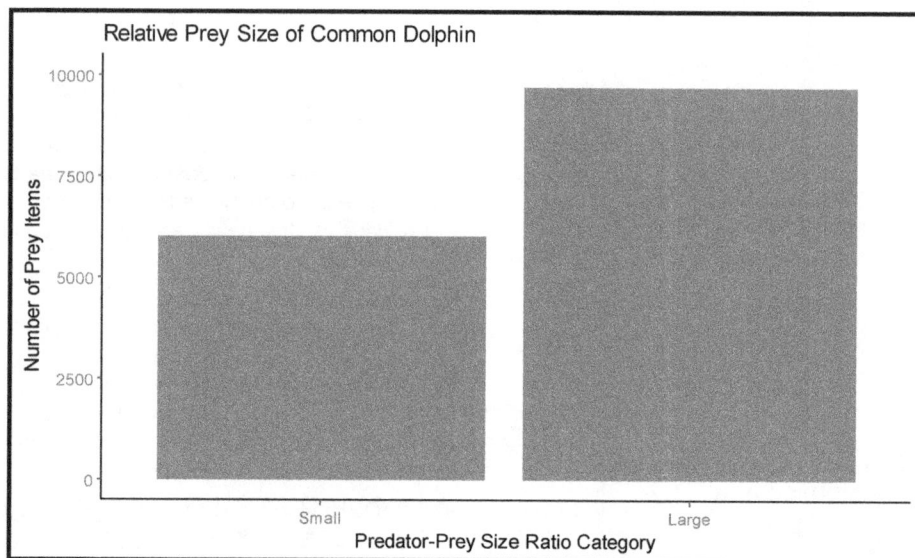

Relative Prey Size of Common Dolphin

Once you have created a bar graph of count data, you can export it from R so that you can include it in a manuscript or presentation. If you are using RGUI, you can do this by clicking on the R GRAPHICS window containing your bar graph to select it, before clicking on FILE on the main menu bar and selecting SAVE AS. This will allow you to save it in a variety of different formats. If you are using RStudio, you can export your graph by clicking on the EXPORT button at the top of the window displaying your bar graph and selecting SAVE AS IMAGE.

When including a bar graph in a manuscript, it is important that you provide an appropriate figure legend for it. This legend should provide all the information required for the reader to interpret the contents of the graph. For the above bar graph, an appropriate legend would be:

Figure 1: The number of prey recorded in the stomach contents of common dolphin (Delphinus sp) in two different relative prey size categories. Small: Prey less than 5% of the dolphin's body length (PPSR < 0.05); Large: Prey equal to or greater than 5% of the dolphin's body length (PPSR => 0.05).

When processing your data to prepare them for making a bar graph, there are two main commands that are used to divide the data into different categories. These are the `ifelse` command and the `case_when` command. Both of these commands are contained in the `dplyr` package. The main difference between these two commands is the number of categories that the data can be divided into. When you use the `ifelse` command, you can only divide your data into two categories. In contrast, when you use the `case_when` command, you can divide your data into as many different categories as you like. In both cases, you would use the workflow outlined above to categorise your data and create a bar graph from them, but you would vary the categorisation command used in step 5. However, regardless of the categorisation command you use, it is usually best to use a numerical code for your categories when using such commands, and then add any text labels to the categories when you create your bar graph using the `scale_x_discrete` command. This is because when you create a bar graph with the GGPlot `geom_bar` command, by default the categories are displayed on the X axis in numerical order (for numeric categories) or in alphabetical order (for text categories). As a result, if you use a numerical code it is much easier to create a graph with the categories in the required order on the X axis.

To explore how you can use different commands to divide your data into categories which can then be used to create bar graphs, in the next part of this exercise you will modify the categorisation command in used in step 5 before creating a new bar graph from the resulting data set. Firstly, you will divide the common dolphin prey size data into four categories, rather than just two as you did in the original flow diagram. This will be done by using a `case_when` command to create a new column called `ppsr_cat_2` with the new categories in it. As with the `ifelse` command, the `case_when` command contains a series of logical statements that are used to define the different categories you wish to apply to the specified column of data in your data set. To do this, edit the categorisation code from step 5 of the above flow diagram (this is CODE BLOCK 31 in the document R_CODE_DATA_VISUALISATION_WORKBOOK.DOC) so that it looks like this (the required modifications are highlighted in **bold**):

```
common_dolphin_ppsr_data$ppsr_cat_2=
case_when(common_dolphin_ppsr_data$ppsr <0.01~'1',
      common_dolphin_ppsr_data$ppsr<0.05~'2',
    common_dolphin_ppsr_data$ppsr<0.10~'3',TRUE~'4')
```

You can modify this code either by editing it in the R CONSOLE window of RGUI or through the SCRIPT EDITOR window of RStudio (depending on which interface you are using). If you are entering commands directly into the R CONSOLE window, you can use the UP arrow on your keyboard to bring commands and code blocks you have previously run during the same session back on to the command line of this window, and then use the LEFT and RIGHT arrows to scroll through and edit them. In this case, use the UP arrow to bring the previous version of the code block used to categorise your data back onto the command line and edit it so that it looks like the one above. Once you have finished modifying this code block, you can run it by pressing the ENTER key on your keyboard. If you are using RStudio, you can copy and paste the original code block in the SCRIPT EDITOR window before editing the new version to include the required modifications. Once you have done this, select the modified version of the code block and click on the RUN button to run it in the R CONSOLE window.

After you have run this new version of the R code for categorising the common dolphin prey size data, you need to view your data set to ensure the categorisation has been carried out correctly. To do this, enter the following command into R:

```
View(common_dolphin_ppsr_data)
```

This will open a DATA VIEWER window where you can examine the data set to ensure that your data have been categorised correctly. In this case, there should be four categories, labelled 1 to 4, based on the logical statements provided in the case_when command. Once you are happy that the categorisation has been carried out to your required specifications, you can then make a new bar graph based on these new categories. This is done by modifying the column specified in the x argument in the aes element of the ggplot command and two of the style commands in the final block of code from step 8 of the above flow diagram (this is CODE BLOCK 34 in the document R_CODE_DATA_ VISUALISATION_WORKBOOK.DOC). For this example, to make new bar graph based on the four categories in the column called ppsr_cat_2 that you have just created, edit the code from this step so that it looks like the code provided at the top of the next page (the required modifications are highlighted in **bold**).

```
ggplot(data=common_dolphin_ppsr_data,aes(x=ppsr_cat_2)) +
geom_bar(stat="count") + scale_x_discrete(labels=c("Very
Small","Small","Medium","Large")) + labs(title="Relative
Prey Size of Common Dolphin",x="Predator-Prey Size Ratio
  Category",y="Number of Prey Items") + ylim(c(0,7500)) +
                     theme_classic()
```

This code contains three modifications. Firstly, the `x` argument in the `aes` element of the `ggplot` command is now set to `ppsr_cat_2` (rather than `ppsr_cat`, as it was originally). Secondly, the labels argument in the `scale_x_discrete` command has been modified to provide names for four categories (`Very Small`, `Small`, `Medium` and `Large`) rather than two, as was the case previously. Thirdly, the maximum value within the `ylim` command has been changed from `10000` to `7500` to account for the fact that the value for the tallest bar on the resulting graph will be smaller. The graph produced when you run this modified R code should look like this:

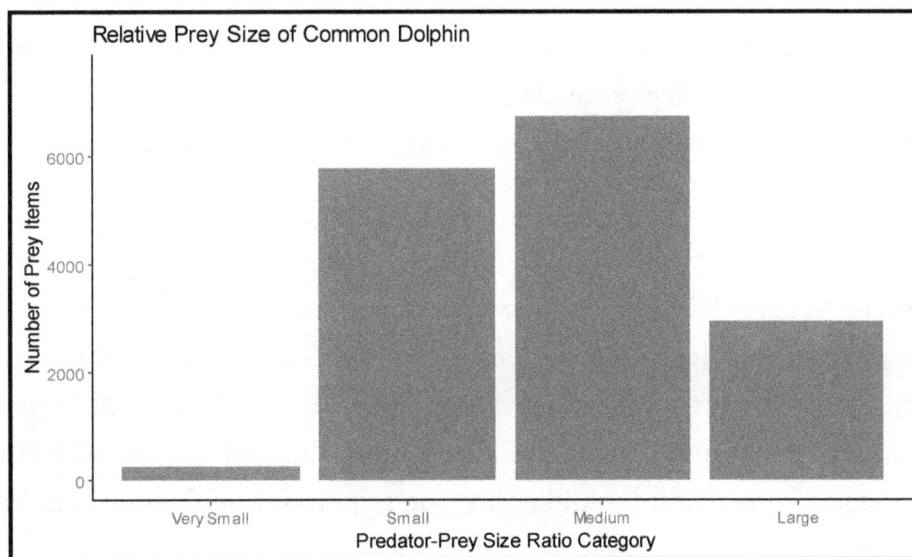

On this graph, the data for the four categories created from the logical statements provided in the `case_when` command have been plotted in numerical order (from left to right, these are 1 to 4), and then labelled with the labels from the `scale_x_discrete` command in the code used to create the graph itself. However, you can also order the bars on your bar graph based on their size, with the tallest bar at the left hand end of the X axis and the shortest one on the right hand end.

The easiest way to do this is to first create a bar graph in the way outlined above. Once you have done this, you can identify which category has the tallest bar, which has the next, and so on. You can then use this information to modify the categorisation command used step 5 to ensure that the bars are plotted based on their relative heights rather than based on the prey size categories they represent. This can be demonstrated by considering the last bar graph that you created (see page 83). On this graph, the category with the tallest bar is Medium, followed by Small, Large and Very Small. This information can then be used to create a new column (called `ppser_cat_3`) containing a new set of categories by editing the categorisation command from page 81 so that it looks like this (the required modifications are highlighted in **bold**):

```
common_dolphin_ppsr_data$ppsr_cat_3=case_when(
common_dolphin_ppsr_data$ppsr<0.01~'4',common_dolphin_ppsr_
    data$ppsr< 0.05~'2',common_dolphin_ppsr_data$
            ppsr<0.10~'1',TRUE~'3')
```

Once you have run this new version of the R code for categorising the common dolphin prey size data, you need to view your data set to ensure the new categorisation has been carried out correctly. To do this, enter the following command into R:

```
View(common_dolphin_ppsr_data)
```

This will open a DATA VIEWER window where you can examine the data set to ensure that your data have been categorised correctly. Once you are happy that the categorisation has been carried out to your required specifications, you can then make a new bar graph based on these new categories. This is done by editing the code you used to make the last bar graph (see page 83) so that it look like this (the required modifications are highlighted in **bold**):

```
ggplot(data=common_dolphin_ppsr_data, aes(x=ppsr_cat_3)) +
    geom_bar(stat="count") + scale_x_discrete(labels=
        c("Medium","Small","Large","Very Small")) +
    labs(title="Relative Prey Size of Common Dolphin",
    x="Predator-Prey Size Ratio Category",y="Number of Prey
        Items") + ylim(c(0,7500)) + theme_classic()
```

The only modifications contained in this code are the column containing the categories specified in the x argument in the aes element of the ggplot command and the labels argument in the scale_x_discrete command. Specifically, the column containing the categories to be used for the X axis has been changed to the new one called ppsr_cat_3 that you have just created, while the order of the labels for the four categories provided in the scale_x_discrete command have been changed so that they are consistent with the order that these new categories will be plotted in. The graph produced when you run this modified R code should look like this.

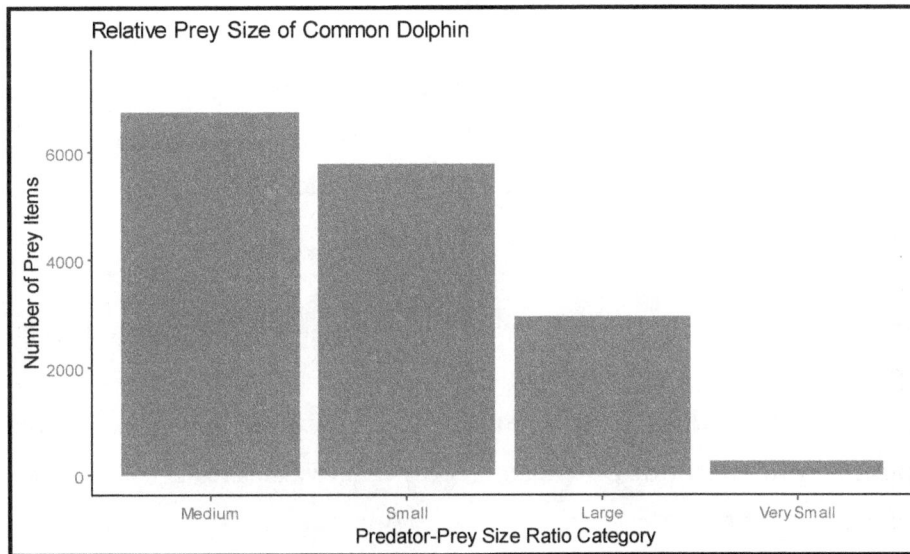

On this graph, the different categories are no longer plotted in order along the X axis based on the prey sizes they contain, but instead are plotted based on the height of each bar, and so the number of records in each category, going from tallest to shortest. The same approach of changing the numerical code given to each category in the categorisation command can be used to create graphs with the bars in any order that you wish them to have.

EXERCISE 2.2: HOW TO CREATE A BAR GRAPH OF COUNT DATA WITH MULTIPLE DATA SERIES ON IT:

In exercise 2.1, you learned how to divide the data in a data set into discrete categories based on the values in one of its columns and create a bar graph which shows the count of data in each of these categories. However, there will be occasions where you wish to create bar graphs which show count data not just for one data series, but for multiple data series. Most, but not all, of the steps for doing this are the same as for making a bar graph with a single data series on it, as was outlined in Exercise 2.1. The main differences are that you need to ensure that your data set contains the values for all the data series you wish to plot on it (which requires modifying the subset command in step 4), and modify the code used to create the bar graph in steps 7 and 8 to define the column which contains the labels that identify which data series each row of data belongs to and how the data from the different data series will be displayed on the graph.

In the first part of this exercise, you will create a multi-series bar graph that contains data on the prey size preferences for two cetaceans species from the data set `cetacean_prey_ sizes_2.csv`, which you imported into R in Exercise 2.1. These species are the common dolphin and the harbour porpoise. To do this, work through the flow diagram that starts at the top of the next page.

NOTE: These instructions assume that you have already installed both the `ggplot2` and the `dplyr` packages in your version of R and that you have loaded their command libraries into your analysis project. You can check if you already have these packages installed in your version of R using the command `library()`. If these packages are already installed they will appear on the list in the R PACKAGES AVAILABLE window that will open when you run this command. If one or both of these packages are not already installed in your version of R, you can find instructions for how to install them in Exercise 2.1. Once you have ensured that these packages are installed in your version of R, you can load their command libraries into your analysis project by entering the following pair of commands into R:

```
library(ggplot2)
library(dplyr)
```

Data set held in an existing R object in your analysis project

For this example, the data set you will use is stored in the R object called `cetacean_prey_sizes_2` created in Exercise 2.1 of this workbook.

The first step in creating a multi-series bar graph is to create a new R object containing just the data for the series you wish to plot on it. This is done using the `subset` command. In order to use this command, you first need to identify the names of the groups of data that you wish to have in your new R object. This can be done by entering the following command into R:

```
unique(cetacean_prey_sizes_2$species)
```

This is CODE BLOCK 35 in the document R_CODE_DATA_VISUALISATION_WORKBOOK.DOC. It will return the names of all the groups present in the column called `species` for the specified data set (`cetacean_prey_sizes_2`), allowing you to identify which data you wish to transfer to a new R object. In this case, it is the rows with the group names `common_dolphin` and `harbour_porpoise`. These group names can now be entered into a `subset` command to create the required new R object. For this example, this can be done by entering the following command into R:

1. Create a new R object containing just the data for the series you wish to plot on your multi-series bar graph

```
two_species_dataset <-
subset(cetacean_prey_sizes_2,species==
   "common_dolphin"|species==
      "harbour_porpoise")
```

This is CODE BLOCK 36 in the document R_CODE_DATA_VISUALISATION_WORKBOOK.DOC, and it will create a new R object called `two_species_dataset` containing just the rows of data with the names `common_dolphin` and `harbour_porpoise` in the column called `species` from the data set called `cetacean_prey_sizes_2`. Once you have done this, you should view the contents of your new R object to check it contains the data you wish it to contain. This can be done by entering the following command into R:

```
View(two_species_dataset)
```

This is CODE BLOCK 37 in the document R_CODE_DATA_VISUALISATION_WORKBOOK.DOC, and it will open a DATA VIEWER window where you can examine the contents of your new R object.

In this example, you will use the `case_when` command to create a new column containing the categories that will be used for the X axis of your bar graph. In this command, you provide a series of logical statements based on a specific column in your data set which define the thresholds for assigning each row to one of the categories provided in them. These statements form a hierarchy and each row is only assigned to a new category if it has not already been assigned one based on an earlier statement in the command. This list of logical statements ends with a `TRUE` statement that provides the category to be used for all rows that do not conform to any of the previous logical statements in the command. To do this for the data set being used in this example, enter the following command into R:

```
two_species_dataset$ppsr_cat=
case_when(two_species_dataset$ppsr<0.01~
'1',two_species_dataset$ppsr<0.05~'2',two_
species_dataset$ppsr<0.10~'3',TRUE~'4')
```

This is CODE BLOCK 38 in the document R_CODE_DATA_VISUALISATION_WORKBOOK.DOC. This command creates a new column called `ppsr_cat` in the data set called `two_species_dataset`. This new column is then filled with values based on the logical statements in the `case_when` command. In this case, this statement provides a series of thresholds (`<0.01`, `<0.05`, `<0.10`) and assigns each row of data with a value of 1, 2, 3 or 4 depending on the value in the `ppsr` column. For example, if the value in this column is 0.016, it would be assigned a value of 2 as it is greater than the first threshold (`<0.01`), but less than the second one (`<0.05`). If the value in the `ppsr` column for a particular row does not conform to any of the thresholds in these logical statements, it is assigned the value from the `TRUE` statement at the end of the `case_when` command. In this case, it is a value of 4.

Once you have done this, you should view the contents of your R object to check that it now contains the new data you wish it to contain. This can be done by entering the following command into R:

```
View(two_species_dataset)
```

This is CODE BLOCK 39 in the document R_CODE_DATA_VISUALISATION_WORKBOOK.DOC, and it will open a DATA VIEWER window where you can examine the contents of your data set to ensure that the categorisation has been carried out correctly.

2. Create a new column which contains the categories that will be used for the X axis of your bar graph

3. Create your initial multi-series bar graph based on the count of records in each category created in step 2

Once your data have been successfully divided into the categories which you will use for its X axis, you are ready to create your initial bar graph based on them. To do this, enter the following block of code into R:

```
ggplot(data=two_species_dataset,aes(x=
ppsr_cat,fill=species)) + geom_bar(stat=
"count",position='dodge',colour="black") +
scale_fill_manual(values=c("blue","red"))
```

This is CODE BLOCK 40 in the document R_CODE_DATA_ VISUALISATION_WORKBOOK.DOC, and it contains three commands separated by + symbols. These are the `ggplot` command, the `geom_bar` command, and the `scale_ fill_manual` command. The `ggplot` command sets the data set which will be used for the graph. This is done using the `data` argument and, in this case, it will be the R object called `two_species_dataset` created in step 1 of this exercise. The column of data that will be plotted on the X axis of the resulting graph is set using the `x` argument of the `aes` element of this `ggplot` command. In this case, it is the column called `ppsr_cat` in this R object (which was created in step 2). Finally, the `fill` argument is use to identify the column containing the identifier that tells R which data series the data in each row belongs to. In this case, it is the column called `species`.

The second command in this code block, `geom_bar`, sets the type of graph that will be created from the data specified in the `ggplot` command. In this case, it will be a bar graph. Within this command, the values to be plotted on the Y axis are set by the `stat` argument. In this case, the argument used is `stat="count"`, meaning that the height of the bar for each category of the X axis will be the count of the number of rows in that category in the data set defined in the `data` argument of the `ggplot` command (in this case, `two_species_dataset`). This command also includes a `position` argument which determines if the multiple data series in the data set are to be plotted side-by-side or stacked on top of each other. In this case, the argument `position="dodge"` is used to create a graph where the multiple data series are potted side-by-side. This command also includes a `colour` argument which sets the outline colour to be used for the bars that will be plotted on it. In this case, the argument is `colour="black"`, which sets the outline colour to black.

The final command, `scale_fill_manual`, is used to set the fill colours to be used for each data series. In this case, these are `red` for the data from the first cetacean species (the common dolphin) and `blue` for the data from the second species (the harbour porpoise).

89

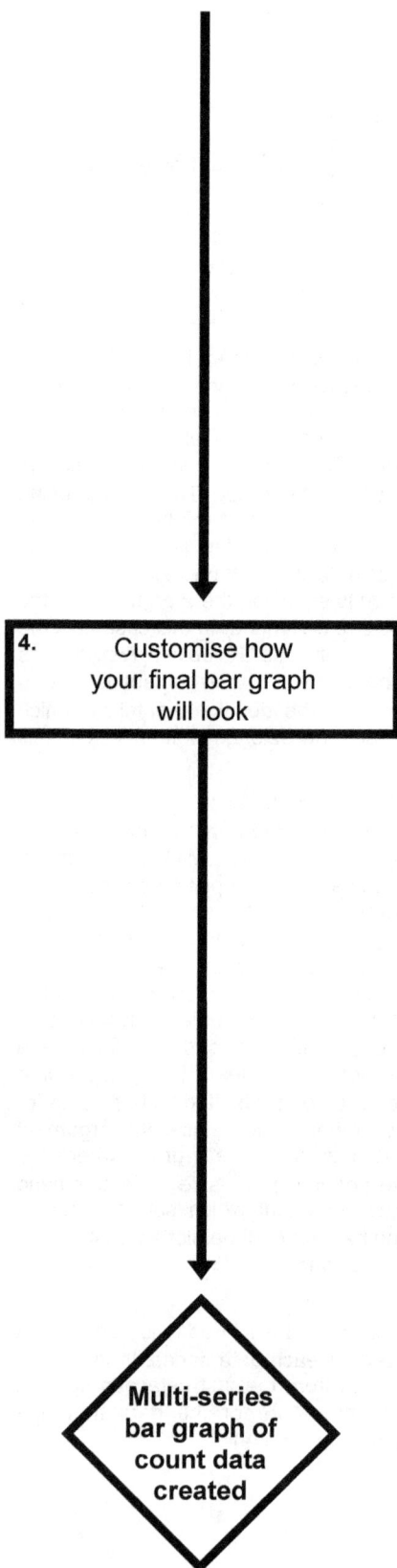

Once you have created your initial bar graph, and you are happy with how the data are displayed on it, you can customise how the final graph will look. This can be done by adding a number of new style commands to the block of code used to produce it. To do this, edit the code from step 3 so that it looks like this (the newly added style commands are highlighted in **bold**):

```
ggplot(data=two_species_dataset,
    aes(x=ppsr_cat,fill=species)) +
geom_bar(stat="count",position='dodge',
    colour="black") + scale_fill_
    manual(values=c("blue","red")) +
scale_x_discrete(labels=c("Very Small",
    "Small","Medium","Large")) +
labs(title="Cetacean Species Prey Size
Preferences",x="Predator-Prey Size Ratio",
y="No. Prey Items") + ylim(c(0,8000)) +
    theme_classic()
```

This is CODE BLOCK 41 in the document R_CODE_DATA_VISUALISATION_WORKBOOK.DOC, and it adds four new commands separated by + symbols to the code block used to create your initial bar graph in step 3. These are the `scale_x_discrete` command, the `labs` command, the `ylim` command, and the `theme_classic` command.

The `scale_x_discrete` command sets the labels that will be used for each group on the X axis of the resulting bar graph. Without this command, each group would be labelled with the values from the `ppsr_cat` column. However, It is much better to include this command to provide more descriptive labels. In this case, these will be `"Very Small"`, `"Small"`, `"Medium"` and `"Large"` respectively.

The `labs` command sets the labels that will be used tor the title of the graph (using the `title` argument), the X axis (using the `x` argument) and the Y axis (using the `y` argument). The `ylim` command sets the minimum and maximum value displayed on the Y axis of the graph. In this case, these will be `0` for the minimum value and `8,000` for the maximum value. Finally, the `theme_classic` command sets the remaining elements of the final graph to those of the pre-existing classic theme. Once you have finished editing this code block, you can run it again to create the final version of your bar graph.

4. Customise how your final bar graph will look

Multi-series bar graph of count data created

Once you have worked through the first part of this exercise, the multi-series bar graph that you created should look like this:

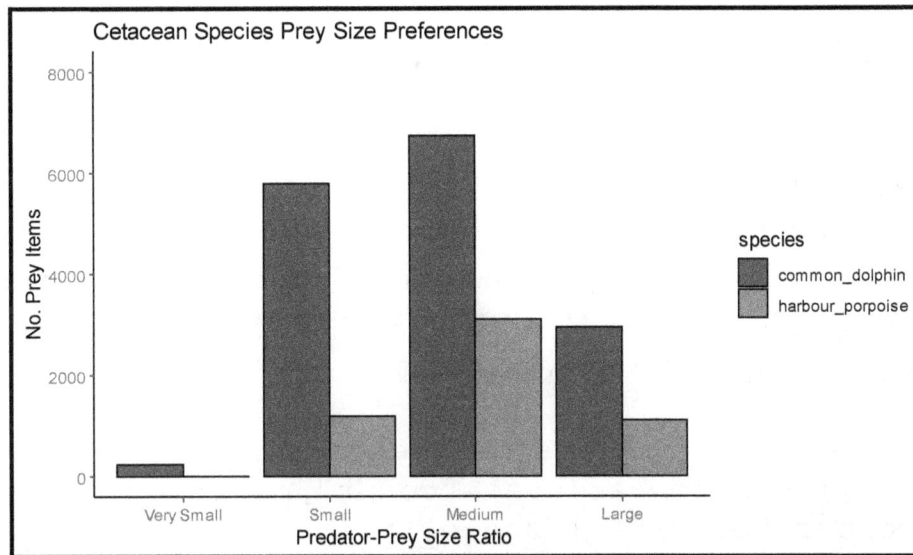

After you have made a multi-series bar graph, you can export it from R so that you can include it in a manuscript or presentation. If you are using RGUI, you can do this by clicking on the R GRAPHICS window containing your bar graph to select it, before clicking on FILE on the main menu bar and selecting SAVE AS. This will allow you to save it in a variety of different formats. If you are using RStudio, you can export your graph by clicking on the EXPORT button at the top of the window displaying your bar graph and selecting SAVE AS IMAGE.

When including a multi-series bar graph in a manuscript, it is important that you provide an appropriate figure legend for it. This legend should provide all the information required for the reader to interpret the contents of the graph. For the above bar graph, an appropriate legend would be:

Figure 1: *A comparison of the relative prey sizes consumed by common dolphin (blue) and harbour porpoise (red) as measured using a predator-prey size ratio (PPSR). Very small: PPSR < 0.01; Small: PPSR 0.01 to 0.05; Medium: PPSR 0.05 to 0.10; Large: PPSR > 0.1.*

The type of multi-series bar graph produced by working through the above flow diagram is just one of a number of different ways that data from multiple data series can be displayed on a single bar graph. Other possible ways include using bars where the data from different series are stacked on top of each other for each category on the X axis (rather than being displayed side-by-side) and using stacked proportional information rather than absolute counts. These types of graphs can be generated by altering the terms used in the `position` argument in the `geom_bar` command. To gain experience in varying the terms in this argument to display multi-series bar graphs in different ways, start by editing the code from step 4 in the above flow diagram (this is CODE BLOCK 41 in the document R_CODE_DATA_VISUALISATION_WORKBOOK.DOC) so that it looks like this (the required modifications are highlighted in **bold**):

```
ggplot(data=two_species_dataset,aes(x=ppsr_cat,fill=
    species)) + geom_bar(stat="count",position='stack',
        colour="black") + scale_fill_manual(values=
  c("blue","red")) + scale_x_discrete(labels=c("Very Small",
   "Small","Medium","Large")) + labs(title="Cetacean Species
Prey Size Preferences",x="Predator-Prey Size Ratio",y="No.
     Prey Items") + ylim(c(0,10000)) + theme_classic()
```

In this modified version of this code, the `position='stack'` argument is used in the `geom_bar` command rather than the `position='dodge'` argument to tell R to create a stacked bar graph rather than a side-by-side bar graph. The only other required modification is to the maximum value in the `ylim` style command. This has been increased from `8000` to `10000`. This has been changed because if this maximum number is not large enough, you may find that the bar segments for some of the data series are not visible on it. **NOTE:** If any bar segments for any of the data series are missing on your own graphs, simply increase the maximum number for the Y axis scale in the `ylim` command until they are all visible on it. When you run this modified code for creating a multi-series bar graph, you should get a graph that looks like the image at the top of the next page.

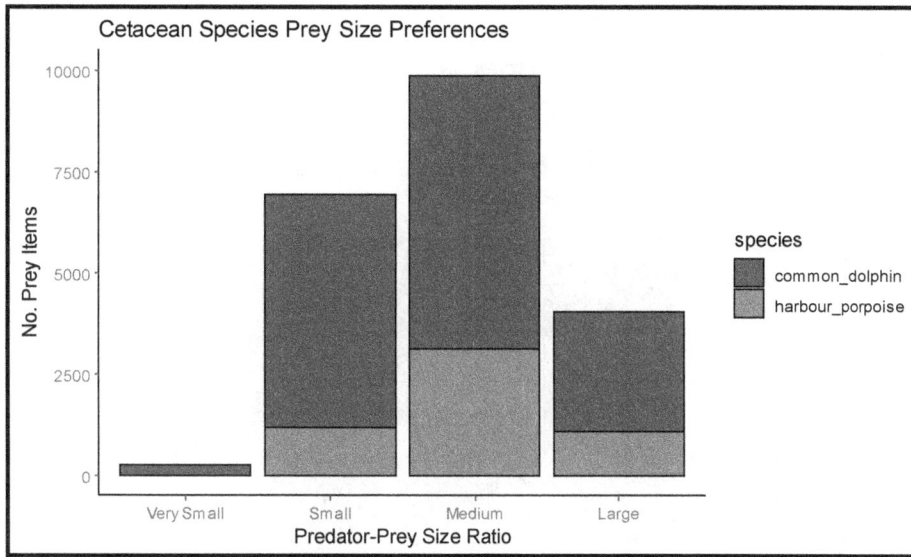

While this type of stacked bar graph allows you to see how the counts in each X axis category compare between the different data series, it is not always easy to tell whether each data series makes up the same or different proportions of the records within each category. If you wish to create a bar graph where you can easily see the proportions of each X axis category that belong to each data series (regardless of the total record count), you can make a proportional stacked graph. Again, this is done by modifying the contents of the `position` argument in the `geom_bar` command. To display the data from the above multi-series stacked graph as proportional bars rather than absolute bars, edit the code from page 92 so that it looks like this (the required modifications are highlighted in **bold**):

```
ggplot(data=two_species_dataset,aes(x=ppsr_cat,fill=
    species)) + geom_bar(stat="count",position='fill',
        colour="black") + scale_fill_manual(values=
c("blue","red")) + scale_x_discrete(labels=c("Very Small",
 "Small","Medium","Large")) + labs(title="Cetacean Species
Prey Size Preferences",x="Predator-Prey Size Ratio",y="No.
        Prey Items") + ylim(c(0,1)) + theme_classic()
```

In this modified version of this code, the `position='fill'` argument is now used in the `geom_bar` command rather than the `position='stack'` argument to tell R to create a proportional stacked bar graph rather than a total count stacked graph. The only other required modification is to the maximum value in the `ylim` style command. This has been changed from `10000` to `1`. This has been done because the bar segments now represent proportions which all add up 1, rather than stacked count data. When you run this modified code for creating a multi-series bar graph, you should get a graph that looks like this:

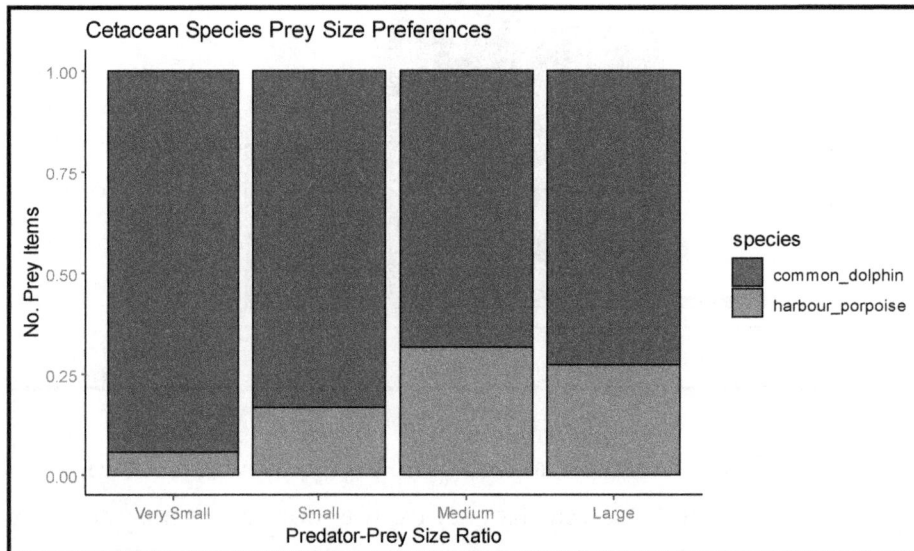

So far in this exercise, you have been working with data from only two out of the four species in the original `cetacean_prey_sizes_2` data set. In the final part of this exercise, you will create a side-by-side multi-series graph with data from all four species (the common dolphin, the harbour porpoise, the pilot whale and the sperm whale) displayed on it. Before you can do this, you first need to create the data set that you will use to make the graph. In the above flow diagram, you did this using the `subset` command in step 1 followed by the `case_when` categorisation command in step 2. However, in this case, you are going to use the whole data set, so you do not need to subset your data first. Instead, you can move straight on to step 2, and run the required categorisation command. As before, you will categorise the data into four groups using the `case_when` command and based on the contents of the `ppsr` column. To do this, edit the categorisation command from step 2 of the above flow diagram (this is CODE BLOCK 38 in the document R_CODE_DATA_VISUALISATION_WORKBOOK.DOC) so that it looks like this (the required modifications are highlighted in **bold**):

`cetacean_prey_sizes_2``$ppsr_cat=case_when(`**`cetacean_prey_sizes_2`**`$ppsr<0.01~'1',`**`cetacean_prey_sizes_2`**`$ppsr<0.05~'2',`**`cetacean_prey_sizes_2`**`$ppsr<0.10~'3',TRUE~'4')`

Once this command has finished running, you can open a DATA VIEWER window to allow you to examine the data and check that the categorisation has been carried out correctly. To do this, enter the following command into R (the required modification to the `View` command from step 2 of the above flow diagram is highlighted in **bold**):

```
View(cetacean_prey_sizes_2)
```

Once you are happy that the data have been correctly divided into the categories you require to create your side-by-side multi-series bar graph, you are now ready to create the graph itself. To do this, edit the final code block from step 4 in the above flow diagram (this is CODE BLOCK 41 in the document R_CODE_DATA_VISUALISATION_ WORKBOOK.DOC) so that it looks like this (the required modifications are highlighted in **bold**):

```
ggplot(cetacean_prey_sizes_2,aes(x=ppsr_cat,fill=
species)) + geom_bar(stat="count",position='dodge',
colour="black") + scale_fill_manual(values=c("blue","red",
"green","yellow")) + scale_x_discrete(labels=c("Very
Small","Small","Medium","Large")) + labs(title="Cetacean
Species Prey Size Preferences",x="Predator-Prey Size
Ratio",y="No. Prey Items") + ylim(c(0,10000)) +
theme_classic()
```

These modifications change the data set on which the graph will be based (`cetacean_prey_sizes_2`), the colours that will be used to fill in the different bars and the maximum value for the Y axis. Specifically, it adds two new colours (`green` and `yellow`) to the `scale_fill_manual` style command to provide colours for the two new data series that will be added to the graph, while the maximum value for the Y axis has been increased to `10000` by modifying the second number in the `ylim` command (this is to account for the different maximum heights of the new bars for the new series that will be added to the graph). When you have finished running this code in R, you should have graph that looks like the image at the top of the next page.

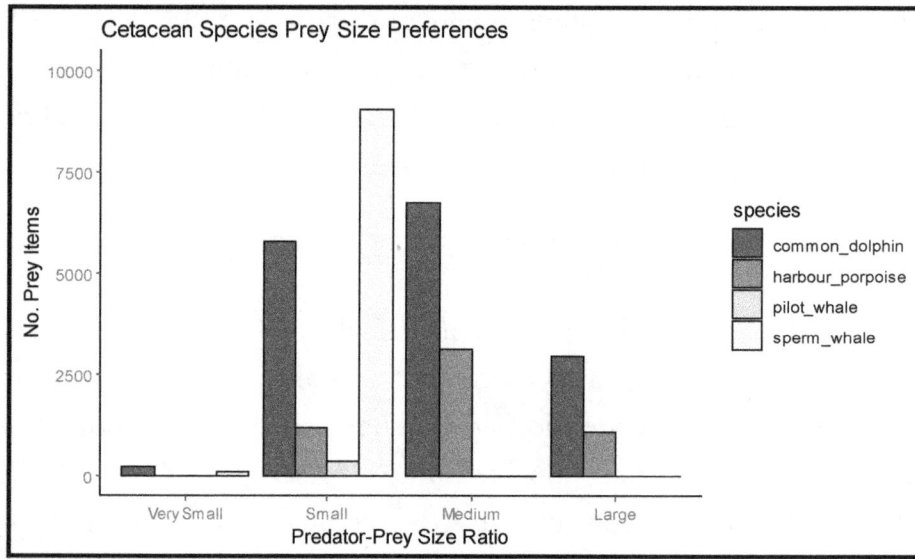

If you examine this graph, you will notice a potential issue with it. This is that the bars representing the pilot whale data are barely visible on it. This is because the counts for each predator-prey size ratio category are much smaller for pilot whales than the common dolphin, the harbour porpoise and the sperm whale. Such issues can make it difficult to display the data for different groups on a single multi-series graph. In these circumstances, it may be better to make a multi-panel figure with each data series being represented on its own bar graph within it. To do this, you first need to separate the data for each data series out into its own R object using the `subset` command. Start by creating a new R object containing just the data for common dolphin by entering the following `subset` command into R:

```
common_dolphin_data <- (subset(cetacean_prey_sizes_2,
           species=="common_dolphin"))
```

Once you have created an R object containing just the common dolphin data, you can edit this `subset` command to do the same for the harbour porpoise data. The edited version of this command should look like this (the required modifications are highlighted in **bold**):

```
harbour_porpoise_data <- (subset(cetacean_prey_sizes_2,
           species=="harbour_porpoise"))
```

This code is then edited to do the same for the pilot whale data, and it should look like this (the required modifications are highlighted in **bold**):

```
pilot_whale_data <- (subset(cetacean_prey_sizes_2,
            species=="pilot_whale"))
```

And finally, the code needs to be edited to do the same for the sperm whale data, and it should look like this (the required modifications are highlighted in **bold**):

```
sperm_whale_data <- (subset(cetacean_prey_sizes_2,
            species=="sperm_whale"))
```

Once you have the data for each species separated out into their own R objects, you can create a separate bar graph for each one and save it as a new R object.

To create a bar graph from the common dolphin data, you need to edit the block of code from step 4 in the above flow diagram (this is CODE BLOCK 41 in the document R_ CODE_DATA_VISUALISATION_WORKBOOK.DOC) so that the graph it creates is based on the R object containing the common dolphin data, and is saved in its own new R object. The edited version of this code should look like this (the required modifications are highlighted in **bold** – __NOTE:__ As this is a single species graph, you can remove the `fill` argument from the `aes` element of the `ggplot` command this code block as well as the `scale_fill_manual` and the `ylim` commands):

```
common_dolphin_bargraph <- ggplot(common_dolphin_data,
  aes(x=ppsr_cat)) + geom_bar(stat="count",colour="black",
  fill="blue") + scale_x_discrete(labels=c("Very Small",
"Small","Medium","Large")) + labs(title="Common Dolphin",
   x="Predator-Prey Size Ratio",y="No. Prey Items") +
                  theme_classic()
```

To view this graph, enter the following command into R:

```
common_dolphin_bargraph
```

The above code is then modified to create individual graphs for the remaining three species. To do this for the harbour porpoise data, edit the above block of code so that it looks like this (the required modifications are highlighted in **bold** – in this case, the code for viewing the resulting graph is included on a new line at the end of the code block to allow it to be viewed automatically):

```
harbour_porpoise_bargraph <- ggplot(harbour_porpoise_data,
  aes(x=ppsr_cat)) + geom_bar(stat="count",colour="black",
    fill="red") + scale_x_discrete(labels=c("Very Small",
"Small","Medium","Large")) + labs(title="Harbour Porpoise",
    x="Predator-Prey Size Ratio",y="No. Prey Items") +
                    theme_classic()
           harbour_porpoise_bargraph
```

To do this for the pilot whale data, edit the above block of code so that it looks like this (the required modifications to the above code are highlighted in **bold**):

```
  pilot_whale_bargraph <- ggplot(pilot_whale_data,
  aes(x=ppsr_cat)) + geom_bar(stat="count",colour="black",
   fill="green") + scale_x_discrete(labels=c("Very Small",
  "Small","Medium","Large")) + labs(title="Pilot Whale",
    x="Predator-Prey Size Ratio",y="No. Prey Items") +
                    theme_classic()
            pilot_whale_bargraph
```

To do this for the sperm whale data, edit the above block of code so that it looks like this (the required modifications to the above code are highlighted in **bold**):

```
  sperm_whale_bargraph <- ggplot(sperm_whale_data,
  aes(x=ppsr_cat)) + geom_bar(stat="count",colour="black",
   fill="yellow") + scale_x_discrete(labels=c("Very Small",
   "Small","Medium","Large"))+ labs(title="Sperm Whale",
    x="Predator-Prey Size Ratio",y="No. Prey Items") +
                    theme_classic()
            sperm_whale_bargraph
```

Once you have created the individual graphs showing the data for each data series, you can use the ggarrange command to create a multi-panel figure containing all four individual bar graphs. To do this, enter the following command into R (**NOTE:** This assumes that the ggpubr package has been installed in your version of R – see pages 65 and 66 for more details):

```
library(ggpubr)
ggarrange(common_dolphin_bargraph,
harbour_porpoise_bargraph,pilot_whale_bargraph,
sperm_whale_bargraph,labels=c("A","B","C","D"),
ncol=2,nrow=2)
```

The resulting multi-panel figure should look like this:

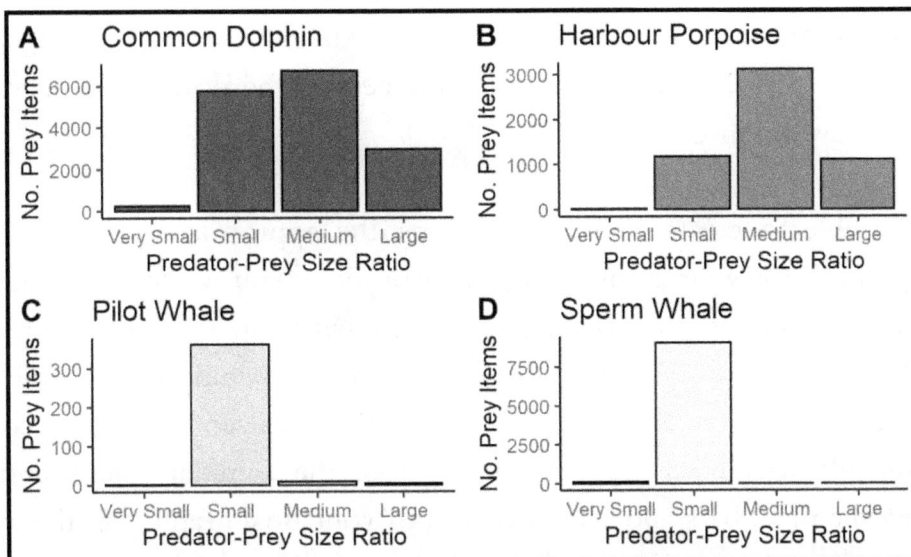

EXERCISE 2.3: HOW TO CREATE A BAR GRAPH OF MEANS/SUMMARY STATISTICS PER GROUP FOR ONE VARIABLE WITH ERROR BARS ON IT:

While bar graphs are commonly used to display count data (see Exercises 2.1 and 2.2), they can also be used to display summary statistics for a data set. For example, bar graphs can be used to display the mean or median for a specific variable to allow you to compare these values for different categories of a second variable in a data set, such as species or sampling location. When plotting this type of graph, you will usually wish to add error bars to show a measure of the variation around the summary statistic for each group that the main bars represent. This variation can be measured as standard deviations, standard errors, interquartile ranges or confidence intervals. This means that the first step in making such graphs is to create a table which contains the values for the summary statistic and measure of variation that you wish to plot on it for each group of data in your data set. This is done using the tools from a variety of R packages including the `Rcpp` package, the `plyr` package and the `plotrix` package. These packages will be downloaded and installed as part of the workflow outlined below.

In order to be able to create a table containing the appropriate summary statistic and measure of variation for your graph using this workflow, you will need to have your data arranged in a spreadsheet or table where each row contains data from a single record in your data set. In this table, there also needs to be a column containing the variable that will be used for the categories on the X axis of your bar graph, and a second column containing the continuous variable which will be used to calculate the summary statistics that will be displayed using the main bars and the error bars of your final graph. For this exercise, you will start by creating a bar graph that shows the mean and standard deviation of the body lengths of angle head lizards sampled from three locations in Malaysia. The instructions for how to do this are provided in the flow diagram that starts at the top of the next page.

NOTE: These instructions assume that you have already installed both the `ggplot2` and the `dplyr` packages in your version of R and that you have loaded their command libraries into your analysis project. You can check if you already have these packages installed in your version of R using the command `library()`. If these packages are already installed they will appear on the list in the R PACKAGES AVAILABLE window that will open when you run this command. If one or both of these packages are not already installed in your version of R, you can find instructions for how to install them in Exercise 2.1. Once you have ensured that these packages are installed in your version of R, you can load their command libraries into your analysis project by entering the following pair of commands into R:

```
library(ggplot2)
library(dplyr)
```

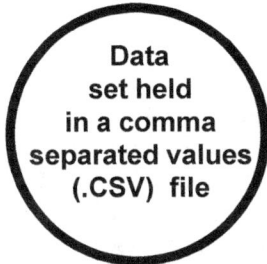

Data set held in a comma separated values (.CSV) file

For this example, the data set you will use is stored in a file called `all_locations.csv` that is located in the WORKING DIRECTORY folder you created during the introduction to this chapter.

1. Set the WORKING DIRECTORY for your analysis project

Before you start any analysis in R, you first need to set the WORKING DIRECTORY. To do this, enter the text `setwd("` and then type the address of your WORKING DIRECTORY, using slashes (/) as the folder separators, before entering a second quotation mark followed by a closing bracket, like this `")`. For example, if your WORKING DIRECTORY has the address C:\STATS_FOR_BIOLOGISTS_TWO, your `setwd` command should look like this:

```
setwd("C:/STATS_FOR_BIOLOGISTS_TWO")
```

If you are using RGUI, enter your `setwd` command in the R CONSOLE window (remembering to use the address of your own WORKING DIRECTORY folder in it) and then press the ENTER key on your keyboard. If you are using RStudio, enter your `setwd` command into the SCRIPT EDITOR window. To run it, select it and then click on the RUN button at the top of this window. You will enter all the remaining commands for this exercise in a similar manner, depending on the user interface you are using.

To check that your WORKING DIRECTORY has been set properly, enter the command `getwd()` and carefully check that the address it returns is the same as the one for the STATS_FOR_BIOLOGISTS_TWO folder you created at the start of this chapter.

Before you move on to step 2, make sure that all the data you wish to use in your analysis project are located in this WORKING DIRECTORY folder. In this case, this is a file called `all_locations.csv`. **NOTE:** If the data you are going to import into R in step 2 are not located in the WORKING DIRECTORY you set in this step, the import code provided in the next step will not work.

The `read.table` command provides the easiest way to load data held in a .CSV file (and stored in the WORKING DIRECTORY you set in step 1) into R so you can analyse it. To do this for the data set being used in this example, enter the following command into R:

```
all_locations <-
read.table(file="all_locations.csv",
   sep=",",as.is=FALSE,header=TRUE)
```

This code has to be entered exactly as it is written here or it will not work. If you wish to use the copy-and-paste approach for entering this command, copy the text directly below CODE BLOCK 42 in the document R_CODE_DATA_ VISUALISATION_WORKBOOK.DOC and paste it into R.

This command will create a new object in R called `all_locations` which will contain the data from the specified .CSV file. To import a different .CSV file into R, all you need to do is change the file name in the `file` argument to the name of the one you wish to import. You can also use whatever name you wish for the R object which will be created by this command. To do this, simply replace `all_locations` at the start of the first line of the above code with the name you wish to use for it. **NOTE:** If your .CSV data set uses a semicolon as the column separator, you would need to replace the `sep=","` argument with `sep=";"`.

Step 2

2. Load your data into R using the `read.table` command

Whenever you import any data into R you need to check that they have loaded correctly. First, you need to check that all the required columns are present in the R object you just created. To do this, enter the following command into R:

```
names(all_locations)
```

This is CODE BLOCK 43 in the document R_CODE_DATA_ VISUALISATION_WORKBOOK.DOC. This command will return the names used for each column in the R object you just created. For this example, the names should be: `id`, `body_length`, `forelimb_length`, `location` and `sex`.

Next, you should view the contents of the whole table using the `View` command. This is done by entering following code into R:

```
View(all_locations)
```

This is CODE BLOCK 44 in the document R_CODE_DATA_ VISUALISATION_WORKBOOK.DOC. This command will open a DATA VIEWER window where you can examine your data set and check that the correct data have been loaded into R.

3. Check the data have loaded into R correctly by checking the names of the columns and by viewing it

4. **Load the required new packages and command libraries into R**

The first step in creating a bar graph which displays summary statistics rather than count data is to create a new table containing the summary statistics you wish to plot on your graph. This is done using tools from a variety of additional packages you have not used before in this workbook, and you need to ensure these packages have been installed in your version of R. These are the `Rcpp`, `plyr` and `plotrix` packages. To do this, you first need to check if you already have these packages installed in your version of R by entering the code `library()`. If none of these packages are listed in the R PACKAGES AVAILABLE window that opens, enter the follow code into R:

```
install.packages(c("Rcpp","plyr","plotrix"))
```

This is CODE BLOCK 45 in the document R_CODE_DATA_ VISUALISATION_WORKBOOK.DOC. This is a variation on the `install.packages` command which you have used previously, with the addition of the `c` element between the command and the list of packages, and it allows you to install multiple packages using a single command. If you already have any of these packages installed in your version of R, remove their name from this command before you run it so that you do not accidently install it again. If you already have all these packages installed, you do not need to run this command at all.

Once you have ensured that the required additional packages have been installed in your version of R, you need to load their command libraries into your analysis project. To do this, enter the following commands into R:

```
libary(Rcpp)
library(plyr)
library(plotrix)
```

This is CODE BLOCK 46 in the document R_CODE_DATA_ VISUALISATION_WORKBOOK.DOC. Unlike the `install. packages` command, you cannot install the command libraries from multiple packages at the same time using the `library` command. Instead, you need to enter a separate command to install the library for each individual package you wish to use. In this case, you need to enter separate `library` commands to install the `Rcpp` command library, the `plyr` command library and the `plotrix` command library to ensure that you have access to all the tools required to complete the rest of this exercise.

103

5. Create a table containing the summary statistics for the data you wish to plot on your bar graph

There are a number of ways that you can created the new table containing the summary statistics which you need to make a bar graph from them. In the first workbook in this series (*An Introduction to Basic Statistics for Biologists using R*) this was done by creating separate tables for each summary statistic before joining them together to create a overall summary table. In this exercise, you will streamline this process by creating a single, overall summary table containing a range of different summary statistics that you might want to use for your bar graph using a single `ddply` command. To do this for the data being used for this example, enter the following code into R:

```
summary_table <- ddply(all_locations,
    c("location"),summarise,n=
    length(body_length),mean_length=
mean(body_length),sd=sd(body_length),
    se=std.error(body_length))
```

This is CODE BLOCK 47 in the document R_CODE_DATA_VISUALISATION_WORKBOOK.DOC. This command will create a summary table from the data held in the data set called `all_locations`. In this summary table, the individual rows are defined by data groupings provided in the `location` column. This is set by the `c("location")` argument. Other arguments are then used to provide a name and a calculation for each summary statistic that will be added to the table based on these groupings. For example, the `n=length(body_length)` argument will create a column in the summary table called `n` that will give the count of the data in each group (which is calculated using the `length` argument). Similarly, the `mean_length=mean(body_length)` argument creates a column in the summary table called `mean_length` that will contain the mean of the `body_length` data for each group, while the `sd=sd(body_length)` argument creates a column in the summary table called `sd` that will contain the standard deviation of the `body_length` data for each group, and the `se=std.error(body_length)` argument creates a column in the summary table called `se` that will contain the standard error of the `body_length` data.

Once your summary table has been created, you should view it to check that it contains the information you need it to contain by entering the following command into R:

```
View(summary_table)
```

This is CODE BLOCK 48 in the document R_CODE_DATA_VISUALISATION_WORKBOOK.DOC. This command will open a DATA VIEWER window where you can review the summary table you have just created to ensure it contains the correct information required to make your intended bar graph.

104

6. Create your initial bar graph with error bars based on data in the summary table created in step 5

Once you have successfully created your summary table and checked that it contains the correct information, you are ready to use it to make an initial bar graph based on the summary statistics it contains. To do this, enter the following block of code into R:

```
ggplot(data=summary_table,
  aes(x=location,y=mean_length)) +
geom_bar(stat="identity",position="dodge") +
  geom_errorbar(aes(ymin=mean_length-sd,
    ymax=mean_length+sd),width=0.2)
```

This is CODE BLOCK 49 in the document R_CODE_DATA_ VISUALISATION_WORKBOOK.DOC, and it contains three commands separated by + symbols. These are the `ggplot` command, the `geom_bar` command and the `geom_ errorbar` command. The `ggplot` command sets the data which will be used for the graph. This is done using the `data` argument and, in this case, it will be the R object called `summary_table` created in step 5 of this exercise. The column of data that will be plotted on the X axis of the resulting graph is set using the `x` argument of the `aes` element of this `ggplot` command. In this case, it is the column called `location` in the `summary_table` data set. The column of data that will be plotted on the Y axis is set using the `y` argument of the `aes` element of this `ggplot` command. In this case, it is the column called `mean_ length` in the `summary_table` data set.

The second command in this code block, `geom_bar`, sets the type of graph which will be created from the data specified in the `ggplot` command. In this case, it will be a bar graph. Within this command, the values to be plotted on the Y axis are set by the `stat` argument. In this case, the argument used is `stat="identity"`, meaning that the height of the bar for each category of the X axis will be set by the values provided in the column defined by the `y` argument in the `aes` element of the `ggplot` command. In this case, this will be the `mean_length` column in the `summary_table` data set).

The third command is `geom_errorbar`. This command creates the error bars that will be added to the resulting graph. The `ymin` and `ymax` arguments in the `aes` element of this command set the upper and lower limits of the error bars. In this case, they are set to be the value from the `mean_length` column in the summary table data set minus the value from the `sd` column (which contains the standard deviation of body length for each group of data) for the `ymin` argument, and `mean_length` plus the value from the `sd` column for the `ymax` argument. The `width=0.2` argument sets the width of the horizontal line at the top of each error bar to `0.2`.

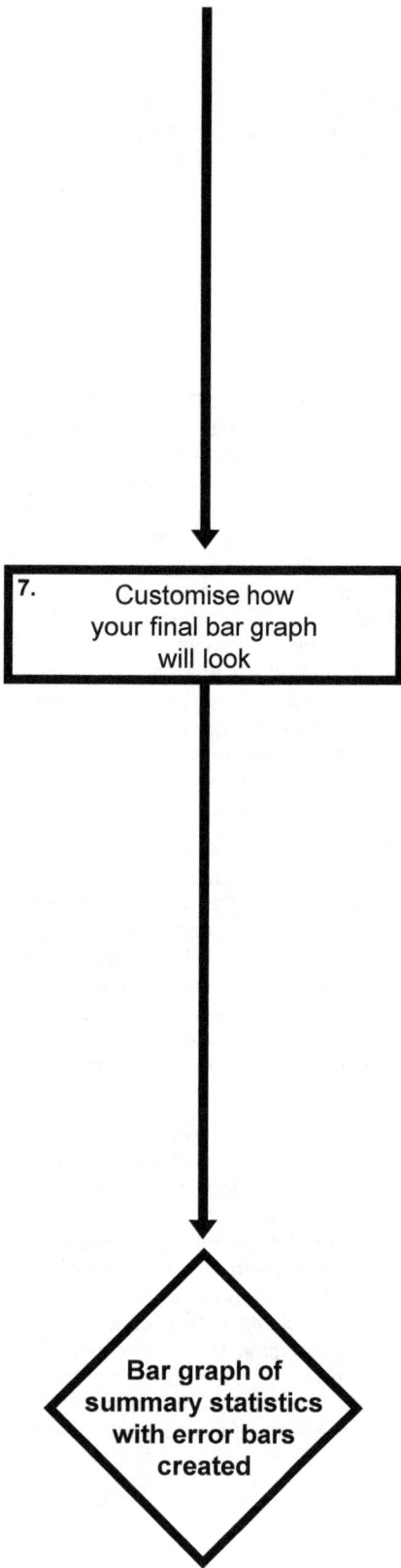

After you have created your initial bar graph, and you are happy with how the summary statistics and error bars are displayed on it, you can customise how your final graph will look. This can be done by adding a number of new style commands to the block of code used to produce it. To do this, edit the code from step 6 so that it looks like this (the newly added style commands are highlighted in **bold**):

```
ggplot(data=summary_table,aes(x=location,
y=mean_length)) + geom_bar(stat="identity",
position="dodge") + geom_errorbar(aes(ymin=
    mean_length-sd,ymax=mean_length+sd),
  width=0.2) + labs(x="Sampling Location",
  y="Body Length (cm)") + ylim(c(0,50)) +
                theme_classic()
```

This is CODE BLOCK 50 in the document R_CODE_DATA_ VISUALISATION_WORKBOOK.DOC, and it adds three new style commands separated by + symbols to the code block used to create your initial bar graph in step 6. These are the `labs` command, the `ylim` command, and the `theme_classic` command. The `labs` command sets the labels that will be used for the X axis (using the `x` argument) and for the Y axis (using the `y` argument). The `ylim` command sets the minimum and maximum value displayed on the Y axis of the graph. In this case, these will be 0 for the minimum value and 50 for the maximum value. Finally, the `theme_classic` command sets the remaining style elements of the graph to those of the pre-existing classic theme. Once you have finished editing this code block, you can run it again to create the final version of your bar graph.

7. Customise how your final bar graph will look

Bar graph of summary statistics with error bars created

After you have worked through this example, you should check the contents of the summary table created in step 5. If it is not already visible, use the command `View(summary_table)` to open a DATA VIEWER window so that you can see it. It should look like this:

	location	n	mean_length	sd	se
1	A	20	34.87302	3.971579	0.8880721
2	B	20	30.13328	5.536410	1.2379790
3	C	20	37.63058	5.992624	1.3399915

The final bar graph with error bars that you created from this table in step 7 should look like this:

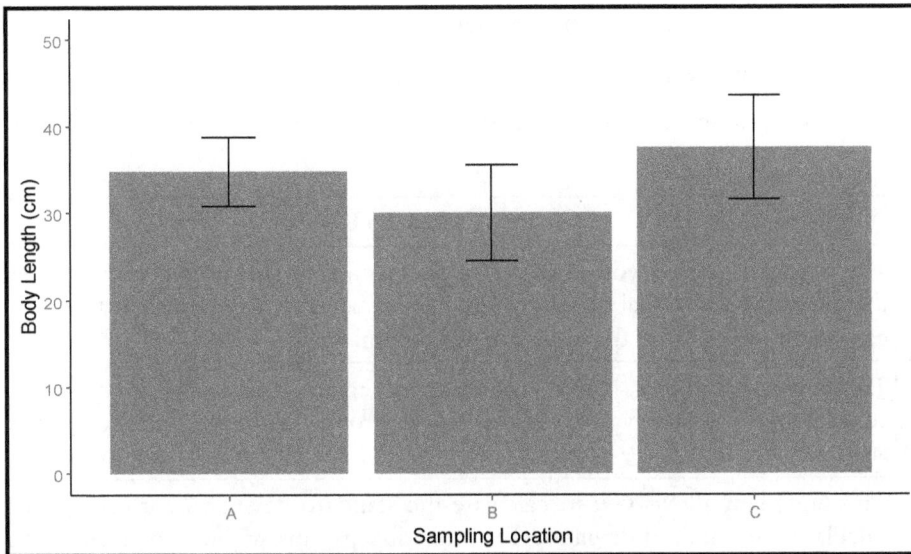

Once you have created a bar graph with error bars, you can export it from R so that you can include it in a manuscript or presentation. If you are using RGUI, you can do this by clicking on the R GRAPHICS window containing your bar graph to select it, before clicking on FILE on the main menu bar and selecting SAVE AS. This will allow you to save it in a variety of different formats. If you are using RStudio, you can export your graph by clicking on the EXPORT button at the top of the window displaying your bar graph and selecting SAVE AS IMAGE.

When including a bar graph with summary statistics and error bars in a manuscript, it is important that you provide an appropriate figure legend for it. This legend should provide all the information required for the reader to interpret the contents of the graph. For the above the bar graph, an appropriate legend would be:

Figure 1: *A comparison of the body lengths of anglehead lizards from three sampling locations in Malaysia. Main bars represent the mean values, while the error bars represent the standard deviation of body length.*

When using the `ddply` command to generate a summary table in order to create a bar graph with error bars based on summary statistics, you can calculate almost any summary statistic you wish to use by varying the arguments that are included in it. The table below provides information about the arguments you can use as part of the `ddply` command to generate different summary statistics in a summary table which you can then use to plot different types of information on a bar graph (or another types of graph – see Exercise 2.4).

Argument	How To Use It
`mean`	This argument allows you to calculate the mean value for each group of data. To calculate the mean for individual groups within a data set, use the argument `mean` in the `ddply` command in step 5 of the above flow diagram.
`median`	This argument allows you to calculate the median value for each group of data. To calculate the median value for individual groups within a data set, use the argument `median` in the `ddply` command in step 5 of the above flow diagram.
`sd`	This argument allows you to calculate the standard deviation for each group of data. To calculate the standard deviation for individual groups within a data set, use the argument `sd` in the `ddply` command in step 5 of the above flow diagram.
`std.error`	This argument allows you to calculate the standard error for each group of data. In order to be able to use this argument in the `ddply` command, you need to have the `plotrix` package installed in your version of R, and have its command library loaded into your analysis project. To calculate the standard error for individual groups within a data set, use the argument `std.error` in the `ddply` command in step 5 of the above flow diagram.
`IQR`	This argument allows you to calculate the interquartile range for each group of data. However, in order to add error bars which show the interquartile range, you will need to divide this value by two (as the error bars above and below the main bar are drawn separately). For example, for a column of data called `body_length`, this can be done by including the following code with the `ddply` command: `half_IQR=(IQR(body_length)/2)`

Argument	How To Use It
group.CI	A confidence interval cannot easily be calculated using the ddply command. Instead, you need to use the group.CI command to calculate the confidence intervals and then use the cbind command to join this information on to the summary table generated by the ddply command in step 5. To use this command, you need to have the Rmisc package installed in your version of R and have its command library loaded into your analysis project. The options for the group.CI command are the column containing the data you wish to generate a confidence interval from followed by the column with the groups you wish to calculate it for, the R object containing these data, and the required confidence interval. For example, using the command group.CI(body_length~location,data=all_locations,0.95) will calculate a table with the 95% confidence intervals for a column called body_length based on the groups in a column called location in an R object called all_locations. **NOTE:** To add these confidence intervals as error bars to your graph, you will need to modify the geom_errorbar command in step 6 so that it uses the correct data for them. These data are held in the columns labelled body_length.upper and body_length.lower in the R object created by this command.

For the next part of this exercise, you will change the information displayed by the error bars from the standard deviation of body length to the standard error. To do this, you will need to edit the final block of code provided in step 7 of the above flow diagram (this is CODE BLOCK 50 from the document R_CODE_DATA_VISUALISATION_WORKBOOK.DOC) so that it looks like this (required modifications are highlighted in **bold**):

```
ggplot(data=summary_table,aes(x=location,y=
mean_length)) + geom_bar(stat="identity",position=
"dodge") + geom_errorbar(aes(ymin=mean_length-se,ymax=
mean_length+se),width=0.2) + labs(x="Sampling
Location",y="Body Length (cm)") + ylim(c(0,50)) +
theme_classic()
```

NOTE: This code assumes that you have already included the calculation of the standard error when using the ddply command in step 5. If you have not done this, you will need to modify the code from step 5 to calculate the standard error before you can run the above code.

Once you have run the above command, you should have a bar graph with error bars that looks like the image at the top of the next page.

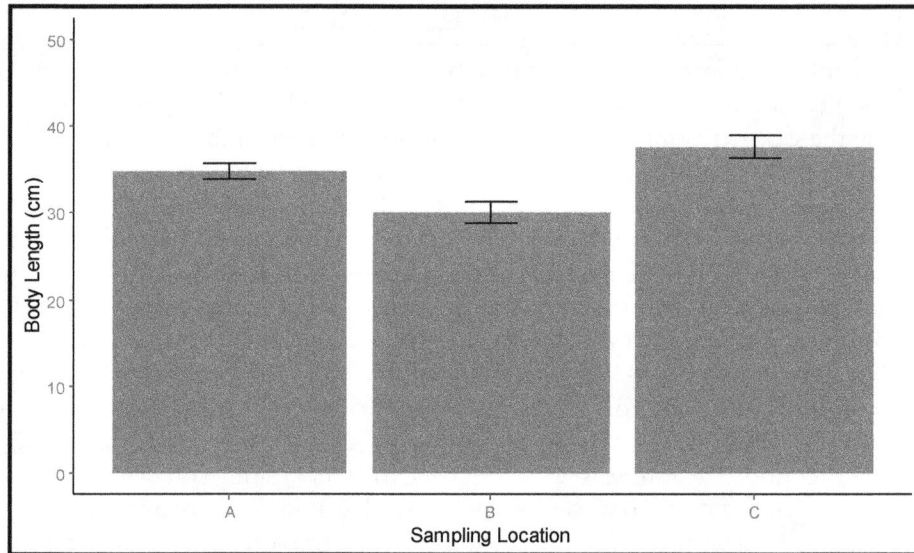

If you only wish to have error bars running in one direction (either only above or only below the main bars on your graph), this can be done by modifying the `ymin` and `ymax` arguments in the block of code used to create your bar graph. In order to have error bars that only extend above the bars displaying the mean values, you would need to modify the `ymin` argument so that it only contains the name of the column which contains the values used for your main bars and no other mathematical functions or column names. Similarly, to have error bars that only extend below your main bars, you would need to modify the `ymax` argument so that it only contains the name of the column which contains the values used for your main bars and no other mathematical functions or column names. As an example of this, you will create a new bar graph of mean body length of anglehead lizards sampled from the three locations in Malaysia with error bars representing the standard error that only extend above the main bars. To do this, edit the block of code used to create the last graph you made so that it look like this (the required modifications are highlighted in **bold**):

```
ggplot(data=summary_table, aes(x=location,
y=mean_length)) + geom_bar(stat="identity",position=
"dodge") + geom_errorbar(aes(ymin=mean_length,ymax=
mean_length+se),width=0.2) + labs(x="Sampling Location",
y="Body Length (cm)") + ylim(c(0,50)) + theme_classic()
```

The graph you obtain when you run this code should look like this:

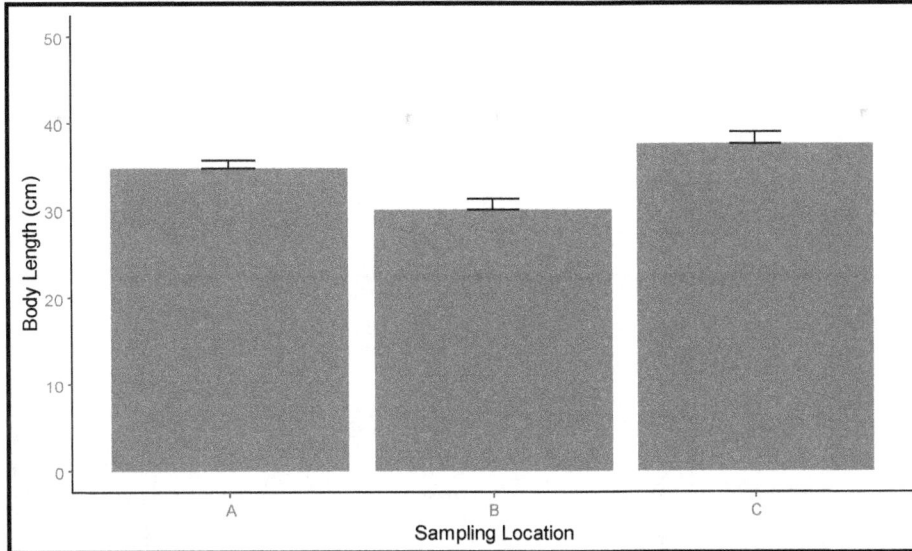

So far, the bar graphs of summary statistics that you have been creating have been based on a single category or grouping, in this case the sampling location. However, there will be times when you wish to compare data not just by the groups in a single variable, but also by sub-groups within them. For example, in a sexually dimorphic species, you may wish to compare the body lengths not only between sampling locations, but also between males and females from each of these locations. This can be done by modifying the ddply command in step 5 of the above flow diagram to create a summary table which includes summary statistics based on the groupings in two columns of the original data set and not just one. For the lizard morphological data being used in this exercise, this can be done using the columns containing the sampling location (called location) and the sex of each lizard that was sample (called sex). To create a table of summary statistics for males and females from each of the three sampling locations, edit the ddply command from step 5 in the above flow diagram (this is CODE BLOCK 47 from the document R_CODE_DATA_ VISUALISATION_WORKBOOK.DOC) so that it looks like this (the required modifications are highlighted in **bold**):

```
summary_table_2 <- ddply(all_locations,c("location",
"sex"),summarise,n=length(body_length),mean_length=
         mean(body_length),sd=sd(body_length),se=
                std.error(body_length))
```

Once you have run this command, you can use the following `View` command to allow you to view the summary table created by it so that you can check it contains the required information:

```
View(summary_table_2)
```

At this stage, the table called `summary_table_2` should look like this:

	location	sex	n	mean_length	sd	se
1	A	female	9	33.33678	3.165843	1.0552810
2	A	male	11	36.12995	4.253188	1.2823843
3	B	female	10	31.04879	7.336319	2.3199478
4	B	male	10	29.21777	3.004209	0.9500141
5	C	female	10	36.72453	4.316991	1.3651524
6	C	male	10	38.53663	7.439930	2.3527125

If you examine this table, you will see that it now contains summary statistics for male and female lizards from each of the three sampling locations. The appropriate summary statistics for each of these six groupings can be identified using the information in the `location` and `sex` columns. Once you have a table of summary statistics for your data with this structure, you can then edit the block of code used to create the final bar graph in the above flow diagram (this is CODE BLOCK 50 from the document R_CODE_DATA_ VISUALISATION_WORKBOOK.DOC) so that separate bars are plotted for the summary statistics for male and females from each location based on the groupings specified in these two columns. This is done by adding a new argument, called `fill`, to the `aes` element of the `ggplot` command. In this case, it specifies that the sub-groupings to be used for the graph are held in the column called `sex`. You also need to add a new command to this code to specify what colours should be used for the bars for groups specified by the `fill` command. This is the `scale_fill_manual` command. In this case, you will use this command to set the colour for females to blue and for males to red by including the argument `values=c("blue","red")`. Finally, you need to add a `position=position_dodge` argument to the `geom_errorbar` command to ensure that separate error bars are added to the bars representing each sub-group of data. The modified version of this code block should look like the code at the top of the next page (the required modifications are highlighted in **bold**).

```
ggplot(data=summary_table_2,aes(x=location,
  y=mean_length,fill=sex)) + geom_bar(stat="identity",
position="dodge") + geom_errorbar(aes(ymin=mean_length-se,
      ymax=mean_length+se),width=0.2,position=
    position_dodge(0.9)) + scale_fill_manual(values=
  c("blue","red")) + labs(x="Sampling Location",y="Body
    Length (cm)") + ylim(c(0,50)) + theme_classic()
```

When you have run this modified block of code, you should have a final graph that looks like this:

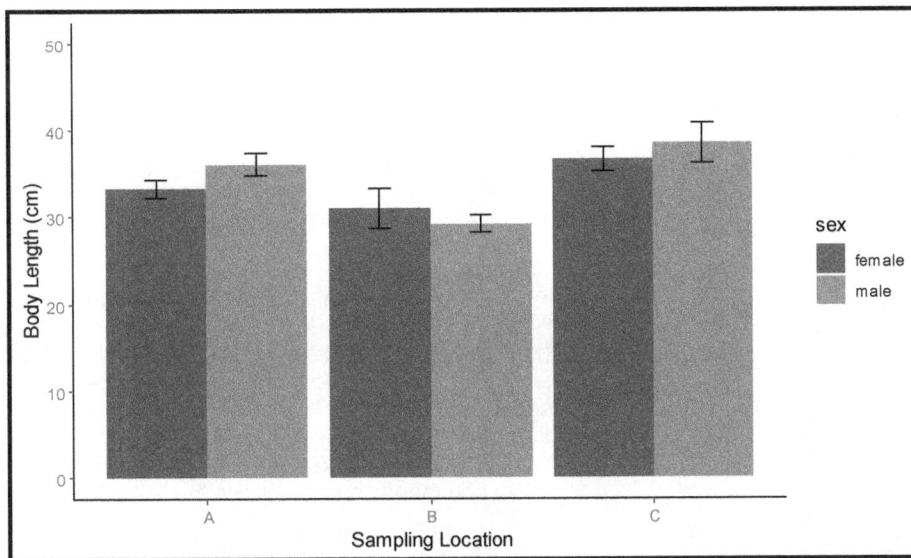

EXERCISE 2.4: HOW TO CREATE A POINT GRAPH OF SUMMARY STATISTICS FOR ONE VARIABLE WITH ERROR BARS:

While the most common type of graph that is used to display summary statistics is a bar graph, you can also plot such data using other graphs types. Of these alternative graph types, the most common one is a point graph showing the summary statistics for individual groups, along with the associated error bars. Such point graphs are most often used by biologists when the groups to be plotted on the X axis form an ordinal scale rather than just being discrete categories. For example, a point graph of summary statistics can be used where the groups to be plotted on the X axis are temporal variables such as year, month, day or hour of the day. Regardless of the type of graph you are using, the workflow for creating graphs of summary statistics for groups of data is very similar to that used to create the bar graphs in Exercise 2.3. To illustrate this, in this exercise, you will create a point graph of the mean body mass for great tits (a small passerine bird from Eurasia) at different hours of the day. For this exercise, the error bars you will include on this graph will display the standard error of the body mass. To create this first graph, work through the flow diagram provided below. **NOTE:** This workflow assumes that you have already downloaded and installed the `ggplot2`, `Rcpp`, `plyr`, `dplyr` and `plotrix` packages, and that the command libraries from these packages have already been loaded into your analysis project (see Exercises 2.1 and 2.3). If you have not already done these steps as part of earlier exercises, you will need to do them before you start working through this flow diagram.

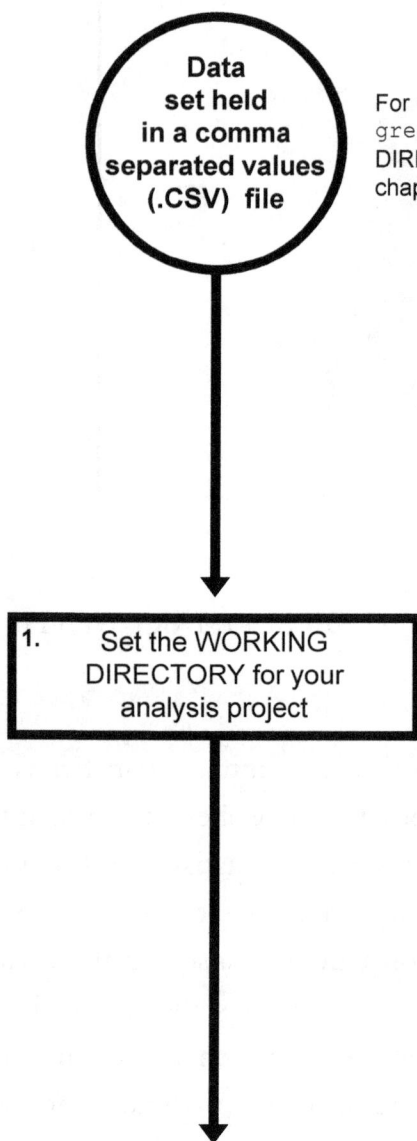

Data set held in a comma separated values (.CSV) file

For this example, the data set you will use is stored in a file called `great_tit_daily_mass.csv` that is located in the WORKING DIRECTORY folder you created during the introduction to this chapter.

Before you start any analysis in R, you first need to set the WORKING DIRECTORY. To do this, enter the text `setwd("` and then type the address of your WORKING DIRECTORY, using slashes (/) as the folder separators, before entering a second quotation mark followed by a closing bracket, like this `")`. For example, if your WORKING DIRECTORY has the address C:\STATS_FOR_BIOLOGISTS_TWO, your `setwd` command should look like this:

```
setwd("C:/STATS_FOR_BIOLOGISTS_TWO")
```

1. Set the WORKING DIRECTORY for your analysis project

If you are using RGUI, enter your `setwd` command in the R CONSOLE window (remembering to use the address of your own WORKING DIRECTORY folder in it) and then press the ENTER key on your keyboard. If you are using RStudio, enter your `setwd` command into the SCRIPT EDITOR window. To run it, select it and then click on the RUN button at the top of this window. You will enter all the remaining commands for this exercise in a similar manner, depending on the user interface you are using.

To check that your WORKING DIRECTORY has been set properly, enter the command `getwd()` and carefully check that the address it returns is the same as the one for the STATS_FOR_BIOLOGISTS_TWO folder you created at the start of this chapter.

Before you move on to step 2, make sure that all the data you wish to use in your analysis project are located in this WORKING DIRECTORY folder. In this case, this is a file called `great_tit_daily_mass.csv`. **NOTE:** If the data you are going to import into R in step 2 are not located in the WORKING DIRECTORY you set in this step, the import code provided in the next step will not work.

2. Load your data into R using the `read.table` command

The `read.table` command provides the easiest way to load data held in a .CSV file (and stored in the WORKING DIRECTORY you set in step 1) into R so you can analyse it. To do this for the data set being used in this example, enter the following command into R:

```
great_tit_daily_mass <-
read.table(file="great_tit_daily_mass.csv",
    sep=",",as.is=FALSE,header=TRUE)
```

This code has to be entered exactly as it is written here or it will not work. If you wish to use the copy-and-paste approach for entering this command, copy the text directly below CODE BLOCK 51 in the document R_CODE_DATA_VISUALISATION_WORKBOOK.DOC and paste it into R.

This command will create a new object in R called `great_tit_daily_mass` which will contain the data from the specified .CSV file. To import a different .CSV file into R, all you need to do is change the file name in the `file` argument to the name of the one you wish to import. You can also use whatever name you wish for the R object which will be created by this command. To do this, simply replace `great_tit_daily_mass` at the start of the first line of the above code with the name you wish to use for it. **NOTE:** If your .CSV data set uses a semicolon as the column separator, you would need to replace the `sep=","` argument with `sep=";"`.

3. Check the data have loaded into R correctly by checking the names of the columns and by viewing it

Whenever you import any data into R you need to check that they have loaded correctly. First, you need to check that all the required columns are present in the R object you just created. To do this, enter the following command into R:

```
names(great_tit_daily_mass)
```

This is CODE BLOCK 52 in the document R_CODE_DATA_VISUALISATION_WORKBOOK.DOC. This command will return the names used for each column in the R object you just created. For this example, the names should be: `x`, `mass`, `wing`, `sex`, `age`, `julday`, `hourday`, `meantemp`, `location`, `daypropo` and `daypart`.

Next, you should view the contents of the whole table using the `View` command. This is done by entering the following code into R:

```
View(great_tit_daily_mass)
```

This is CODE BLOCK 53 in the document R_CODE_DATA_VISUALISATION_WORKBOOK.DOC. This command will open a DATA VIEWER window where you can examine your data set and check that the correct data have been loaded into R.

The first step in creating a point graph which displays summary statistics is to create a new table containing the summary statistics for the variable that you wish to plot on your graph. To do this for the data being used in this example, enter the following code into R:

```
great_tit_summary_table <-
ddply(great_tit_daily_mass,c("hourday"),
summarise,n=length(mass),mean_mass=
mean(mass),se_mass=std.error(mass))
```

This is CODE BLOCK 54 in the document R_CODE_DATA_ VISUALISATION_WORKBOOK.DOC. This command will create a summary table from data held in the data set called `great_tit_summary_table`. In this summary table, the individual rows are defined by data groupings provided in the `hourday` column. This is set by the `c("hourday")` argument. Other arguments are then used to provide a name and a calculation for each summary statistic that will be added to the table based on these groupings. For example, the `n=length(mass)` argument will create a column in the summary table called `n` that will give the count of the data in each group (which is calculated using the `length` argument). Similarly, the `mean_mass= mean(mass)` argument creates a column in the summary table called `mean_mass` which will contain the mean of the `mass` data for each group, while the `se_mass= std.error(mass)` argument creates a column in the summary table called `se_mass` that will contain the standard error of the `mass` data. **NOTE:** If this command doesn't work, check that you have all the required packages and command libraries installed and loaded into R (see page 114 for more details).

Once your summary table has been created, you should view it to check that it contains the information you need it to contain. To do this, enter following code into R:

```
View(great_tit_summary_table)
```

This is CODE BLOCK 55 in the document R_CODE_ BASIC_STATS_WORKBOOK.DOC. This command will open a DATA VIEWER window where you can review the summary table that you have just created to ensure it contains the correct information required to make your intended point graph.

4. Create a table containing the summary statistics for the data that you wish to plot on your point graph

Once you have successfully created your summary table and checked that it contains the correct information, you are ready to use it to create your initial point graph based on the summary statistics it contains. To do this, enter the following block of code into R:

```
ggplot(data=great_tit_summary_table,
    aes(x=hourday,y=mean_mass)) +
geom_errorbar(aes(ymin=mean_mass-se_mass,
    ymax=mean_mass+se_mass),width=0.2) +
    geom_point(stat="identity",size=3)
```

This is CODE BLOCK 56 in the document R_CODE_DATA_ VISUALISATION_WORKBOOK.DOC, and it contains three commands separated by + symbols. These are the `ggplot` command, the `geom_errorbar` command, and the `geom_point` command. The `ggplot` command sets the data set which will be used for the graph. This is done using the `data` argument and, in this case, it will be the R object called `great_tit_summary_table` created in step 4 of this exercise. The column of data from this data set that will be plotted on the X axis of the resulting graph is set using the `x` argument of the `aes` element of this `ggplot` command. The column of data that will be plotted on the Y axis is set using the `y` argument in the `aes` element.

The second command is `geom_errorbar`. This command creates the error bars that will be added to the resulting graph, and it needs to come before the `geom_point` command to ensure the error bars are drawn underneath the points rather than on top of them. The `ymin` and `ymax` arguments in the `aes` element of the `geom_errorbar` command set the upper and lower limits of the error bars. In this case, they are set to be the value from the `mean_mass` column in the summary table data set minus the value from the `se_mass` column (which contains the standard error for body mass for each group of data) for the `ymin` argument, and `mean_mass` plus the value from the `se_mass` column for the `ymax` argument. The `width=0.2` argument sets the width of horizontal line at the end of each error bar to 0.2.

The third command in this code block, `geom_point`, sets the type of graph that will be created from the data specified in the `ggplot` command. In this case, it will be a point graph. Within this command, the values to be plotted on the Y axis are set by the `stat` argument. In this case, the argument used is `stat="identity"`, meaning that the position of the point for each category of the X axis will be set by the values provided in the column defined by the `y` argument in the `aes` element of the `ggplot` command (in this case, the `mean_mass` column in the `great_tit_summary_table` data set). The size of the points representing the mean values for each group on the graph are set by the `size` argument in the `geom_point` command. In this case, the size is set to 3.

5. Create your initial point graph with error bars based on the data in the summary table created in step 4

117

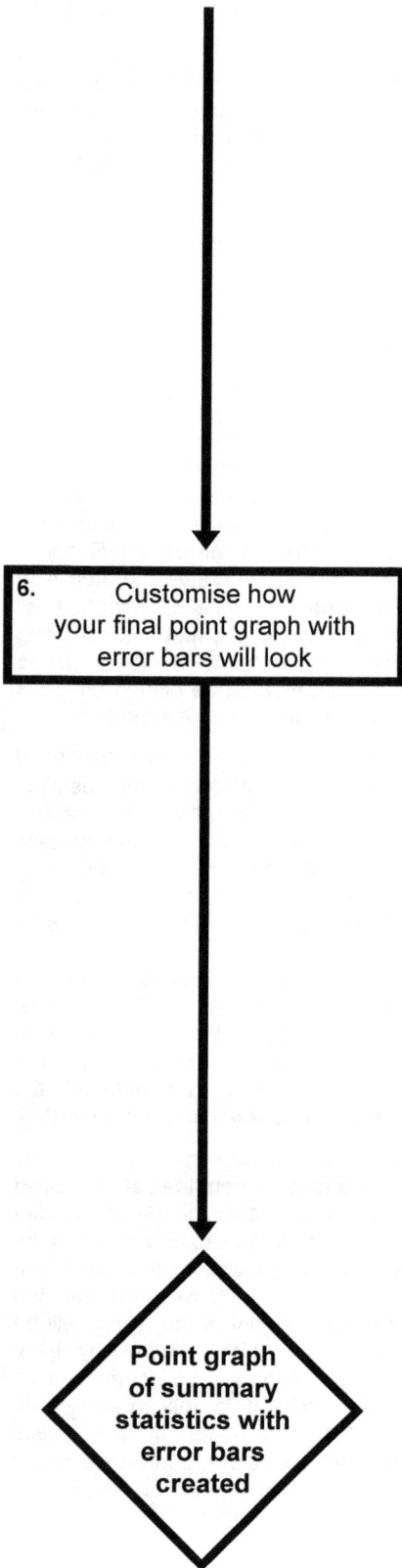

Once you have created your initial point graph with error bars, and you are happy with how the summary statistics and error bars are displayed on it, you can customise how your final graph will look. This can be done by adding a number of new style commands to the block of code used to produce it. To do this, edit the code from step 5 so that it looks like this (the newly added style commands are highlighted in **bold**):

```
ggplot(data=great_tit_summary_table,
    aes(x=hourday, y=mean_mass)) +
geom_errorbar(aes(ymin=mean_mass-se_mass,
    ymax=mean_mass+se_mass),width=0.2) +
    geom_point(stat="identity",size=3) +
labs(x="Hour of Day",y="Body Mass (g)") +
    ylim(c(17,22)) + theme_classic()
```

```
6.   Customise how
your final point graph with
     error bars will look
```

This is CODE BLOCK 57 in the document R_CODE_DATA_VISUALISATION_WORKBOOK.DOC, and it adds three new style commands separated by + symbols to the code block used to create your initial point graph in step 5. These are the `labs` command, the `ylim` command, and the `theme_classic` command. The `labs` command sets the labels that will be used for the X axis (using the `x` argument) and the Y axis (using the `y` argument). The `ylim` command sets the minimum and maximum value displayed on the Y axis of the graph. In this case, these will be `17` for the minimum value and `22` for the maximum value. Finally, the `theme_classic` command sets the remaining style elements of the final graph to those of the pre-existing classic theme. Once you have finished editing this code block, you can run it again to create the final version of your bar graph.

```
Point graph
of summary
statistics with
error bars
created
```

After you have worked through this example, you should check the contents of the summary table created in step 4. If it is not already visible, use the command `View(great_tit_summary_table)` to open a DATA VIEWER window so that you can see it. It should look like this:

	hourday	n	mean_mass	se_mass
1	0	341	18.72287	0.05750562
2	1	452	18.95708	0.05049581
3	2	457	19.29519	0.05450674
4	3	480	19.54792	0.05162874
5	4	509	19.63261	0.04752587
6	5	462	19.76342	0.05414572
7	6	551	19.92523	0.05113331
8	7	521	20.04511	0.05537578
9	8	412	19.90777	0.05178499
10	9	303	19.79670	0.05850003
11	10	205	19.83610	0.07281537
12	11	72	19.64583	0.12311072
13	12	6	19.65000	0.53400999

The final point graph with error bars that you created from this table in step 6 should look like this:

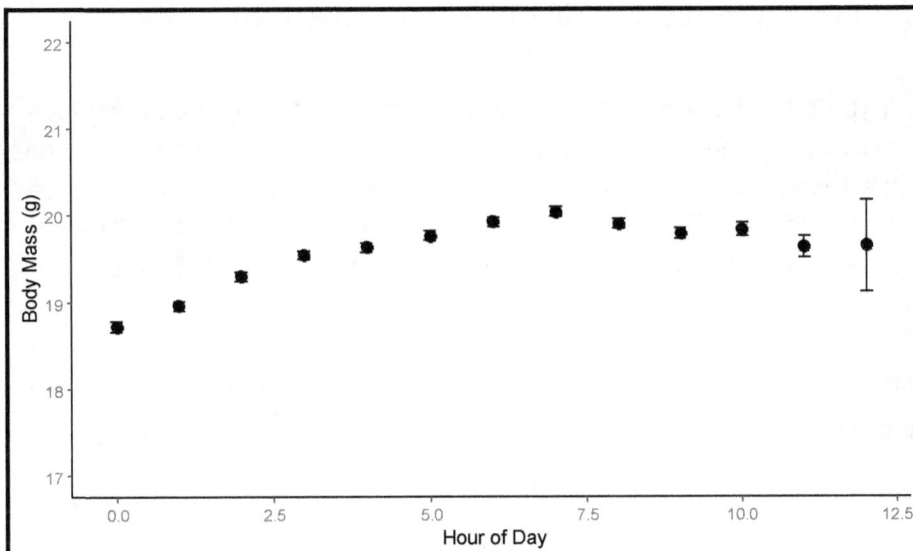

When you examine this graph, you will see that the bars showing the standard errors around the mean for the final hour of the day (hour 12) are much larger than those for the other hours of the day. If you go back to the great_tit_summary_table, you will see that this is because the sample size for this time period (n=6) is much smaller than for any of the other time periods. This is probably too few records to allow you to accurately estimate the standard error of a mean. As a result, it may be better to exclude these data from the final graph. To do this, you first need to use the subset command to remove the data from hour 12 from the summary table by entering the following line of code into R:

```
great_tit_summary_table=subset(great_tit_summary_table,
                    hourday<12)
```

Once you have done this, you need to check that your data have been processed correctly by entering the following View command into R:

```
View(great_tit_summary_table)
```

After you are happy that the data you wish to remove from the summary table have been correctly removed from it, you can then edit the code used to create your point graph (this is CODE BLOCK 57 from the document R_CODE_DATA_VISUALISATION_WORKBOOK.DOC) to create a new graph based on this subset of data. In this edited version of the command, the values in the ylim command have been changed to reflect the different ranges of data that are now displayed on the graph. This edited block of code should look like this (the required modifications are highlighted in **bold**):

```
ggplot(data=great_tit_summary_table,aes(x=
    hourday, y=mean_mass)) + geom_errorbar(aes(ymin=
  mean_mass-se_mass,ymax=mean_mass+se_mass),width=0.2) +
  geom_point(stat="identity",size=3) + labs(x="Hour of the
        Say",y="Body Mass (g)") + ylim(c(18,21)) +
              theme_classic()
```

The graph that is produced when you run this block of code should look like the image at the top of the next page.

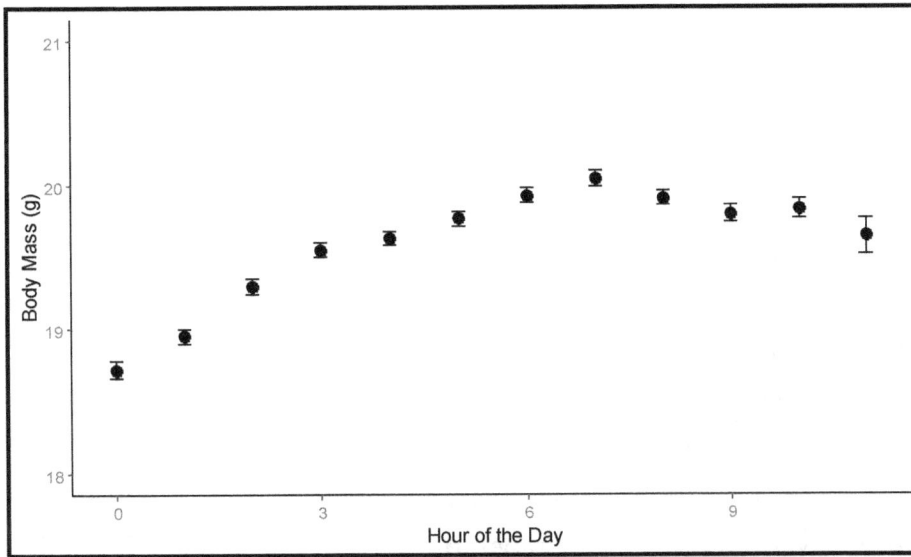

Once you have created a point graph with error bars that you are happy with, you can export it from R so that you can include it in a manuscript or presentation. If you are using RGUI, you can do this by clicking on the R GRAPHICS window containing your point graph to select it, before clicking on FILE on the main menu bar and selecting SAVE AS. This will allow you to save it in a variety of different formats. If you are using RStudio, you can export your graph by clicking on the EXPORT button at the top of the window displaying your point graph and selecting SAVE AS IMAGE.

When including a point graph of summary statistics with error bars in a manuscript, it is important that you provide an appropriate figure legend for it. This legend should provide all the information required for the reader to interpret the contents of the graph. For the above point graph, an appropriate legend would be:

Figure 1: *A comparison of the body mass of great tits at different hours of the day. Circles represent the mean values, while the error bars show the standard error of the mean. Only hourly samples with measurements from more than 50 individuals are shown on this graph.*

As the groupings on the X axis of this graph represent values on an ordinal scale (in this case, the hours of the day), you could choose to connect the individual mean values represented by the points to help illustrate how mean body mass changes throughout the hours of daylight. This can be done by adding a new graphing command to the code block used to create the graph (which can be found at the bottom of page 120). This new graphing command is `geom_line`, and it should be inserted into the block of code between the `ggplot` command and the `geom_errorbar` command so that the line is plotted underneath both the error bars and the points, rather than on top of them. At least three arguments should be specified in this new `geom_line` command. These are the `stat` argument, which sets the data that will be connected by the line, the `linetype` argument that specifies the type of line, and the `size` argument, which determines the width of the line. In this case, you will use `stat="identity"` (which tells the command to use the values in the `y` argument in the `aes` element of the `ggplot` command as the values to connect with the line), `linetype="dashed"` (to connect the points with a dashed line), and `size=0.5` to set the width of the line. To create a graph with a line connecting the mean values in this way, edit the above block of code so that it looks like this (the required modifications are highlighted in **bold**):

```
ggplot(data=great_tit_summary_table,aes(x=hourday,
  y=mean_mass)) + geom_line(stat="identity",linetype=
"dashed",size=0.5) + geom_errorbar(aes(ymin=mean_mass-
      se_mass,ymax=mean_mass+se_mass),width=0.2)+
      geom_point(stat="identity",size=3) + labs(x=
  "Hour of the Day",y="Body Mass (g)") + ylim(c(18,21)) +
                  theme_classic()
```

When you run this code block, it should produce a graph that looks like this:

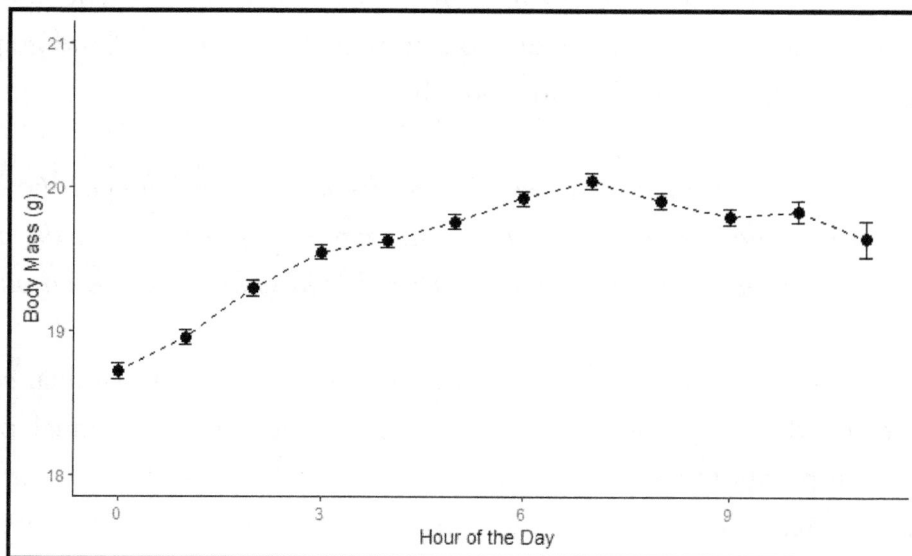

EXERCISE 2.5: HOW TO CREATE A POINT GRAPH OF SUMMARY STATISTCIS FOR TWO VARIABLES WITH HORIZONTAL AND VERTICAL ERROR BARS:

So far the graphs you have created in this chapter have plotted summary statistics for a single variable. However, there will be some situations where you wish the plot summary statistics for two variables against each other in order to visualise any relationship there may be between them. For example, you may wish to know whether there is a relationship between the average mass of individuals from different sampling locations and their average size (as measured by their body length). These types of relationships can be visualised by creating an X-Y scatter plot where the mean or median values for different groups in your data set for one variable are plotted on the X axis and the mean or median values for the same groups for the other variable are plotted on the Y axis. On such graphs, you may also want to include a measure of the variance in each group by adding two sets of error bars to it, one running horizontally (showing the variation in the variable being plotted on the X axis) and a second running vertically (showing the variation in the variable being plotted on the Y axis).

In this exercise, you will learn how to create a point graph of the summary statistics for two variables in different groups in a data set along with error bars to show the variation in each group for each variable. This will be done using the same data on the morphometrics of anglehead lizards sampled from three locations in Malaysia used in Exercise 2.3. You will start by creating an X-Y scatter plot of the average body length and the average forelimb length for the three locations, with error bars that represent the standard error of the measurements from each location. To create this graph, work through the flow diagram that starts at the top of the next page.

NOTE: This workflow assumes that you have already downloaded and installed the ggplot2, Rcpp, dplyr, plyr and plotrix packages and that the command libraries from these packages have already been loaded into your analysis project (see Exercises 2.1 and 2.3). If you have not already done these steps as part of earlier exercises in this chapter, you will need to do them before you start working through this flow diagram.

Data set held in an existing R object in your analysis project

For this example, the data set you will use is stored in the R object called `all_locations` created in Exercise 2.3 of this workbook.

The first step in making a point graph of summary statistics for two variables with horizontal and vertical error bars is to create a table of summary statistics for the variables that you wish to plot on it. In this example, the variables are the body lengths and forelimb lengths of anglehead lizards sampled from three different locations in Malaysia. To do this, enter the following command into R:

```
lizard_summary_table <-
ddply(all_locations,c("location"),
 summarise,n=length(body_length),
  mean_length=mean(body_length),
se_body_length=std.error(body_length),
mean_forelimb_length=mean(forelimb_length),
   se_forelimb_length=std.error(
        forelimb_length))
```

1. Create a table containing the summary statistics that you wish to display on your point graph

This is CODE BLOCK 58 in the document R_CODE_DATA_VISUALISATION_WORKBOOK.DOC. This `ddply` command will create a summary table from the data held in the data set called `all_locations`. In this summary table, the individual rows are defined by data groupings provided in the `location` column. This is set by the `c("location")` argument. Other arguments are then used to provide a name and a calculation for each summary statistic for each variable that will be added to the table based on these groupings. For example, the `n=length(body_length)` argument will create a column in the summary table called `n` that will give the count of the data in each group (which is calculated using the `length` argument). Similarly, the `mean_length=mean(body_length)` argument creates a column in the summary table called `mean_length` that will contain the mean of the `body_length` data for each group, while the `se_body_length=std.error(body_length)` argument creates a column in the summary table called `se_body_length` that will contain the standard error of the `body_length` data for each group. Similar arguments are used to create columns in the summary table with the required summary statistics for the data on forelimb length held in the column called `forelimb_length`.

Once your summary table has been created, you should view it to check that it contains the information you need it to contain. To do this, enter the following code into R:

```
View(lizard_summary_table)
```

This is CODE BLOCK 59 in the document R_CODE_DATA_VISUALISATION_WORKBOOK.DOC. This command will open a DATA VIEWER window where you can review the summary table that you have just created to ensure it contains the information required to make your graph.

Once you have successfully created your summary table and checked that it contains the correct information, you are ready to use it to make your initial X-Y scatter plot based on the summary statistics it contains. To do this, enter the following block of code into R:

```
ggplot(data=lizard_summary_table,
aes(x=mean_length,y=mean_forelimb_length)) +
    geom_point(stat="identity",size=3,
        shape=21,fill="black")
```

This is CODE BLOCK 60 in the document R_CODE_DATA_VISUALISATION_WORKBOOK.DOC, and it contains two commands separated by a + symbol. These are the `ggplot` command and the `geom_point` command. The `ggplot` command sets the data set which will be used for the graph. This is done using the `data` argument and, in this case, it will be the R object called `lizard_summary_table` created in step 1 of this exercise. The column of data that will be plotted on the X axis of the resulting graph is set using the `x` argument of the `aes` element of this `ggplot` command. In this case, it is the column called `mean_body_length` in the `lizard_summary_table` data set. The column of data that will be plotted on the Y axis is set using the `y` argument of the `aes` element. In this case, it is the column called `mean_forelimb_length`.

2. Create your initial point graph from the summary table created in step 1

The second command in this code block, `geom_point`, sets the type of graph that will be created from the data specified in the `ggplot` command. In this case, it will be a point graph. Within this command, the values to be plotted on the X and Y axes are set by the `stat` argument. In this case, the argument used is `stat="identity"`, meaning that position of each point on the scatter plot is determined by the values found in the columns named in the `x` and `y` arguments in the `aes` element of the `ggplot` command. The size of the points representing the mean values for each group on the graph are set by the `size` argument in the `geom_point` command, while the shape and the fill colour are set by the `shape` and `fill` arguments in it. In this case, the `size` is set to 3, the `shape` to 21 and the `fill` argument to `black`. **Note:** The number included in the `shape` argument is a code (called pch) that refers to a specific shape. Details of which codes refer to which shapes can be found at *www.sthda.com/english/wiki/r-plot-pch-symbols-the-different-point-shapes-available-in-r.*

125

After you have created your basic scatter plot where each point represents the mean values for the two variables for each different group in your data set, you can then add the horizontal and vertical error bars to it. This is done by adding two new graphing commands to the code block used to create the scatter plot. To do this, edit the code from step 2 so that it looks like this (the newly added graphing commands are highlighted in **bold**):

```
ggplot(data=lizard_summary_table,aes(x=
    mean_length,y=mean_forelimb_length)) +
    geom_errorbar(aes(xmin=mean_length-
        se_body_length,xmax=mean_length+
se_body_length),width=0.2) + geom_errorbar(
        aes(ymin=mean_forelimb_length-
            se_forelimb_length,ymax=
    mean_forelimb_length+se_forelimb_length),
    width=1) + geom_point(stat="identity",
        size=3,shape=21,fill="black")
```

This is CODE BLOCK 61 in the document R_CODE_DATA_ VISUALISATION_WORKBOOK.DOC. The two newly added commands are both `geom_errorbar` commands, one which creates the horizontal error bars and another which creates the vertical error bars. **NOTE:** These `geom_ errorbar` commands need to be added before the `geom_ point` command. This ensures that the error bars plot underneath the points representing the mean values.

The first `geom_errorbar` command creates the horizontal error bars for the graph. The `xmin` and `xmax` arguments in the `aes` element of this command set the upper and lower limits of the error bars. In this case, they are set to be the value from the `mean_body_length` column in the summary table data set minus the value from the `se_body_length` column (which contains the standard error for body length for each group of data) for the `xmin` argument, and `mean_body_length` plus the value from the `se_body_length` column for the `xmax` argument. The width of the line at the end of the horizontal error bar is set by the `width` argument in this command. This is in the same units as the variable being plotted on the X axis. In this case, the argument used is `width=0.2`.

The second `geom_errorbar` command creates the vertical error bars for the graph. The `ymin` and `ymax` arguments in the `aes` element of this command set the upper and lower limits of the error bars. In this case, they are set to be the value from the `mean_forelimb_length` column in the summary table data set minus the value from the `se_forelimb_length` column (which contains the standard error for forelimb length for each group of data) for the `ymin` argument, and `mean_forelimb_length` plus the value from the `se_forelimb_length` column for the `ymax` argument. The width of the line at the top of the error bar is set by the `width` argument in this command. This is in the same units as the variable being plotted on the Y axis. In this case, the argument used is `width=1`. Once you have finished editing this code block, you can run it again to create an updated version of your graph.

3. Add horizontal and vertical error bars to your graph based on the data from the summary table created in step 1

Once you have created your initial X-Y scatter plot and added horizontal and vertical error bars to it, and you are happy with how the summary statistics and error bars are displayed on it, you can customise how the final graph will look. This can be done by adding a number of new style commands to the block of code used to produce it. To do this, edit the code from step 3 so that it looks like this (the newly added style commands are highlighted in **bold**):

```
ggplot(data=lizard_summary_table,aes(x=
   mean_length,y=mean_forelimb_length)) +
   geom_errorbar(aes(xmin=mean_length-
      se_body_length,xmax=mean_length+
se_body_length),width=0.2) + geom_errorbar(
      aes(ymin=mean_forelimb_length-
         se_forelimb_length,ymax=
   mean_forelimb_length+se_forelimb_length),
   width=1) + geom_point(stat="identity",
size=3,shape=21,fill="black") + labs(x="Body
   Length (cm)",y="Forelimb Length (cm)") +
      xlim(c(28,40)) + ylim(c(6.8,7.8)) +
                 theme_classic()
```

4. Customise how your final scatter plot will look

This is CODE BLOCK 62 in the document R_CODE_DATA_ VISUALISATION_WORKBOOK.DOC, and it adds four new style commands separated by + symbols to the code block used to create your initial scatter plot with two sets of error bars in step 3. These are the `labs` command, the `xlim` command, the `ylim` command, and the `theme_classic` command. The `labs` command sets the labels that will be used tor the X axis (using the `x` argument) and the Y axis (using the `y` argument). The `xlim` command sets the minimum and maximum value displayed on the X axis of the graph. In this case, these will be `28` for the minimum value and `40` for the maximum value. The `ylim` command sets the minimum and maximum value displayed on the Y axis of the graph. In this case, these will be `6.8` for the minimum value and `7.8` for the maximum value. Finally, the `theme_classic` command sets the remaining style elements of the final graph to those of the pre-existing classic theme. Once you have finished editing this code block, you can run it again to create an updated version of your graph.

Finally, you need to add a way to allow the reader to identify which point on your scatter plot represents the summary statistics for which group of data. This can be done in a number of different ways, including adding labels, using different symbols for each group of data or using different colours for each group. This is done by modifying the block of code used to create the customised graph in step 4. For this example, you will add labels based on the name of each group provided in the summary table created in step 1. To do this, edit the code from step 4 so that it looks like this (the newly added style command is highlighted in **bold**):

```
ggplot(data=lizard_summary_table,aes(x=
mean_length,y=mean_forelimb_length))  +
    geom_errorbar(aes(xmin=mean_length-
       se_body_length,xmax=mean_length+
se_body_length),width=0.2) + geom_errorbar(
      aes(ymin=mean_forelimb_length-
           se_forelimb_length,ymax=
mean_forelimb_length+se_forelimb_length),
     width=1) + geom_point(stat="identity",
size=3,shape=21,fill="black") + labs(x="Body
  Length (cm)",y="Forelimb Length (cm)") +
     xlim(c(28,40)) + ylim(c(6.8,7.8)) +
     theme_classic() + geom_text(aes(label=
        location,hjust=-0.55,vjust=1.5))
```

This is CODE BLOCK 63 in the document R_CODE_DATA_ VISUALISATION_WORKBOOK.DOC, and it adds one new style command to the code block used to create your customised graph in step 4. This is the `geom_text` command. In this command, you can specify the column containing the labels you wish to use for each point (using the `label` argument in the `aes` element). In this case, this is the `location` column in the `lizard_summary_table` data set specified in the `data` argument in the `ggplot` command. In addition, you can specify the offset required for these labels using the `hjust` and the `vjust` arguments. The former adjusts the position of the label horizontally, and the latter adjusts it vertically. The units for these arguments are the same as the units for the X axis (for `hjust`) and the Y axis (for `vjust`). In this case, the arguments included in the `geom_text` command are `hjust=-0.55` and `vjust=1.5`. Once you have finished editing this code block, you can run it again to create the final version of your graph.

5. Add a way to identify which group each point on your graph represents

Point graph of summary statistics for two variables with horizontal and vertical error bars created

At the end of this exercise, you should have an X-Y scatter plot showing the mean values for the two variables, with both horizontal and vertical error bars, that looks like this:

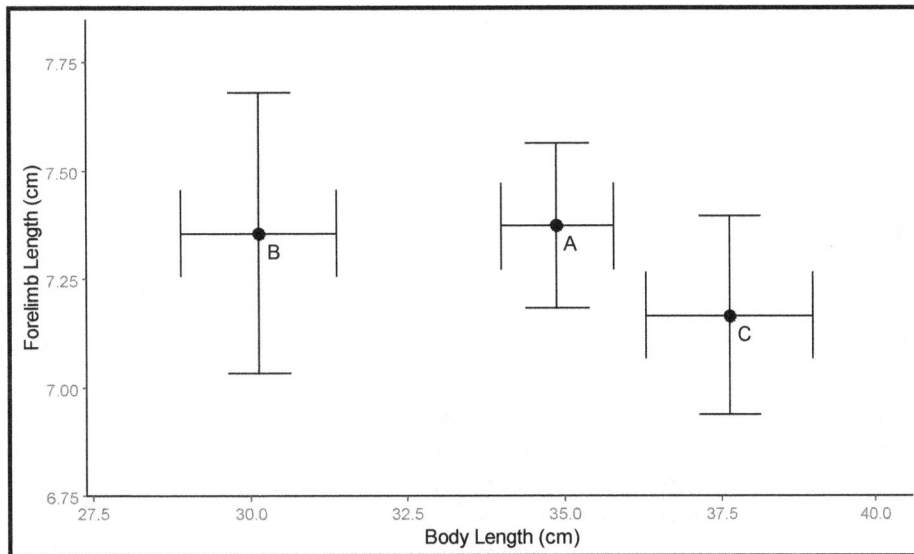

Once you have created a point graph showing summary statistics for two different variables and associated error bars, you can export it from R so that you can include it in a manuscript or presentation. If you are using RGUI, you can do this by clicking on the R GRAPHICS window containing your scatter graph to select it, before clicking on FILE on the main menu bar and selecting SAVE AS. This will allow you to save it in a variety of different formats. If you are using RStudio, you can export your graph by clicking on the EXPORT button at the top of the window displaying your scatter plot and selecting SAVE AS IMAGE.

When including such a point graph of summary statistics in a manuscript, it is important that you write an appropriate figure legend for it. This legend should provide all the information required for the reader to interpret the contents of the graph. For the above graph, an appropriate legend would be:

Figure 1: A comparison of the morphometrics of anglehead lizards sampled from three locations (A, B and C) in Malaysia. The points represent the mean values for body length (X axis) and forelimb length (Y axis), while the error bars represent the standard errors for these two variables.

The symbols representing the different groups of data on this type of graph can be identified in a number of ways. This includes using the groupings in a column from the summary table on which the scatter plot is based, using custom text labels, using symbols with different shapes and/or using symbols with different colours. The commands and arguments required to label the different groups on an X-Y scatter plot of summary statistics in these ways are provided in the table below.

Method for identifying the data from different groups	Commands and arguments required to apply it
Using text labels based on a column in the summary table	To identify the data from different groups on your X-Y scatter plot using a text label from a column in the summary table your graph is based on, you need to include the `geom_text` command at the end of your code block, and include the `label` argument in its `aes` element, along with the `hjust` and `vjust` arguments to tell R exactly where you wish the labels to be positioned. This `label` argument should contain a term that identifies the column with the groupings you wish to use as your labels. For example, to create labels for the data on a scatter plot based on a column called `location` that are positioned directly beside each symbol, you would include the following command in your final code block: `geom_text(aes(label=location,hjust=0,vjust=0))` **NOTE:** The values for the `hjust` argument are in the units of the X axis, while the values for the `vjust` argument are in the units of the Y axis. Negative values will move the position of the labels right and up, while positive values will move them left and down.

Using custom text labels	To identify the data from different groups on your X-Y scatter plot using custom text labels, you need to include the `geom_text` command at the end of your code block, and include the `label` argument in its `aes` element, along with the `hjust` and `vjust` arguments to tell R exactly where you wish the labels to be positioned. This `label` argument should contain a term that identifies the custom text labels you wish to use for the data from each group. For example, to create labels for the data on your scatter plot called `Loc. A`, `Loc. B` and `Loc. C` that are positioned directly beside each symbol, you would include the following command in your final code block:
	
	`geom_text(aes(label=c("Loc. A","Loc. B","Loc.`
	`C"),hjust=0,vjust=0)))`
	
	NOTE: The values for the `hjust` argument are in the units of your X axis, while the values for the `vjust` argument are in the units of your Y axis. Negative values will move the position of the labels right and up, while positive values will move them left and down.

Method for identifying the data from different groups	Commands and arguments required to apply it
Using different symbols	To identify the data from different groups on your X-Y scatter plot using different symbols, you need to include a `shape` argument in the `aes` element of your `ggplot` command, and use it to identify the column from your summary table that contains the information about which group each data point represents. For example, to use different shapes to display the summary statistics from the groupings provided in a column called `location`, you would include the argument `shape=location` in the `aes` element of your `ggplot` command. **NOTE:** You would also need to remove any other `shape` arguments you have anywhere else in your code block or they will over-ride the `shape` argument in the `aes` element of the `ggplot` command. If you wish to manually specify which symbols should be used for the data from each group, you would need to also include a `scale_shape_manual` command to tell R which shape should be used to plot the data from each group using pch codes (see *www.sthda.com/english/wiki/r-plot-pch-symbols-the-different-point-shapes-available-in-r* for more details). For example, to use a circle, a square and a triangle as the shapes to plot the summary statistics from three different groups, you would need to include the following command in your code block: `scale_shape_manual(values=c(19,15,17))`
Using different colours	To identify the data from different groups on your X-Y scatter plot using different colours, you need to include a `colour` argument in the `aes` element of your `ggplot` command, and use it to identify the column from your summary table that contains the information about which group each data point represents. For example, to use different colours to display the summary statistics based on the groupings provided in a column called `location`, you would include the argument `colour=location` in the `aes` element of your `ggplot` command. **NOTE:** You would also need to remove any other `colour` arguments you have anywhere else in your code block or they will over-ride the `colour` argument in the `aes` element of the `ggplot` command. If you wish to manually specify which colour should be used for the data from each group, you would need to also include a `scale_colour_manual` command to tell R which colour should be used to plot the data from each group. For example, to use blue, red and green as the colours to plot the summary statistics from three different groups, you would need to include the following command in your code block: `scale_colour_manual(values=c("blue","red","green"))` **NOTE:** This option will only work for symbols with pch codes 1 to 20. For symbols with codes 21 to 25, the `colour` argument and the `scale_colour_manual` command will only affect the outline of the shape. To change the fill colour, you would also need to add a `fill` argument to the `aes` element of the `ggplot` command, and to manually determine what these fill colours should be, you would include the `scale_fill_manual` command at the end of your code block.

For the next part of this exercise, you will change the labels for the data points on your scatter plot from being based on the contents of a specific column of your summary table to using custom text. This is done by editing the `label` argument in the `aes` element of the `geom_text` command of the block of code used to create your final graph in step 5 of the above flow diagram (this is CODE BLOCK 63 in the document R_CODE_DATA_ VISUALISATION_WORKBOOK.DOC). To use the labels Loc. A., Loc. B and Loc. C, rather than just A, B and C (as was the case when you base the labels on the contents of the column called `location`), you should edit this `label` argument so that it says `label=c("Loc. A","Loc. B","Loc. C")`. When you do this, you also need to change the values in the `hjust` and `vjust` arguments to ensure that the new labels do not overlap with the error bars for each point. The modified block of code should then look like this (the required modifications are highlighted in **bold**):

```
ggplot(data=lizard_summary_table,
 aes(x=mean_length,y=mean_forelimb_length)) +
 geom_errorbar(aes(xmin=mean_length-se_body_length,
  xmax=mean_length+se_body_length),width=0.2) +
  geom_errorbar(aes(ymin=mean_forelimb_length-
   se_forelimb_length,ymax=mean_forelimb_length+
   se_forelimb_length), width=1) + geom_point(stat=
  "identity",size=3,shape=21,fill="black")+labs(x="Body
 Length (cm)",y="Forelimb Length (cm)") + xlim(c(28,40)) +
 ylim(c(6.8,7.8)) + theme_classic() + geom_text(aes(label=
  c("Loc. A","Loc. B","Loc. C"),hjust=-0.2,vjust=4))
```

This should produce a new scatter plot that looks like the image at the top of the next page.

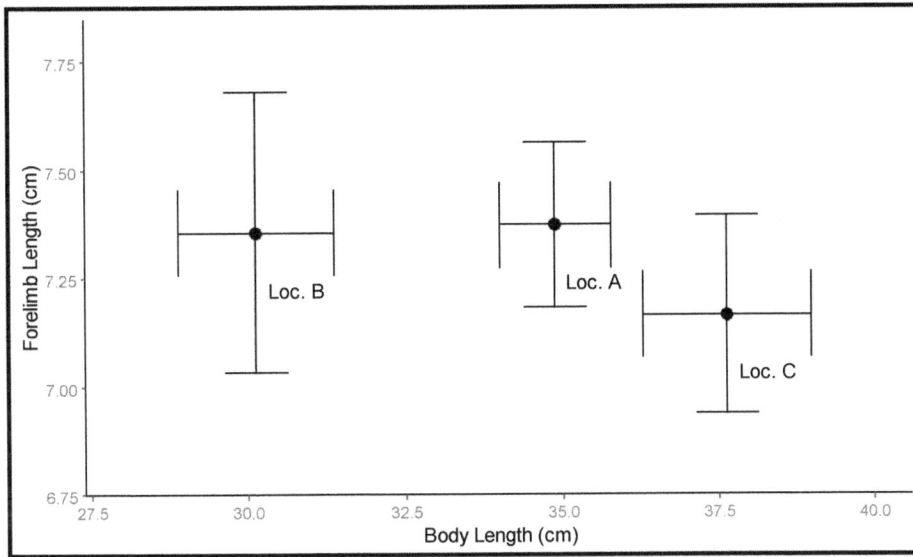

Next, you will alter the symbols used for your data points. This is done using a `shape` argument in the `aes` element of the `ggplot` command to define the column which contains the data groupings, and then adding the `scale_shape_manual` command to allow you to set the specific shape that should be used to represent the summary statistics from each group of data. To create a graph that uses a circle for summary statistics for group A, a square for the summary statistics for group B and a triangle for the summary statistics from group C, add the argument `shape=location` to the `aes` element of the `ggplot` command of the above code block, and then add the `scale_shape_manual` command with the argument `values=c(19,15,17)` at the end of the code block. You will also need to remove the `shape` argument from the `geom_point` command so that it does not over-ride the new `shape` argument in the `ggplot` command. You can also remove the `geom_text` command from the end of the code block as you no longer need to have text labels if you are using different shapes to identify the data from different groups in your data set. The modified block of code should then look like the code at the top of the next page (the required modifications are highlighted in **bold**).

```
       ggplot(data=lizard_summary_table,
      aes(x=mean_length,y=mean_forelimb_length,
  shape=location)) + geom_errorbar(aes(xmin=mean_length-
     se_body_length,xmax=mean_length+se_body_length),
 width=0.2) + geom_errorbar(aes(ymin=mean_forelimb_length-
      se_forelimb_length, ymax=mean_forelimb_length+
 se_forelimb_length), width=1) + geom_point(stat="identity",
      size=3,fill="black") + labs(x="Body Length
    (cm)",y="Forelimb Length (cm)") + xlim(c(28,40))+
        ylim(c(6.8,7.8)) + theme_classic() +
      scale_shape_manual(values=c(19,15,17))
```

Once you have finished modifying this command, you can run it. This should produce a new scatter plot that looks like this:

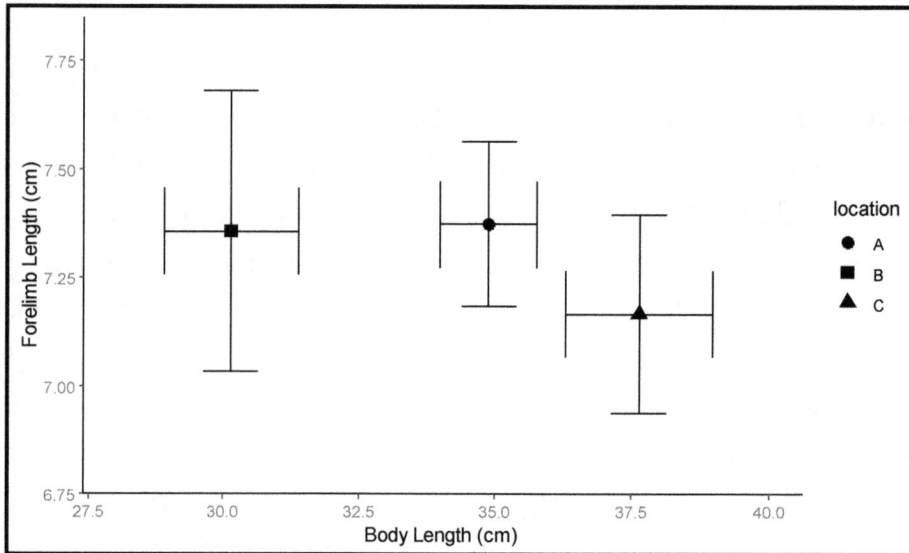

EXERCISE 2.6: HOW TO CREATE A BOX PLOT TO SUMMARISE THE DATA FROM DIFFERENT GROUPS IN A DATA SET:

The bar and point graphs you created in the earlier exercises in this chapter all have one thing in common. This is that they use a combination of central values, such as means, and a measure of variance, such as standard deviation or standard error, to represent and compare the data from different groups. However, sometimes it is more useful to be able to compare more detailed information about the range of values for a particular variable found in each group. While this could be done by plotting and comparing separate frequency distribution histograms for each group of data, as was done in the exercises in Chapter 3, an alternative way of doing this is to use box plots. A box plot can be used to show a number of characteristics of the distribution of data for different groups in a data set in a format that is easy to understand and compare. This is because all the information for the different groups is plotted side-by-side on a single graph. The distribution characteristics which can be shown on box plots include the median value, the interquartile range, the full range of values, outliers, the confidence interval around the median, the mean value and the distribution of all the individual data points within each group. In this exercise you will produce a series of box plots for a data set consisting of the body lengths of individuals from three species of deep water benthic sharks. You will start by producing a relatively simple box plot that just compares the three species in terms of their median values, their interquartile ranges, their range of data that falls within 1.5 times the interquartile range above and below it, and any outliers in their distributions. To do this, work through the flow diagram that starts at the top of the next page.

NOTE: These instructions assume that you have already installed the `ggplot2` package in your version of R and have its command library loaded into your analysis project. You can check if you already have the `ggplot2` package installed by using the command `library()`. If you find you need to install the this package, instructions for how to do it can be found in Exercise 2.1. Once you have ensured you have the `ggplot2` package installed in your version of R, you can load its command library into your analysis project by entering the following command into R:

```
library(ggplot2)
```

Data set held in a comma separated values (.CSV) file

For this example, the data set you will use is stored in a file called `shark_data.csv` that is located in the WORKING DIRECTORY folder you created during the introduction to this chapter.

1. Set the WORKING DIRECTORY for your analysis project

Before you start any analysis in R, you first need to set the WORKING DIRECTORY. To do this, enter the text `setwd("` and then type the address of your WORKING DIRECTORY, using slashes (/) as the folder separators, before entering a second quotation mark followed by a closing bracket, like this `")`. For example, if your WORKING DIRECTORY has the address C:\STATS_FOR_BIOLOGISTS_TWO, your `setwd` command should look like this:

```
setwd("C:/STATS_FOR_BIOLOGISTS_TWO")
```

If you are using RGUI, enter your `setwd` command in the R CONSOLE window (remembering to use the address of your own WORKING DIRECTORY folder in it) and then press the ENTER key on your keyboard. If you are using RStudio, enter your `setwd` command into the SCRIPT EDITOR window. To run it, select it and then click on the RUN button at the top of this window. You will enter all the remaining commands for this exercise in a similar manner, depending on the user interface you are using.

To check that your WORKING DIRECTORY has been set properly, enter the command `getwd()` and carefully check that the address it returns is the same as the one for the STATS_FOR_BIOLOGISTS_TWO folder you created at the start of this chapter.

Before you move on to step 2, make sure that all the data you wish to use in your analysis project are located in this WORKING DIRECTORY folder. In this case, this is a file called `shark_data.csv`. **NOTE:** If the data you are going to import into R in step 2 are not located in the WORKING DIRECTORY you set in this step, the import code provided in the next step will not work.

The `read.table` command provides the easiest way to load data held in a .CSV file (and stored in the WORKING DIRECTORY you set in step 1) into R so you can analyse it. To do this for the data set being used in this example, enter the following command into R:

```
shark_data <-
read.table(file="shark_data.csv",sep=",",
as.is=FALSE,header=TRUE)
```

This code has to be entered exactly as it is written here or it will not work. If you wish to use the copy-and-paste approach for entering this command, copy the text directly below CODE BLOCK 64 in the document R_CODE_DATA_VISUALISATION_WORKBOOK.DOC and paste it into R.

This command will create a new object in R called `shark_data` which will contain the data from the specified .CSV file. To import a different .CSV file into R, all you need to do is change the file name in the `file` argument to the name of the one you wish to import. You can also use whatever name you wish for the R object which will be created by this command. To do this, simply replace `shark_data` at the start of the first line of the above code with the name you wish to use for it. **NOTE:** If your .CSV data set uses a semicolon as the column separator, you would need to replace the `sep=","` argument with `sep=";"`.

2. Load your data into R using the `read.table` command

Whenever you import any data into R you need to check that they have loaded correctly. First, you need to check that all the required columns are present in the R object you just created. To do this, enter the following command into R:

```
names(shark_data)
```

This is CODE BLOCK 65 in the document R_CODE_DATA_VISUALISATION_WORKBOOK.DOC. This command will return the names used for each column in the R object you just created. For this example, the names should be: `id`, `body_length`, `species` and `sex`.

Next, you should view the contents of the whole table using the `View` command. This is done by entering following code into R:

```
View(shark_data)
```

This is CODE BLOCK 66 in the document R_CODE_DATA_VISUALISATION_WORKBOOK.DOC. This command will open a DATA VIEWER window where you can examine your data set and check that the correct data have been loaded into R.

3. Check the data have loaded into R correctly by checking the names of the columns and by viewing it

4. Create your initial box plot with separate boxes for each group of data in your data set

Once you have loaded your data into your analysis project, you are ready to use it to create your initial box plot. To do this, enter the following block of code into R:

```
ggplot(data=shark_data,aes(x=species,
y=body_length)) + geom_boxplot(coef=1.5,
    notch=FALSE,outlier.colour="black",
    outlier.shape=16,outlier.size=2)
```

This is CODE BLOCK 67 in the document R_CODE_DATA_ VISUALISATION_WORKBOOK.DOC, and it contains two commands separated by a + symbol. These are the ggplot command and the geom_boxplot command. The ggplot command sets the data set which will be used for the graph. This is done using the data argument and, in this case, it will be the R object called shark_data created in step 2 of this exercise. The column of data which contains the groupings that will be plotted on the X axis of the resulting graph is set using the x argument of the aes element of this ggplot command. In this case, it is the column called species in the shark_data data set. The column containing the data that will be used to create the 'boxes' for each group of data is set using the y argument of the aes element of this ggplot command. In this case, it is the column called body_length.

The second command in this code block, geom_boxplot, sets the type of graph that will be created from the data specified in the ggplot command. In this case, it will be a box plot. Within this command, you can use a variety of different arguments to determine what data will be included in the 'boxes' for each species and how they will be displayed. The coef argument determines the length of the whiskers above and below the box representing the interquartile range. In this case, the argument used is coef=1.5, which means the whiskers will be 1.5 times the interquartile range (i.e. the height of the main box). Using a larger value in this argument will result in longer whiskers and potentially fewer outliers, while using a smaller value will result in shorter whiskers and potentially more outliers. The notch argument determines whether there will be a notch in the main box representing the 95% confidence interval around the median. In this case, the argument used is notch=FALSE, meaning no notches will be present on the boxes on the graph. The outlier.colour, outlier. shape and outlier.size arguments set the colour, shape and size used to plot any outliers that are not included in the range of values covered by the main box and the whiskers. In this case, the terms in these arguments will create outliers which are black in colour, have a shape determined by the code 16 (which is a solid circle) and have a size of 2.

138

5. Customise how your final box plot will look

Once you have created your initial box plot, and you are happy with how the distribution of data for each group is being displayed, you can customise how the final graph will look. This can be done by adding a number of new style arguments and commands to the block of code used to produce it. To do this, edit the code from step 4 so that it looks like this (the newly added style arguments and commands are highlighted in **bold**):

```
ggplot(data=shark_data,aes(x=species,
y=body_length)) + geom_boxplot(coef=1.5,
  notch=FALSE,outlier.colour="black",
  outlier.shape=16,outlier.size=2,fill=
"grey",colour="black") + labs(x="Species",
y="Body Length (m)") + ylim(c(1.5,5.5)) +
          theme_classic()
```

This is CODE BLOCK 68 in the document R_CODE_DATA_ VISUALISATION_WORKBOOK.DOC. This code block adds arguments to the `geom_boxplot` command to set the colours to be used for the boxes themselves, and then adds three new style commands separated by + symbols to the code block used to create your initial box plot. The new arguments added to the `geom_boxplot` command are `fill`, which sets the fill colour for the central boxes for each group, and `colour`, which sets the colour used for their outlines. In this case, the arguments set the fill colour to `grey` and the outline to `black`

The newly added commands are the `labs` command, the `ylim` command, and the `theme_classic` command. The `labs` command sets the labels that will be used tor the X axis (using the `x` argument) and the Y axis (using the `y` argument). The `ylim` command sets the minimum and maximum value displayed on the Y axis of the graph. In this case, these will be `1.5` for the minimum value and `5.5` for the maximum value. Finally, the `theme_classic` command sets the remaining style elements of the final graph to those of the pre-existing classic theme. Once you have finished editing this code block, you can run it again to create the final version of your graph.

Box plot created showing the distribution of data for different groups in a data set

At the end of this exercise, you should have a box plot showing the range of data for each of the three species in the `shark_data` data set that looks like this:

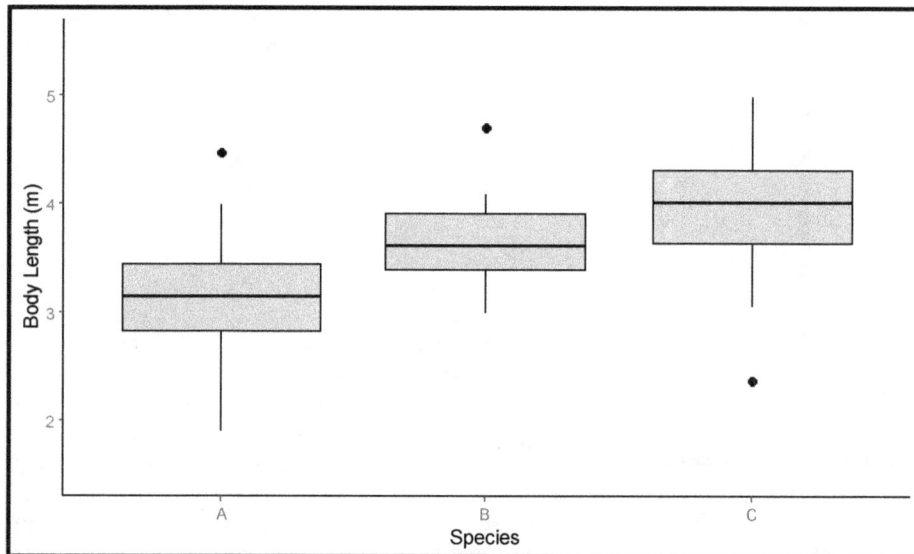

Once you have created a box plot, you can export it from R so that you can include it in a manuscript or presentation. If you are using RGUI, you can do this by clicking on the R GRAPHICS window containing your box plot to select it, before clicking on FILE on the main menu bar and selecting SAVE AS. This will allow you to save it in a variety of different formats. If you are using RStudio, you can export your graph by clicking on the EXPORT button at the top of the window displaying your box plot and selecting SAVE AS IMAGE.

When including a box plot in a manuscript, it is important that you write an appropriate figure legend for it. This legend should provide all the information required for the reader to interpret the contents of the graph. For the above box plot, an appropriate legend would be:

Figure 1: *A comparison of the body lengths of three different species of deep water shark. Horizontal bar: Median value; Box: Interquartile range; Whiskers: 1.5 times the interquartile range; Points: Data points beyond the range of the whiskers.*

The basic box plots created using GGPlot can be customised in a number of different ways so that they show exactly what you wish them to show. This includes changing the range of values covered by the whiskers, adding a notch to represent the confidence intervals around the median value, adding a symbol to mark the mean value of the data and adding data points to show the distribution of all the data on which the boxes are based. Information on how to make these customisations is provided in the table below.

Required customisation	How to achieve it for a box plot created with the `geom_boxplot` command
Changing the fill and outline colours of the boxes representing the interquartile range	The colours used to draw the boxes representing the interquartile ranges can be customised by using the `fill` and `colour` arguments in the `geom_boxplot` command. The setting for these two arguments are the names of the colour(s) you wish to use for the fill and the outline, respectively. For example, to create a graph with boxes representing the interquartile range that are outlined in black and filled with grey, you would include the arguments `fill="grey", colour="black"` in the `geom_boxplot` command.
Adding a notch representing the 95% CI around the median value	To add a notch to the boxes of the interquartile range which indicates the 95% confidence interval around the median value for the data they represent, you need to change the `notch` argument in the `geom_box` command from FALSE to TRUE.
Changing the range of values covered by the whiskers above and below the interquartile range boxes	The length of the whiskers on a box plot are controlled by the `coef` argument in the `geom_boxplot` command. The value used in the argument is a multiplier for the interquartile range. To change the range of values that are represented by these whiskers, you simply need to change the multiplier value in this argument. For example, if you include the argument `coef=1.5`, the whiskers will extend above and below the boxes to a range of 1.5 times the interquartile range (represented by the boxes themselves). Similarly, if you use the argument `coef=2.0`, the whiskers will cover a range of values that is 2 times the interquartile range.
Changing the symbols used to represent any outliers	The symbols used for the outliers (the points that lie beyond the range of values covered by the boxes and the whiskers) are determined by the arguments `outlier.shape`, `outlier.colour`, `outlier.fill`, and `outlier.size` in the `geom_boxplot` command. The `outlier.shape` argument uses pch codes to determine the shape of the symbol used for the outliers (see *www.sthda.com/english/wiki/r-plot-pch-symbols-the-different-point-shapes-available-in-r*), while the `outlier.colour` and `outline.fill` arguments use the desired colour names (**NOTE:** The `outlier.fill` argument is only required for symbols with pch codes from 21 to 25) and the `outlier.size` argument uses a numerical value. For example, including the arguments `outlier.colour="black", outlier.shape=16` and `outlier.size=2`, will result in a box plot with round, black symbols of size 2 for any outliers present in the data set being used to make it. If you wish to create a box plot with no outliers on it, this can be done by setting the `outlier.shape` argument to NA.

Required customisation	How to achieve it for a box plot created with the `geom_boxplot` command

Adding a marker to represent the mean value for the data for each box	The line in the middle of the interquartile range box represents the median value of the data on which it is based. However, you can also add a marker to represent the mean value (in case this differs). To do this, you would need to add a new command to your code block. This is a `stat_summary` command. In this command, you need to include the arguments `fun` (which determine the summary statistic the marker will represent), `geom` (which determines the type of symbol), `shape` (which determines the shape of the marker), `size` (which determines its size), `colour` (which determines its colour) and `fill` (which determines its fill colour – this argument is only needed when you use pch codes 21 to 25 in the `shape` argument). For example, to add a round, black marker of size 4 to represent the mean values of the data for each group, you would include the following command in the bock of code used to create your box plot:
	
	`stat_summary(fun=mean,geom="point",shape=16,size=4,`
	`colour="black")`

Adding symbols to represent individual data points	You can add markers representing the values for the individual data points on which the boxes on your box plot are based by including additional commands in your block of code. This can be done in a number of ways, but the most useful one is to use the `geom_jitter` command. This produces a cloud of symbols representing the individual data points. Within this command, you can include various arguments to customise their appearance. For example, if you include the command `geom_jitter(shape=21, size=2, colour="black", fill="white", position=position_jitter(0.1))` in the code used to create your box plot, the individual data points will be represented by symbols with the pch code 21 that will have a size of 2, a black outline and a white fill. In addition, they will be spread out to the left and the right of the central axis of each box (marked by the position of the whiskers) by a distance of 0.1 (as set by the `position=position_jitter(0.1)` argument).
Changing the order of the boxes	If you wish to change the order of the boxes on a box plot, this can be done by adding a `scale_x_discrete` command to the block of code used to create it. Within this command, the `limits` argument can be used to set the list of the groups from the original data set that will be represented by boxes on the final graph, and their order. For example, if you have groups in your data set called A, B and C, and you wish to plot them in reverse alphabetical order, you would include the following command in the code block you are using to make it: `scale_x_discrete(limits=c("C","B","A"))`

For the next part of this exercise, you will customise your box plot in a number of ways to show different elements of the distribution of body lengths for the different species in the shark data set. Firstly, you will start by adding a notch to the central boxes which show the interquartile range for the distribution of body sizes for each species of shark. This notch will represent the 95% confidence interval around the median value, and it can be used as an indicator of whether there are likely to be significant differences in the central values between the different groups displayed on your box plot. To do this, edit the final command in the above flow diagram (this is CODE BLOCK 68 in the document R_CODE_DATA_VISUALISATION_WORKBOOKDOC) so that the notch=FALSE argument in the geom_boxplot command is replaced with notch=TRUE. It should now look like this (the required modification is highlighted in **bold**):

```
ggplot(data=shark_data,aes(x=species,y=body_length)) +
geom_boxplot(coef=1.5,notch=TRUE,outlier.colour="black",
        outlier.shape=16,outlier.size=2,fill="grey",
colour="black") + labs(x="Species",y="Body Length (m)") +
            ylim(c(1,6)) + theme_classic()
```

When you run this modified block of code, it should produce a new box plot that looks this:

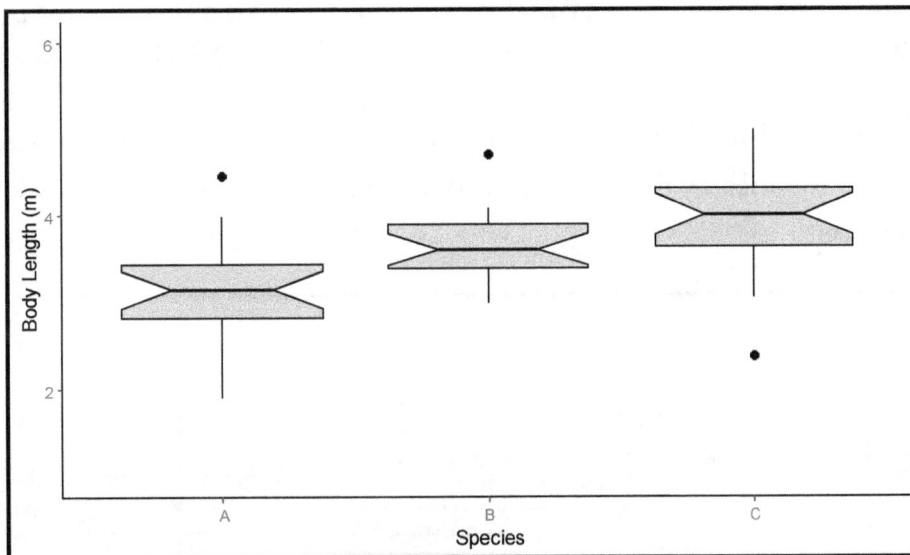

Next, you will add a marker to indicate the position of the mean for each species. This allows you to compare the mean value with the median value (as indicated by the central line in each box). If the data for a group have a more-or-less normal distribution, the mean and the median values should be similar. In contrast, in the data for a group have a clearly non-normal distribution, the mean and median values will differ considerably from each other. Thus, by adding the mean for each group on to a box plot graph, you can get an idea of whether or not the distribution of data in each group is likely to be normal. The mean value for each group can be shown on a box plot by adding a new command to the code block used to make it. This is the `stat_summary` command. Within this command, you can use the `fun=mean` argument to tell R to plot the mean value (rather than any other summary statistic), the `geom="point"` argument to set the code to plot the mean value as a point, and the `shape`, `size` and `colour` arguments to set the shape, size and colour of the symbols used to display the mean values (**NOTE:** Depending on the symbol you are using, you may also need to include a `fill` argument to set the fill colour of the symbol you are using to represent your mean value). In order to add a marker indicating the value of the mean for each group to your box plot, you will need to edit the above block of code so that it looks like this (the required modifications are highlighted in **bold**):

```
ggplot(data=shark_data,aes(x=species,y=body_length)) +
geom_boxplot(coef=1.5,notch=TRUE,outlier.colour="black",
     outlier.shape=16,outlier.size=2,fill="grey",
colour="black") + stat_summary(fun=mean,geom="point",
  shape=16,size=4,colour="black") + labs(x="Species",
y="Body Length (m)") + ylim(c(1,6)) + theme_classic()
```

Once you have finished modifying this code block, you can run it. This should produce a new box plot with the mean value for each group marked on it as a large black circle and it should looks like this:

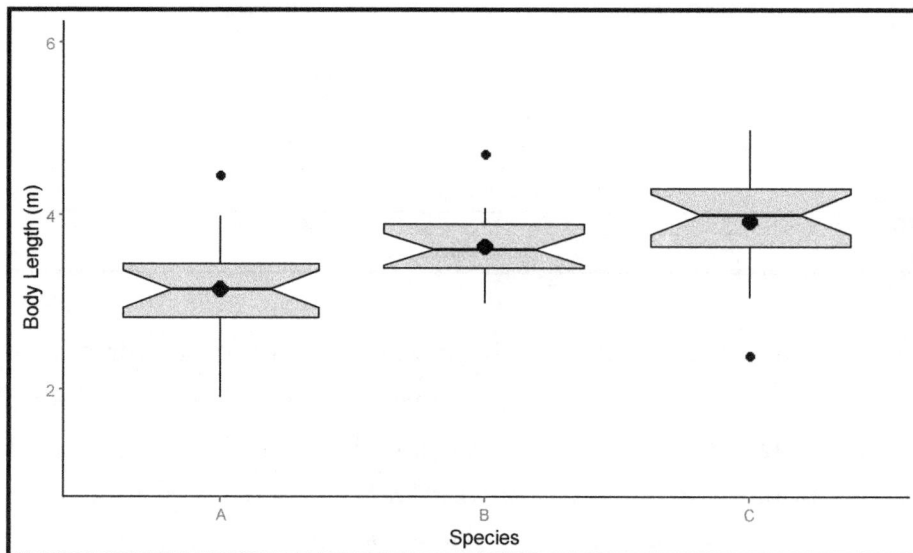

You can now add another command to the code block used to create your box plot which will allow you to plot all the data points for each group on top of the boxes themselves. This allows you to see the actual distribution of the data within each group. This is the `geom_jitter` command. In this command, you can use the `shape`, `size`, `colour` and `fill` arguments to set how the symbols used for these data points will look, while the `position` argument is used to determine whether all the points are plotted along the central line for each box (using the argument `position=position_jitter(0)`), or whether they are spread out across the box (by using a non-zero value at the end of this argument). For this exercise, you will set the `shape` argument for the additional `geom_jitter` command to `21`, the `size` argument to `2`, the `colour` argument to `"black"`, the `fill` argument to `"white"` and the `position` argument to `position_jitter(0.1)`. This command needs to be placed before the `stat_summary` command in the code block so that these points are plotted underneath the symbols for the mean values. Finally, you will need to change the `outlier.colour`, `outlier.shape` and `outlier.size` arguments in the `geom_boxplot` command to NA. If you do not do this, any outliers will be plotted on the box plot twice (once using the symbols set in the `geom_boxplot` command and once using the symbols set by the `geom_jitter` command). When you have made these changes, the modified block of code should look like this (the required modifications are highlighted in **bold**):

```
ggplot(data=shark_data,aes(x=species,y=body_length)) +
geom_boxplot(coef=1.5,notch=TRUE,outlier.colour="NA",
    outlier.shape=NA,outlier.size=NA,fill="grey",
colour="black") + geom_jitter(shape=21,size=2,colour=
"black",fill="white",position=position_jitter(0.1)) +
  stat_summary(fun=mean,geom="point",shape=16,size=4,
colour="black") + labs(x="Species",y="Body Length (m)") +
          ylim(c(1,6)) + theme_classic()
```

Once you have finished modifying this command, you can run it. This should produce a new box plot that looks like the image at the top of the next page.

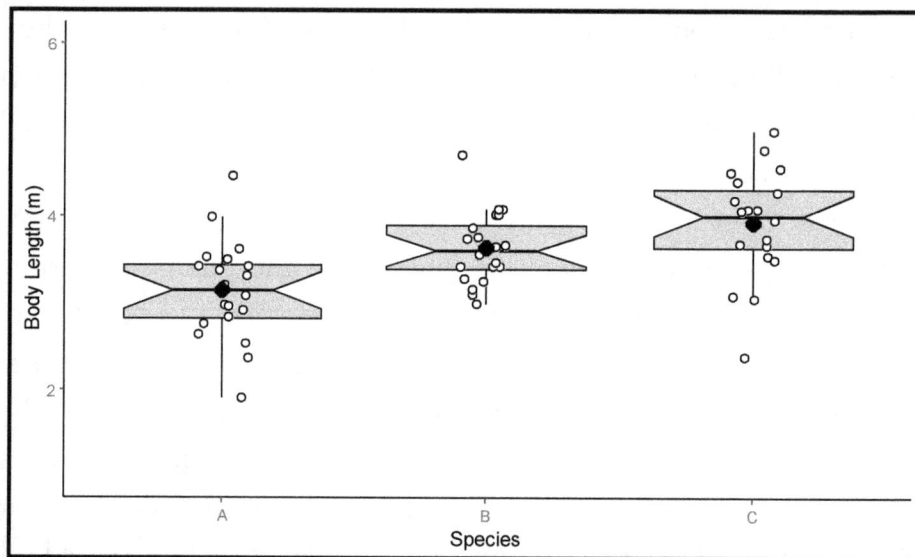

This will add white points with a black outline indicating the value for each data point in each group to your box plot. These will be spread out to the left and right of the central line of each box to allow you to clearly see data points with similar values that might otherwise have overlapped with each other. It also makes it easier to see the whiskers for each of the boxes on your box plot.

As well as customising the different types of information which are presented on a box plot, you can also customise the number of different groups that are plotted on it, and the colours used for them. For example, you can use the `scale_x_discrete` command to determine which groups from your data set will be plotted, and in what order they will appear. In this example, you will use this command to change the order that the boxes representing the data from the three shark species so they appear in reserve alphabetical order (with C on the left hand side, B in the middle and A on the right). To do this, edit the above block of code so that it looks like the code provided at the top of the next page (the required modifications are highlighted in **bold**).

```
ggplot(data=shark_data,aes(x=species,y=body_length)) +
geom_boxplot(coef=1.5,notch=TRUE,outlier.colour="NA",
    outlier.shape=NA,outlier.size=NA,fill="grey",
    colour="black") + geom_jitter(shape=21,size=2,
        colour="black",fill="white",position=
    position_jitter(0.1)) + stat_summary(fun=mean,
        geom="point",shape=16,size=4,colour="black") +
labs(x="Species",y="Body Length (m)") + ylim(c(1,6)) +
theme_classic() + scale_x_discrete(limits=c("C","B","A"))
```

Once you have finished modifying this command, you can run it. This should produce a new box plot that looks like this:

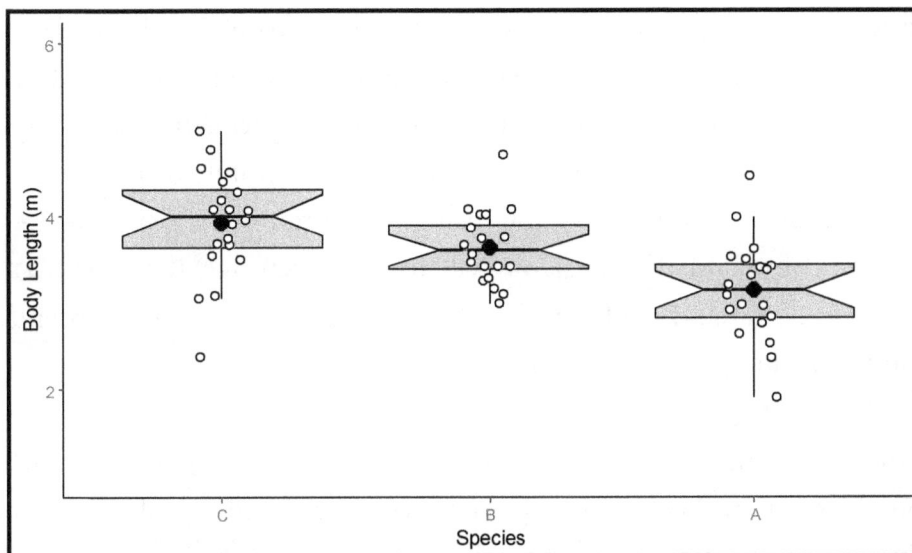

As you can see, the order that the boxes for the three species appear on this graph has been changed to the order specified in the newly added `scale_x_discrete` command. If you wanted to produce a graph with only the boxes for some of these species, you can do this by only including the names of the groupings you wish to include on your graph in the `scale_x_discrete` command. For example, to create a box plot with only the data for species A and B on it (in that order, you would include the command `scale_x_discrete(limits=c("A","B"))` in your code block.

Finally, you can add separate boxes for different sub-groups within each of the main groups in your data set. This is done by including `fill` and `colour` arguments in the `aes` element of the `ggplot` command rather than in the `geom_box` command and setting them to the name of the column containing the sub-groups within each of these main groups. For example, to create a box plot with separate boxes for males and females of each of the three shark species in the data set being used in this exercise, you would need to set the term in these arguments to `sex`. You would then need to specify the colours to be used for each sub-group by adding `scale_fill_manual` and `scale_colour_manual` commands to your block of code. Two colours would need to be specified in each of these commands (one for each sex). In this case, you will use red shades for males and blue shades for females. For this specific data set, you will also need to change the term in the `notch` argument in the `geom_boxplot` command from `TRUE` to `FALSE`. This is because the confidence intervals of the mean for some of the subgroups within each species are greater than the interquartile range, and this causes problems when trying to draw notched boxes. As a result, you will need to use un-notched boxes to display the interquartile ranges and this is done by using the argument `notch=FALSE` in the `geom_boxplot` command. Finally, you will need to remove the `colour` and `fill` arguments from this command so that they do not override the same arguments which have now been included in the `ggplot` command.

If you make these modifications to the above code block and then run it, you will notice that while you get separate boxes for the data from male and females from each species, neither the shapes representing the mean values for each group or the points representing the individual data points will plot in the correct positions. To deal with this for the individual data points values, you need to change the term in the `position` argument of the `geom_jitter` command from `position_jitter(0.1)` to `position_jitterdodge(jitter.width=0.1,dodge.width=0.75)`. In this new argument, the `jitter.width` term determines the left-right offset for the points within each subset (in this case `0.1`), while the `dodge.width` term determines the left-right offset between each subset (in this case, `0.75`). This means that together they can be used to correctly align the points with the boxes for each subset. For the mean values, you will need to add a similar argument to the `stat_summary` command. However, in this case you will use a value of `0` for `jitter.width` (since you wish the shape representing the mean to be plotted in the middle of each box), and a value of `0.75` for `dodge.width`,

This is the same as the offset between the subsets used for the individual data points. **NOTE:** The best way to determine the most appropriate values for these two terms for your own box plots for both the `geom_jitter` and the `stat_summary` commands is by trying different values until you find the ones that position the points in the exact locations you wish them to be.

Finally, since it will not be intuitively obvious which colour represents females and which represents males, you may wish to include a legend on your box plot. This is done by adding a `theme` command containing the argument `legend.position=right`. This will add a legend to the right hand side of your box plot which will provide information about the colours used for the different sub-groups that will be displayed on your graph.

Thus to create a box plot from the shark data set with separate boxes for males and females from each of the three species, you would need to edit the block of code you used on page 147 so that it looks like this (the required modifications are highlighted in **bold**):

```
ggplot(data=shark_data,aes(x=species,
y=body_length,colour=sex,fill=sex))+
geom_boxplot(outlier.colour="white",
outlier.shape=16,notch=FALSE) + geom_jitter(shape=21,
position=position_jitterdodge(jitter.width=0.1,
dodge.width=0.75)) + scale_colour_manual(values=
c("red","blue")) + scale_fill_manual(values=c("coral",
"cornflowerblue")) + stat_summary(fun=mean,geom="point",
shape=16,size=4,position=position_jitterdodge(
jitter.width=0,dodge.width=0.75)) + labs(x="Species",
y="Body Length (m)") + ylim(c(1,6)) + theme_classic() +
scale_x_discrete(limits=c("C","B","A")) +
theme(legend.position="right")
```

Once you have finished modifying this command, you can run it. This should produce a new box plot that looks like the image at the top of the next page.

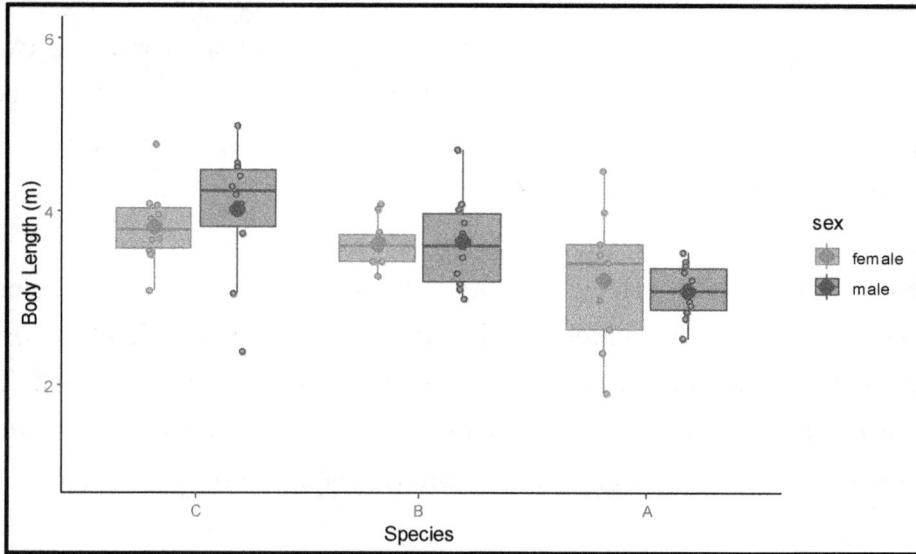

--- Chapter Five ---

How To Create Graphs Displaying Individual Data Points With GGPlot

So far, the graphs which you have learned how to make in this workbook have been based on summaries of information in specific data sets. However, in some situations you will want to present the underlying data rather than such summaries. In this chapter, you will learn how to make a variety of different types of graphs which display individual data points in a data set. In each case, you will learn about all the steps you need to carry out to create a specific type of graph, starting with a data set held in a spreadsheet or table, and finishing with how to write a suitable figure legend for it.

If you have not already done so, before you start the exercises in this chapter, you first need to create a WORKING DIRECTORY folder on your computer and load the necessary data into it (**NOTE:** If you have already created this folder and downloaded data for a previous chapter in this workbook, you do not need to do this again). To do this on a computer with a Windows operating system, open Windows Explorer and navigate to the location where you would like to create the folder (such as your C:\ drive or your DOCUMENTS folder). Next, right click anywhere in this location and select NEW> FOLDER. Now call this folder STATS_FOR_BIOLOGISTS_TWO by typing this into the folder name section to replace what it is currently called (which will most likely be NEW FOLDER). To create a WORKING DIRECTORY folder on a computer running a Mac operating system, open Finder and navigate to the location where you would like to create the folder (such as your DOCUMENTS folder or your DESKTOP). Next, click on FILE> NEW FOLDER, and then type the name STATS_FOR_BIOLOGISTS_TWO before pressing the ENTER key on your keyboard.

Once you have created your WORKING DIRECTORY folder, you are ready to download the data sets you will use for the exercises in this workbook from *www.gisinecology.com/stats-for-biologists-2*. After you have downloaded the compressed folder containing the required data by following the instructions provided on that page, you need to extract all the data files from it and copy them into the folder called STATS_FOR_BIOLOGISTS_TWO that you have just created.

Next, you need to check that the required data have been extracted to the correct folder. If you are using a computer with a Windows operating system, you can use Windows Explorer to open your newly created WORKING DIRECTORY folder and examine its contents. If all the files from the compressed folder are present in it (there should be a total of 90 of them), you can click on the folder icon at the left hand end of the ADDRESS BAR at the top of the WINDOWS EXPLORER window to reveal its full address. Write this address down as you will need it to set this folder as your WORKING DIRECTORY during the exercises provided in this workbook (see pages 12 and 13 for details of how to modify folder addresses so they will be recognised by R).

If you are using a computer with a Mac operating system, you can use Finder to open your newly created WORKING DIRECTORY folder and examine its contents. If all the required data files are present in it (there should be a total of 90 of them), press the CMD and I keys on your keyboard at the same time. This will open the GET INFO window where you will find its address (which is also called the pathway). Write this address down somewhere as you will need it to set this folder as your WORKING DIRECTORY during the exercises provided in this workbook (see pages 12 and 13 for details of how to modify folder addresses so they will be recognised by R).

After you have loaded the required data into your WORKING DIRECTORY folder, you can open RGUI or RStudio, depending on which option you wish to use (see Chapter 2 for more details). Once you have opened your preferred R user interface, you need to create a file called CHAPTER_FIVE_EXERCISES where you will save the results of your analyses from your R CONSOLE window as you work through this chapter. To do this using RGUI, click on the FILE menu and select SAVE WORKSPACE. To do this in RStudio, click on SESSION and select SAVE WORKSPACE AS. In both cases, save it as a WORKSPACE file with the name CHAPTER_FIVE_EXERCISES.RDATA in your WORKING DIRECTORY folder (this will be the one called STATS_FOR_BIOLOGISTS_TWO that you have just created). If you are using RStudio, you will also want to save the contents of your SCRIPT EDITOR window (where you will enter and edit the R code you will use to carry out specific commands). To do this, click on the FILE menu and select SAVE AS.

Save your file as an R SCRIPT file with the name CHAPTER_FIVE_EXERCISES.R in your WORKING DIRECTORY folder. As you work through the exercises in this chapter, remember to regularly save the contents of your R CONSOLE window (which will contain the R objects you have created up to that point) to your WORKSPACE file and, if you are using RStudio, the contents of your SCRIPT EDITOR window to your R SCRIPT file.

Finally, you need to remove any data that are currently held in R's temporary memory. To do this, enter the following command into R:

```
rm(list=ls())
```

If you are using RGUI, you can simply type this code after the command prompt at the bottom of the R CONSOLE window (it looks like this: >) and then press the ENTER key on your keyboard to run it. If you are using RStudio, you can type this command into the SCRIPT EDITOR window (the upper left hand window). To run this command, select it and then click on the RUN button at the top of this window. This will run it in the R CONSOLE window (the lower left hand one in the main RStudio user interface). You are now ready to start the exercises in this chapter.

EXERCISE 3.1: HOW TO CREATE AN X-Y SCATTER PLOT DISPLAYING ONE DATA SERIES USING GGPLOT:

An X-Y scatter plot is the most common type of graph used by biologists to display the values for individual data points in a data set. On such graphs, one point is plotted for each row of data, with their positions being determined by the values that are plotted on the X and Y axes. In order to be able to do this, you need to have your data arranged in a spreadsheet or table where each row contains data from a single record in your data set. In this table, there also needs to be separate columns containing the values for the variables you wish to plot on the X and Y axes of your graph. These will usually be continuous variables, but one or both of them can also be ordinal variables (that is, ones measured on an ascending scale). In general, you should avoid plotting data on categorical scales on an X-Y scatter plot. Instead, it would be better to use a bar graph for such data.

In this exercise, you will create an X-Y scatter plot which shows how relative investment in reproductive tissues in male bats species with complex echolocation (known as microchiropteran bats) varies with the average body mass of each species. The relative investment in reproduction will be measured as the percentage of the total body mass that is made up of the testes during the breeding season (see MacLeod and MacLeod 2009. *Oikos*. 118: 903-916). In both cases, the body mass data and the percentage testes mass (or PTM for short) data have been log-transformed. To do this, work through the following flow diagram (**NOTE:** If you have not already done so for an earlier exercise, you will need to download the `ggplot2` package and install it in your version of R before you start working through these instructions – see Exercise 1.1 for details of how to do this):

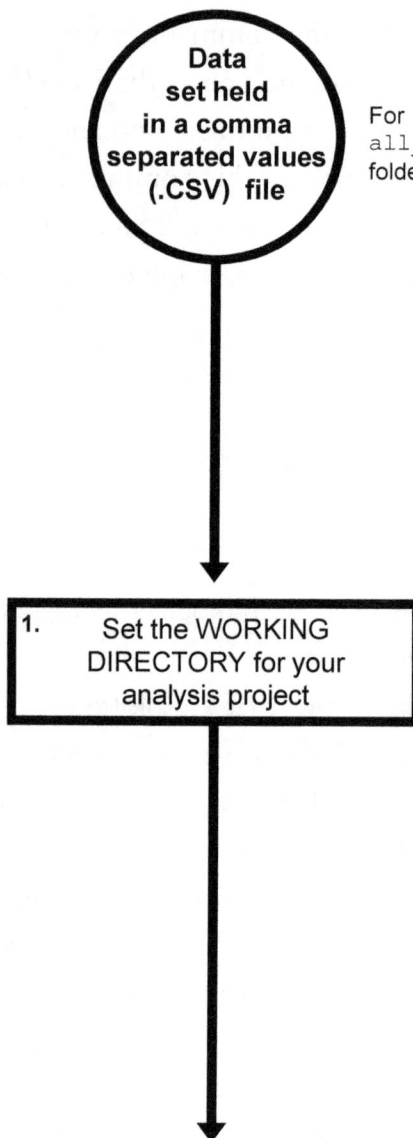

Data set held in a comma separated values (.CSV) file

For this example, the data set you will use is stored in a file called `all_bat_data.csv` that is located in the WORKING DIRECTORY folder you created during the introduction to this chapter.

Before you start any analysis in R, you first need to set the WORKING DIRECTORY. To do this, enter the text `setwd("` and then type the address of your WORKING DIRECTORY, using slashes (/) as the folder separators, before entering a second quotation mark followed by a closing bracket, like this `")`. For example, if your WORKING DIRECTORY has the address C:\STATS_FOR_BIOLOGISTS_TWO, your `setwd` command should look like this:

```
setwd("C:/STATS_FOR_BIOLOGISTS_TWO")
```

If you are using RGUI, enter your `setwd` command in the R CONSOLE window (remembering to use the address of your own WORKING DIRECTORY folder in it) and then press the ENTER key on your keyboard. If you are using RStudio, enter your `setwd` command into the SCRIPT EDITOR window. To run it, select it and then click on the RUN button at the top of this window. You will enter all the remaining commands for this exercise in a similar manner, depending on the user interface you are using.

1. Set the WORKING DIRECTORY for your analysis project

To check that your WORKING DIRECTORY has been set properly, enter the command `getwd()` and carefully check that the address it returns is the same as the one for the STATS_FOR_BIOLOGISTS_TWO folder you created at the start of this chapter.

Before you move on to step 2, make sure that all the data you wish to use in your analysis project are located in this WORKING DIRECTORY folder. In this case, this is a file called `all_bat_data.csv`. **NOTE:** If the data you are going to import into R in step 2 are not located in the WORKING DIRECTORY you set in this step, the import code provided in the next step will not work.

2. Load your data into R using the `read.table` command

3. Check the data have loaded into R correctly by checking the names of the columns and by viewing it

The `read.table` command provides the easiest way to load data held in a .CSV file (and stored in the WORKING DIRECTORY you set in step 1) into R so you can analyse it. To do this for the data set being used in this example, enter the following command into R:

```
all_bat_data <- read.table(file=
"all_bat_data.csv",sep=",",as.is=FALSE,
header=TRUE)
```

This code has to be entered exactly as it is written here or it will not work. If you wish to use the copy-and-paste approach for entering this command, copy the text directly below CODE BLOCK 69 in the document R_CODE_DATA_VISUALISATION_WORKBOOK.DOC and paste it into R.

This command will create a new object in R called `all_bat_data` which will contain the data from the specified .CSV file. To import a different .CSV file into R, all you need to do is change the file name in the `file` argument to the name of the one you wish to import. You can also use whatever name you wish for the R object which will be created by this command. To do this, simply replace `all_bat_data` at the start of the first line of the above code with the name you wish to use for it. **NOTE:** If your .CSV data set uses a semicolon as the column separator, you would need to replace the `sep=","` argument with `sep=";"`.

Whenever you import any data into R you need to check that they have loaded correctly. First, you need to check that all the required columns are present in the R object you just created. To do this, enter the following command into R:

```
names(all_bat_data)
```

This is CODE BLOCK 70 in the document R_CODE_DATA_VISUALISATION_WORKBOOK.DOC. This command will return the names used for each column in the R object you just created. For this example, the names should be: `id`, `log_body_mass`, `log_ptm` and `type`.

Next, you should view the contents of the whole table using the `View` command. This is done by entering following code into R:

```
View(all_bat_data)
```

This is CODE BLOCK 71 in the document R_CODE_DATA_VISUALISATION_WORKBOOK.DOC. This command will open a DATA VIEWER window where you can examine your data set and check that the correct data have been loaded into R.

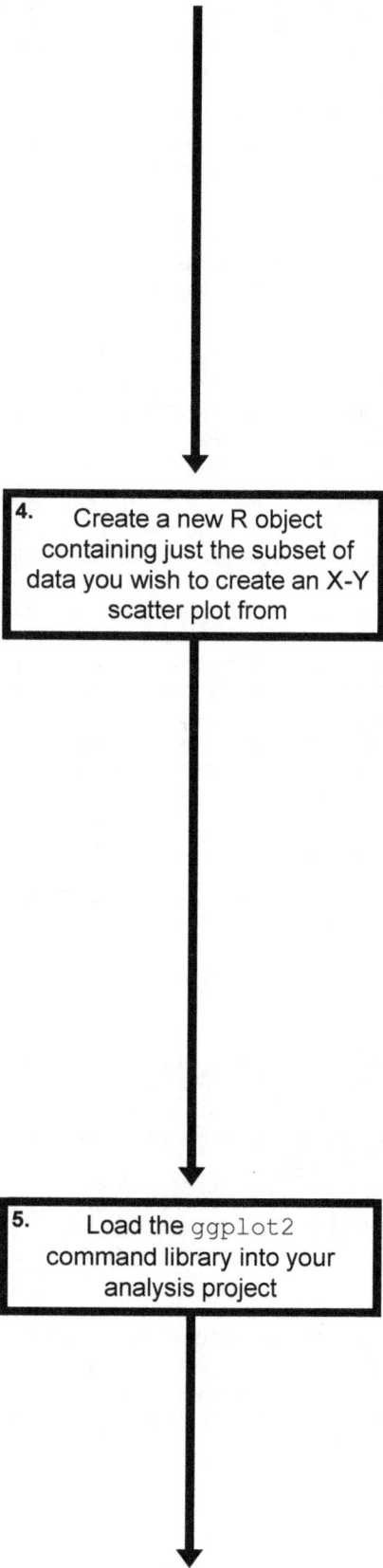

In the data set being used in this example, there are data from two different types of bat: Mircochiropteran bat species, which have complex echolocation, and macrochiropteran bat species, which do not. However, for this example, you only wish to create an X-Y scatter plot of data from the microchiropteran bat species. This means that before you can do anything else, you will need to create a new R object that only contains the data from the microchiropteran bat species. In R, this is done using the `subset` command. To do this for the data set being used in this example, enter the following code into R:

```
micro_bat_data <- subset(all_bat_data,
            type=="micro")
```

4. Create a new R object containing just the subset of data you wish to create an X-Y scatter plot from

This is CODE BLOCK 72 in the document R_CODE_DATA_ VISUALISATION_WORKBOOK.DOC. This command will create a new R object called `micro_bat_data` which will only contain the rows from the `all_bat_data` data set that have the label `micro` in the `type` column. This indicates these data are from microchiropteran bat species

Next, you should view the contents of the subset of data you have just created using the `View` command. This is done by entering following code into R:

```
View(micro_bat_data)
```

This is CODE BLOCK 73 in the document R_CODE_DATA_ VISUALISATION_WORKBOOK.DOC. This command will open a DATA VIEWER window where you can examine your data set and check that it contains the required subset of data. In this example, it should only contain data from microchiropteran bat species and not from any macrochiropteran bat species (which have the label `macro` in the `type` column).

5. Load the `ggplot2` command library into your analysis project

Once you have successfully subsetted your data, you are ready to create your initial scatter plot. However, before you can do this, you need to load the `ggplot2` command library into your analysis project. To do this, enter the following command into R:

```
library(ggplot2)
```

This is CODE BLOCK 74 in the document R_CODE_DATA_ VISUALISATION_WORKBOOK.DOC.

```
6.  Create your initial scatter
    plot based on the subset of
    data created in step 4
```

After you have loaded the `ggplot2` command library into your R project, you are ready to use it to create your initial scatter plot based on the log body mass and log percentage testes mass (PTM) data from the microchiropteran bat species. To do this, enter the following block of code into R:

```
ggplot(data=micro_bat_data,
aes(x=log_body_mass,y=log_ptm)) +
        geom_point()
```

This is CODE BLOCK 75 in the document R_CODE_DATA_ VISUALISATION_WORKBOOK.DOC, and it contains two commands separated by a + symbol. These are the `ggplot` command and the `geom_point` command. The `ggplot` command sets the data set which will be used for the graph. This is done using the `data` argument and, in this case, it will be the R object called `micro_bat_data` created in step 4 of this exercise. The data that will be plotted on the X axis of the resulting graph is set using the `x` argument in the `aes` element of this `ggplot` command. In this case, it is the column called `log_body_mass` in the `micro_bat_data` data set, which contains log-transformed body mass data for each bat species. The data that will be plotted on the Y axis of the resulting graph is set using the `y` argument in the `aes` element of this `ggplot` command. In this case, it is the column called `log_ptm`, which contains log-transformed percentage testes mass data for each bat species.

The second command in this code block, `geom_point`, sets the type of graph that will be created from the data specified in the `ggplot` command. In this case, it will be a scatter plot containing one point for each row of data in the data set specified in the `data` argument of the `ggplot` command. Initially, no arguments will be included in this command.

Once you have created your initial scatter plot, and you are sure that it is displaying the intended data on each axis, you can customise how your final graph will look. This can be done by adding a number of new style commands to the block of code used to produce it. To do this, edit the code from step 6 so that it looks like this (the newly added style commands are highlighted in **bold**):

```
ggplot(data=micro_bat_data,
  aes(x=log_body_mass,y=log_ptm)) +
geom_point() + labs(x="Log Body Mass (g)",
    y="Log Percentage Testes Mass")+
  xlim(c(0.5,2.5)) + ylim(c(-1.0,1.0)) +
            theme_classic()
```

7. **Customise how your final scatter plot will look**

This is CODE BLOCK 76 in the document R_CODE_DATA_ VISUALISATION_WORKBOOK.DOC, and it adds four new style commands separated by + symbols to the code block used to create your initial scatter plot in step 6. These are the `labs` command, the `xlim` command, the `ylim` command, and the `theme_classic` command. The `labs` command sets the labels that will be used tor the X axis (using the `x` argument) and the Y axis (using the `y` argument). The `xlim` command sets the minimum and maximum values for the X axis of the graph. In this case, these will be `0.5` for the minimum value and `2.5` for the maximum value. The `ylim` command sets the minimum and maximum values for the Y axis of the graph. In this case, these will be `-0.1` for the minimum value and `1.0` for the maximum value. Finally, the `theme_classic` command sets the remaining style elements of the final graph to those of the pre-existing classic theme. Once you have finished editing this code block, you can run it again to create the final version of your graph.

X-Y scatter plot created from your data set

At the end of the first part of this exercise, the X-Y scatter plot created by the code block provided in step 7 should look like this:

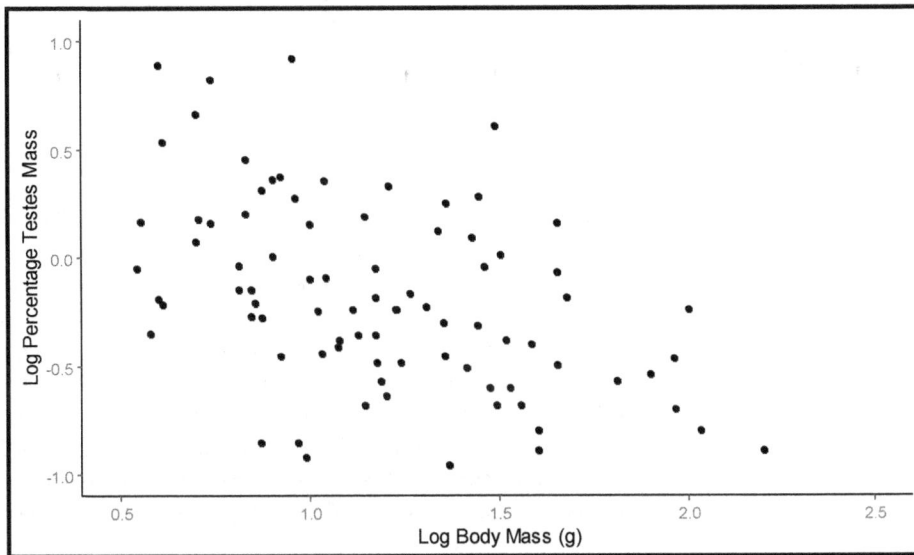

Once you have created a scatter plot, you can export it from R so that you can include it in a manuscript or presentation. If you are using RGUI, you can do this by clicking on the R GRAPHICS window containing your scatter plot to select it, before clicking on FILE on the main menu bar and selecting SAVE AS. This will allow you to save it in a variety of different formats. If you are using RStudio, you can export your graph by clicking on the EXPORT button at the top of the window displaying your scatter plot and selecting SAVE AS IMAGE.

When including an X-Y scatter plot in a manuscript, it is important that you provide an appropriate figure legend for it. This legend should provide all the information required for the reader to interpret the contents of the graph. For the above scatter plot, an appropriate legend would be:

Figure 1: *The relationship between body mass and relative investment in testes mass (as measured by percentage testes mass) in microchiropteran bats. Both variables have been log-transformed prior to being plotted on this graph.*

When creating an X-Y scatter plot, there are a number of different ways that you can customise how the data on it are displayed. This can include changing the shape of the symbol used to represent each point, the colours used to plot it, its size, whether there are any labels associated with each one, and whether a line of best fit is plotted on the graph to show the trend in the relationship between the variables plotted on the X and Y axes. The commands and arguments used to customise the symbols on a scatter plot in these ways are provided in the table below.

Required customisation	How to achieve it for an X-Y scatter plot created with the `geom_point` command
Changing the symbol used to represent each point	The symbols used to represent each point on a scatter plot can be customised by including a `shape` argument in the `geom_point` command. The setting for this argument is the pch code for the specific shape you wish to use. Solid shapes are specified by pch codes 1 to 20 For example, the pch code for a solid square is 15, for a solid triangle is 17 and for a solid circle is 19. Open shapes are specified by pch codes 21 to 25. For example, the pch code for an open circle is 21, and for an open square is 22. A full list of the available pch codes and the symbol shapes they represent can be found at *www.sthda.com/english/wiki/r-plot-pch-symbols-the-different-point-shapes-available-in-r.*
Changing the size of the symbols representing each point	The size of the symbols on a scatter plot can be customised by including a `size` argument in the `geom_point` command. The setting for this argument is a number representing the size of the shape. For example, including the argument `size=3` would create a graph with larger symbols than including the argument `size=2`.
Changing the fill and outline colours of the symbols representing each point	The colours for the symbols on a scatter plot can be customised by including a `colour` and/or a `fill` argument in the `geom_point` command. The setting for these two arguments are the names of the desired colours you wish to use for the fill and the outline colours, respectively. **NOTE:** For symbols with pch codes 1 to 20, you can only specify the colour of the symbol using the `colour` argument. For symbols with codes pch codes 21 to 25, you can specify both the outline colour using the `colour` argument and the fill colour using the `fill` argument. For example, to create a scatter plot with symbols that are outlined in black and filled with blue, you would include the arguments `colour="black"` and `fill="blue"` in the `geom_point` command.
Adding labels to allow you to identify each point on your scatter plot	A label can be added to each point on a scatter plot by including a `geom_text` command in the block of code used to create it. This command should include a `label` argument (in its `aes` element), an `hjust` argument and a `vjust` argument. The `label` argument is used to specify the column in the data set used to create the graph which contains the labels you wish to appear beside each point, while the `hjust` and `vjust` arguments include numbers that allow you to specify exactly where the labels will be plotted in relation to the individual data points. For example, if you include the command `geom_text(aes(label="id"),hjust=-0.3,vjust=1.3)` in a block of code, it will add a label to each data point based on the contents of a column called `id` that are offset horizontally by `-0.3` and vertically by `1.3`, with these values being in the units of the X and Y axes respectively.

Required customisation	How to achieve it for an X-Y scatter plot created with the `geom_point` command
Adding a line of best fit representing the relationship between the X and Y variables	A line of best fit representing the relationship between the X and Y variables can be added to a scatter plot by including a `geom_smooth` command in the block of code used to create it. In this command, you can specify the method used to fit this line to the data using a `method` argument, its colour using a `colour` argument, the type of line using an `lty` argument, and its thickness using a `size` argument. For example, to add a solid black linear line of best fit to a scatter plot, you would add the command `geom_smooth(method="lm", colour="black", lty="solid", size=1)` to the code block used to create it.
Adding confidence intervals around a line of best fit	Confidence intervals around a line of best fit can be added to a scatter plot by including four additional arguments to the `geom_smooth` command used to create the line itself. These are an `se` argument, a `level` argument, a `fill` argument and an `alpha` argument. The `se` argument sets whether confidence intervals around the line of best fit will be displayed or not. To display the confidence intervals, this argument should be set to TRUE. The `level` argument sets the width of the confidence intervals and should be a value between 0 and 1. To add 95% confidence intervals, the `level` argument should be set to 0.95. The `fill` argument sets the colour of the shaded area representing the confidence intervals, while the `alpha` argument will contain a value between 0 and 1 which represents the level of transparency this shaded area should have. For example, adding the arguments `se=TRUE, level=0.95, fill="black", alpha=0.3` to a `geom_smooth` command will add a dark, semi-transparent shaded area representing the 95% confidence intervals around the line of best fit created by this command.

To explore how these commands and arguments can be used to customise how data are displayed on X-Y scatter plots, you will now customise your graph comparing log body mass and log percentage testes mass (PTM) in microchiropteran bats. Firstly, you will change the symbols used to plot each data point from the default settings (which are small solid black circles) to larger, open circles with a black outline and a blue fill. This is done by adding `shape`, `size`, `colour` and `fill` arguments to the `geom_point` command (**NOTE:** You can also use the argument `pch` in this command to change the symbol used on your scatter plot). In this case, you will set the `shape` argument to 21 (the code for an open circle), the `size` argument to 3, the `colour` argument to `black` and the `fill` argument to `blue`. To do this, edit the final code from step 7 (this is CODE BLOCK 76 from the document R_CODE_DATA_VISUALISATION_WORKBOOK.DOC) so that it looks like the block of code at the top of the next page (the newly added arguments are highlighted in **bold**).

```
      ggplot(data=micro_bat_data,aes(x=log_body_mass,
  y=log_ptm)) + geom_point(shape=21,size=3,colour="black",
fill="blue") + labs(x="Log Body Mass (g)",y="Log Percentage
    Testes Mass") + xlim(c(0.5,2.5)) + ylim(c(-1.0,1.0)) +
                        theme_classic()
```

You can modify this code either by editing it in the R CONSOLE window of RGUI or through the SCRIPT EDITOR window of RStudio (depending on which interface you are using). If you are entering commands directly into the R CONSOLE window, you can use the UP arrow on your keyboard to bring commands and code blocks you have previously run during the same session back on to the command line of this window, and then use the LEFT and RIGHT arrows to scroll through and edit them. Once you have finished modifying this code block, you can run it by pressing the ENTER key on your keyboard. If you are using RStudio, you can copy and paste the original code block in the SCRIPT EDITOR window before editing the new version to include the required modifications. After you have done this, select the modified version of the code block and click on the RUN button to run it in the R CONSOLE window.

Once you have run this new version of the R code for creating your X-Y scatter plot of the relationship between log body mass and log PTM in microchiropteran bats, you should have a graph that looks like this:

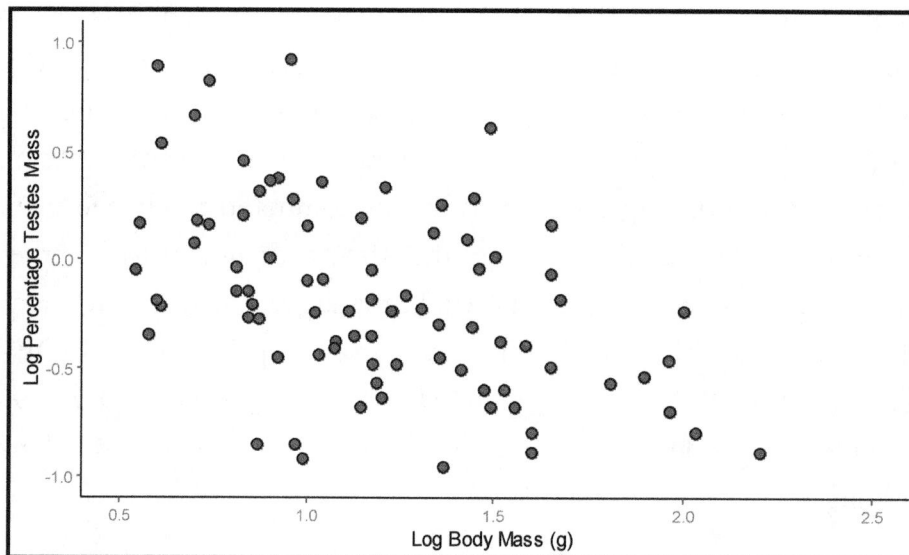

As well as changing the characteristics of the symbols used to represent each point on your X-Y scatter plot, you can also add labels to each data point so that you can tell which point is which. This is done by adding a `geom_text` argument to the R code used to create the above graph. In this command, you can specify a column in your original data set which contains the labels you wish to add to each data point using the `label` argument in the `aes` element, while you can determine the position of the labels relative to the symbols themselves using the `hjust` (which sets the horizontal position) and the `vjust` (which sets the vertical position) arguments in this `geom_text` command. For example, to add labels to the above graph based on the species identification number contained in the column called `id` in the `micro_bat_data` data set that is offset to the bottom right of each symbol, you would include the command `geom_text(aes(label="id")`, `hjust=-0.3,vjust=1.3)` in the R code used to create the above scatter plot. This can be done by editing the above code block so that it looks like this (the newly added command is highlighted in **bold**):

```
ggplot(data=micro_bat_data,aes(x=log_body_mass,y=
  log_ptm)) + geom_point(shape=21,size=3,colour="black",
fill="blue") + labs(x="Log Body Mass (g)",y="Log Percentage
  Testes Mass") + xlim(c(0.5,2.5)) + ylim(c(-1.0,1.0)) +
  theme_classic() + geom_text(aes(label=id),hjust=-0.3,
                    vjust=1.3)
```

The graph produced when you run this modified R code should look like this:

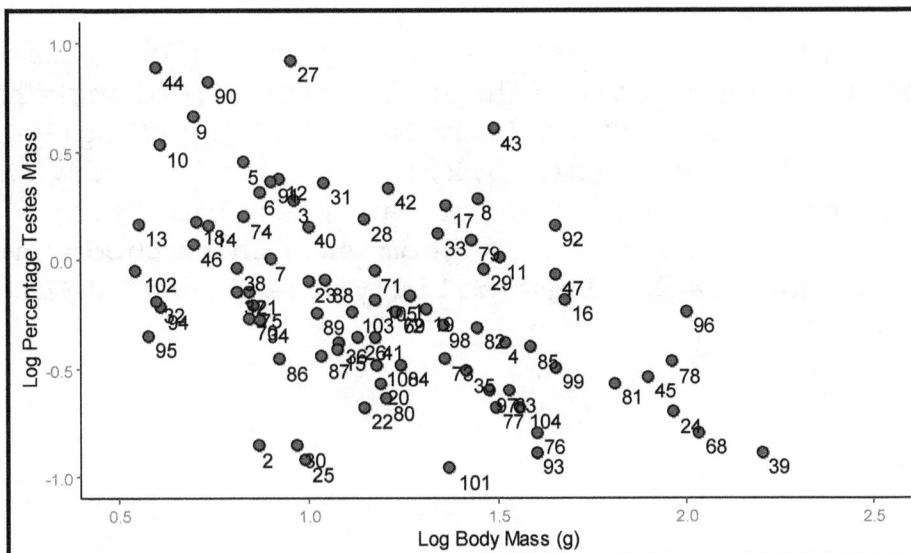

When you examine this graph, you will see that each point on it is now labelled with the species identification number specified in the `id` column of the `micro_bat_data` data set.

The final modification to the way that the data are displayed on your X-Y scatter plot that will be covered in this exercise is to add a line showing the 'best fit' relationship between the two variables displayed on it. This is done by adding a `geom_smooth` command to the code used to generate the above graph. In this command, you can use a variety of arguments to determine how the line of best fit will be calculated and how the line itself will be displayed. These include the `method` argument (which determines the method used to calculate the line of best fit and so whether it is linear or non-linear), the `colour` argument (which sets the colour of the line of best fit), the `lty` argument (which sets the style of the line, that is whether it is solid or dashed), the `size` argument (which sets the thickness of the line), and the `se` argument (which determines whether or not a confidence interval is plotted around the line). In order to demonstrate how these arguments can be used in the `geom_smooth` command to add a line of best fit to an X-Y scatter plot, you will use them to add a thin, solid, black, linear trend line with no confidence intervals to the above graph of the relationship between log body mass and log testes mass in microchiropteran bats. To do this, set the `method` argument to `lm` (which is the setting needed to create a linear line of best fit), the `colour` argument to `black`, the `lty` argument to `solid`, the `size` argument to `1` and the `se` argument to `FALSE`. This can be done by editing the above code block so that it looks like this (the newly added command is highlighted in **bold**):

```
ggplot(data=micro_bat_data, aes(x=log_body_mass,
y=log_ptm)) + geom_point(shape=21,size=3,colour="black",
fill="blue") + labs(x="Log Body Mass (g)",y="Log Percentage
Testes Mass") + xlim(c(0.5,2.5)) + ylim(c(-1.0,1.0)) +
theme_classic() + geom_text(aes(label=id),
hjust=-0.3, vjust=1.3) + geom_smooth(method="lm",
colour="black",lty="solid",size=1,se=FALSE)
```

The graph produced when you run this modified R code should look like this:

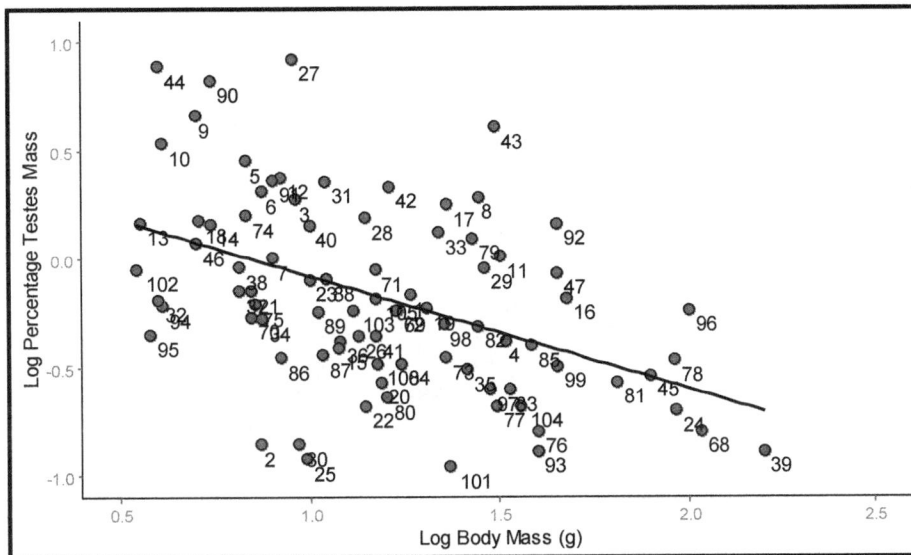

If you wish to add a confidence interval to your trend line, this can be done by changing the `se` argument in the `geom_smooth` command to `TRUE`. However, when you do this, you will also need to add three new arguments to this command to allow you to set exactly how the confidence intervals are displayed. These are a `level` argument, which sets the size of the confidence interval that will be displayed, a `fill` argument, which sets the colour that will be used to display the confidence intervals, and an `alpha` argument that will set how transparent this fill will be. For example, to add a black, semi-transparent shaded area to represent the 95% confidence interval around the above trend line, you would need to add the following arguments to the `geom_smooth` command in the above R code block: `se=TRUE, level=0.95, fill="black", alpha=0.3`. This can be done by editing the above code block so that it looks like this (the newly added arguments are highlighted in **bold**):

```
ggplot(data=micro_bat_data, aes(x=log_body_mass,
 y=log_ptm)) + geom_point(shape=21,size=3,colour="black",
fill="blue") + labs(x="Log Body Mass (g)",y="Log Percentage
   Testes Mass") + xlim(c(0.5,2.5))+ ylim(c(-1.0,1.0)) +
       theme_classic() + geom_text(aes(label=id),
     hjust=-0.3,vjust=1.3)+ geom_smooth(method="lm",
colour="black",lty="solid",size=1,se=TRUE,level=0.95,
              fill="black",alpha=0.3)
```

The graph produced when you run this modified R code should look like this:

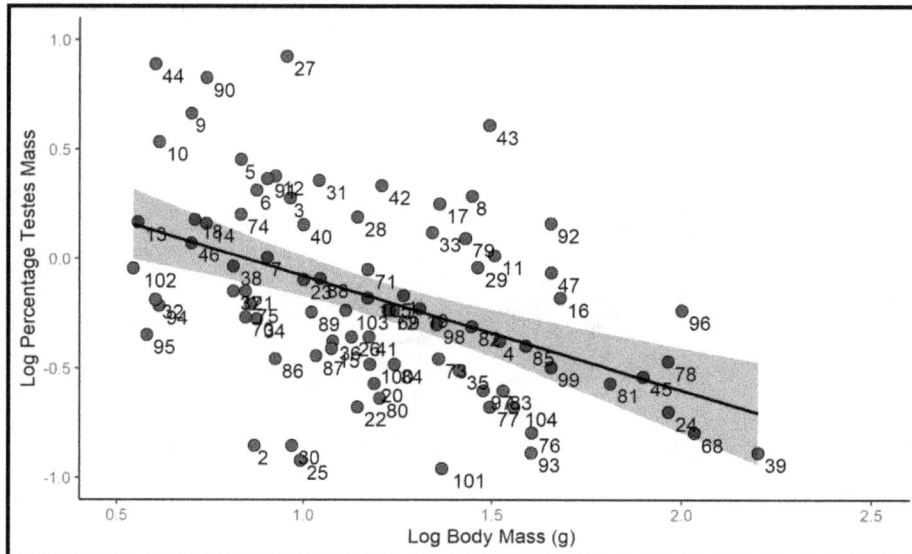

EXERCISE 3.2: HOW TO CREATE AN X-Y SCATTER PLOT WITH MULTIPLE DATA SERIES ON IT:

In exercise 3.1, you learned how to create an X-Y scatter plot to show the relationship between a pair of variables for the data in a single data series. In this exercise, you will learn how to create a similar graph with data from multiple data series on it. The easiest way to do this using GGPlot is to include multiple `geom_point` commands within the same code block, each of which contains a `data` argument with the name of a different R object with a different data set in it. In addition, different terms can be included in the style arguments, such as `shape`, `size`, `colour` and `fill`, in each `geom_point` command to ensure that each data series is represented by a different symbol and so can easily be identified on the final multi-series scatter plot.

In the first part of this exercise, you will create an X-Y scatter plot with two data series displayed on it. These will be the data for log body mass and log percentage testes mass for microchiropteran bats plotted on the single series X-Y scatter plot in Exercise 3.1, and data for the same two variables for macrochiropteran bats (which are bats that lack complex echolocation). To do this, work through the flow diagram that starts at the top of the next page.

NOTE: These instructions assume that you have already installed the `ggplot2` package in your version of R and that you have loaded its command library into your analysis project. If you have not already done this, you will need to do so before you start this exercise. Instructions for how to install this package and load its command library into your analysis project can be found in Exercise 1.1. It also assumes that you already have an R object in your analysis project called `micro_bat_data` which contains the data on body mass and percentage testes mass for the microchiropteran bat species, and an R object called `all_bat_data` that contains the same data for all bat species. Instructions on how to create these R objects can be found in Exercise 3.1.

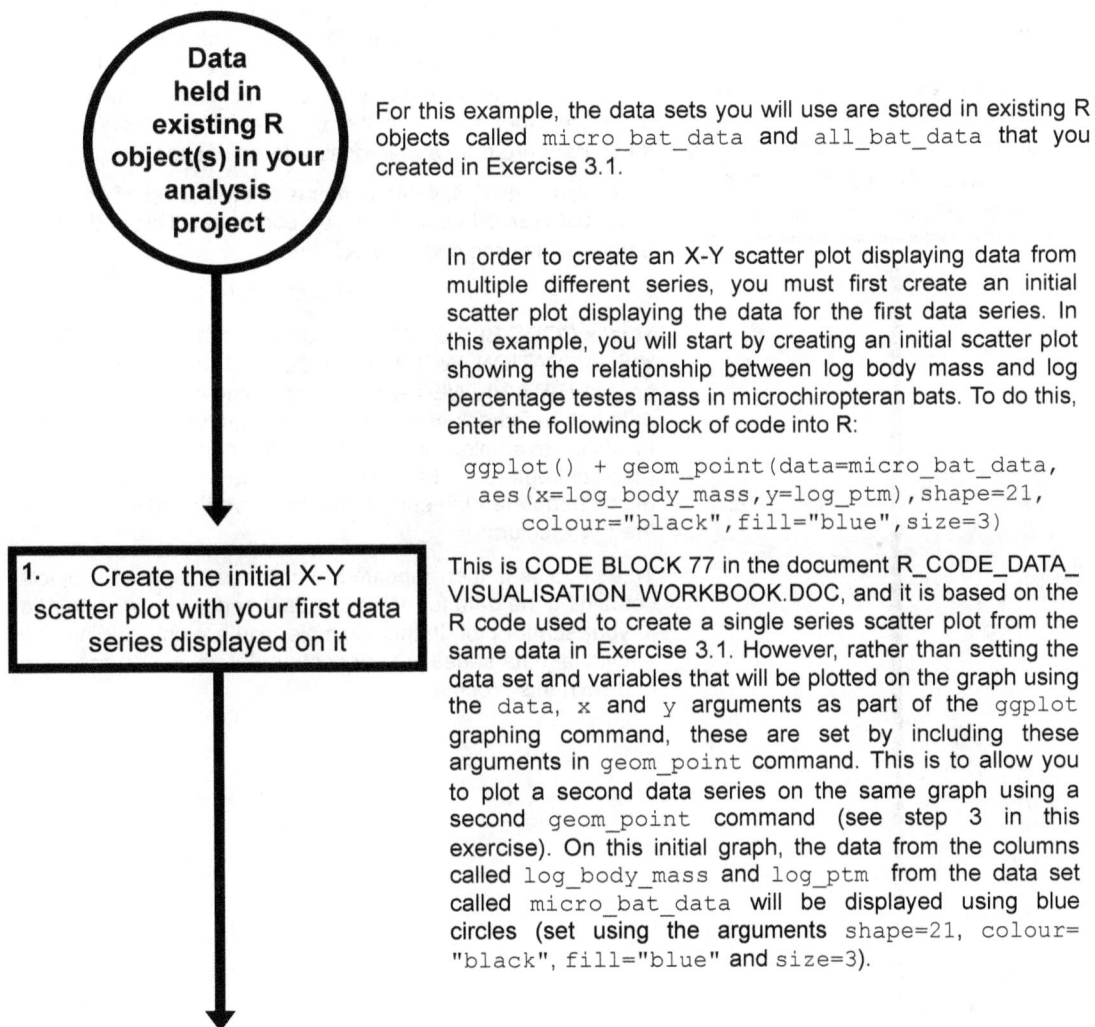

> ## Data held in existing R object(s) in your analysis project

For this example, the data sets you will use are stored in existing R objects called `micro_bat_data` and `all_bat_data` that you created in Exercise 3.1.

In order to create an X-Y scatter plot displaying data from multiple different series, you must first create an initial scatter plot displaying the data for the first data series. In this example, you will start by creating an initial scatter plot showing the relationship between log body mass and log percentage testes mass in microchiropteran bats. To do this, enter the following block of code into R:

```
ggplot() + geom_point(data=micro_bat_data,
  aes(x=log_body_mass,y=log_ptm),shape=21,
    colour="black",fill="blue",size=3)
```

1. Create the initial X-Y scatter plot with your first data series displayed on it

This is CODE BLOCK 77 in the document R_CODE_DATA_VISUALISATION_WORKBOOK.DOC, and it is based on the R code used to create a single series scatter plot from the same data in Exercise 3.1. However, rather than setting the data set and variables that will be plotted on the graph using the `data`, `x` and `y` arguments as part of the `ggplot` graphing command, these are set by including these arguments in `geom_point` command. This is to allow you to plot a second data series on the same graph using a second `geom_point` command (see step 3 in this exercise). On this initial graph, the data from the columns called `log_body_mass` and `log_ptm` from the data set called `micro_bat_data` will be displayed using blue circles (set using the arguments `shape=21`, `colour="black"`, `fill="blue"` and `size=3`).

```
2.   Create new R object(s)
     containing the subset(s) of
     data you wish to add to your
     graph as additional data series
```

In order to add a second data series to your X-Y scatter plot, you first need to create a new R object containing just these data. This can be done by importing a new data set or by creating a new R object based on a subset of data from an existing R object. In this example, you will extract a subset of data from the R object called `all_bat_data` created in step 2 of Exercise 3.1. This subset will contain the data from the macrochiropteran bats (those species which lack complex echolocation). To do this for the data set being used in this example, enter the following code into R:

```
macro_bat_data <- subset(all_bat_data,
                type=="macro")
```

This is CODE BLOCK 78 in the document R_CODE_DATA_ VISUALISATION_WORKBOOK.DOC. This command will create a new R object called `macro_bat_data` which will only contain the rows from the `all_bat_data` data set that have the label `macro` in the `type` column. This indicates these data are from a macrochiropteran bat species.

Next, you should view the contents of the subset of data you have just created using the `View` command. This is done by entering following code into R:

```
View(macro_bat_data)
```

This is CODE BLOCK 79 in the document R_CODE_DATA_ VISUALISATION_WORKBOOK.DOC. This command will open a DATA VIEWER window where you can examine your data set and check that contains the required subset of data. In this example, it should only contain data from macrochiropteran bat species and not from any microchiropteran bat species (which have the label `micro` in the `type` column).

This process is then repeated to create additional R objects containing the data for any other data series you wish to add to your scatter plot. In this example, you are only adding one additional data series to your scatter plot so you do not need to repeat this process.

To add a second data series to your initial X-Y scatter plot, you need to add a second `geom_point` command to the R code you used to create it. To do this for the graph created in step 1, edit the code used to create it so that it looks like this (the newly added graphing command is highlighted in **bold**):

```
ggplot() + geom_point(data=micro_bat_data,
 aes(x=log_body_mass,y=log_ptm),shape=21,
   colour="black",fill="blue",size=3) +
     geom_point(data=macro_bat_data,
 aes(x=log_body_mass,y=log_ptm),shape=21,
     colour="black",fill="red",size=3)
```

This is CODE BLOCK 80 in the document R_CODE_DATA_ VISUALISATION_WORKBOOK.DOC. The newly added `geom_point` command will create a second data series based on the `log_body_mass` and `log_ptm` columns from the data set called `macro_bat_data` (created in step 2 of this exercise). This second data series will be displayed using red circles (set using the arguments `shape=21`, `colour="black"`, `fill="red"` and `size=3`). To add more data series to your X-Y scatter plot, you would simply add additional `geom_point` commands to your code block. Once you have finished editing this code block, you can run it again to create an updated version of your graph.

After you have added all the data series to scatter plot, you can add a line of best fit for the data in your first data series. This is done by adding a `geom_smooth` command to the R code used to create your scatter plot. To do this for the `micro_bat_data` data series, edit the code from step 3 so that it looks like this (the newly added `geom_smooth` command is highlighted in **bold**):

```
ggplot() + geom_point(data=micro_bat_data,
 aes(x=log_body_mass,y=log_ptm),shape=21,
   colour="black",fill="blue",size=3) +
   geom_smooth(data=micro_bat_data,aes(x=
     log_body_mass,y=log_ptm),method="lm",
 colour="blue",lty="solid",size=1,se=TRUE,
     level=0.95,fill="blue",alpha=0.3) +
       geom_point(data=macro_bat_data,
 aes(x=log_body_mass,y=log_ptm),shape=21,
     colour="black",fill="red",size=3)
```

This is CODE BLOCK 81 in the document R_CODE_DATA_ VISUALISATION_WORKBOOK.DOC. In the newly added `geom_smooth` command, the `data` argument is used to tell R to create a line of best fit based on the `micro_bat_data` data set, while the `x` and `y` arguments in the `aes` element set the variables which will be used as the X and Y variables. The `method` arguments determines how the line will be created. In this case it is set to `lm` so that a linear line of best fit is added to the graph. Finally, the `colour`, `lty` and `size` arguments determine the characteristics of the line, and the `se`, `level`, `fill` and `alpha` arguments determine the characteristics of its confidence intervals. Once you have finished editing this code block, you can run it again to create an updated version of your graph.

3. Add the additional data series to your initial X-Y scatter plot

4. Add a line of best fit to your X-Y scatter plot for your first data series

169

After the line of best fit has been added to the scatter plot for the first series, further `geom_smooth` commands can be added to the code block to add lines of best fit for the other data series displayed on the graph. For this example, you will add a second line of best fit for the macrochiropteran bat data series. To do this, edit the code block from step 4 so that it looks like this (the newly added `geom_smooth` command is highlighted in **bold**):

```
ggplot() + geom_point(data=micro_bat_data,
 aes(x=log_body_mass,y=log_ptm),shape=21,
   colour="black",fill="blue",size=3) +
  geom_smooth(data=micro_bat_data, aes(x=
    log_body_mass,y=log_ptm),method="lm",
 colour="blue",lty="solid",size=1,se=TRUE,
    level=0.95, fill="blue",alpha=0.3) +
   geom_point(data=macro_bat_data,aes(x=
 log_body_mass,y=log_ptm),shape=21,colour=
      "black",fill="red",size=3) +
   geom_smooth(data=macro_bat_data,
 aes(x=log_body_mass,y=log_ptm),method="lm",
  colour="red",lty="solid",size=1,se=TRUE,
    level=0.95,fill="red",alpha=0.3)
```

This is CODE BLOCK 82 in the document R_CODE_DATA_ VISUALISATION_WORKBOOK.DOC. The second `geom_ smooth` command includes the same arguments as the `geom_smooth` used to create the first line of best fit. However, it specifies a different R object in its `data` argument (in this case, the R object called `macro_bat_data`), and uses a different colour for the `colour` and `fill` arguments (in this case, `red` rather than `blue`). This matches it to the colours used to plot the data series that it represents a line of best fit for.

To add lines of best fit for additional data series on your X-Y scatter plot, you would simply add additional `geom_smooth` commands to your code block, with each one set to use data from a different data set and to display the line of best fit in a different way. In each case, the new `geom_smooth` command should be added to the code block immediately after the `geom_point` command for the same data series. This will help you keep track of which commands relate to which data series. Once you have finished editing this code block, you can run it again to create an updated version of your graph.

5. Add lines of best fit for the other data series that you have plotted on your scatter plot

170

After you have created your initial scatter plot, and you are sure that it is displaying all the required data series and lines of best fit, you can customise how your final graph will look. This can be done by adding a number of new style commands to the block of code used to produce it. To do this, edit the code block from step 5 so that it looks like this (the newly added style commands are highlighted in **bold**):

```
ggplot() + geom_point(data=micro_bat_data,
 aes(x=log_body_mass,y=log_ptm),shape=21,
   colour="black",fill="blue",size=3) +
  geom_smooth(data=micro_bat_data,aes(x=
   log_body_mass,y=log_ptm),method="lm",
 colour="blue",lty="solid",size=1,se=TRUE,
   level=0.95,fill="blue",alpha=0.3) +
   geom_point(data=macro_bat_data aes(x=
    log_body_mass,y=log_ptm),shape=21,
    colour="black",fill="red",size=3) +
   geom_smooth(data=macro_bat_data,aes(x=
    log_body_mass,y=log_ptm),method="lm",
   colour="red",lty="solid",size=1,se=TRUE,
    level=0.95,fill="red",alpha=0.3) +
labs(x="Log Body Mass (g)",y="Log Percentage
    Testes Mass") + xlim(c(0.5,3.1)) +
   ylim(c(-1.0,1.0)) + theme_classic()
```

6. Customise how your final scatter plot will look

This is CODE BLOCK 83 in the document R_CODE_DATA_VISUALISATION_WORKBOOK.DOC, and it adds four new style commands separated by + symbols to the code block used to create your scatter plot with lines of best fit on it in step 5. These are the `labs` command, the `xlim` command, the `ylim` command, and the `theme_classic` command. The `labs` command sets the labels that will be used tor the X axis (using the `x` argument) and the Y axis (using the `y` argument). The `xlim` command sets the minimum and maximum values for the X axis of the graph. In this case, these will be `0.5` for the minimum value and `3.1` for the maximum value. The `ylim` command sets the minimum and maximum values for the Y axis of the graph. In this case, these will be `-0.1` for the minimum value and `1.0` for the maximum value. Finally, the `theme_classic` command sets the remaining style elements of the final graph to those of the pre-existing classic theme. Once you have finished editing this code block, you can run it again to create the final version of your graph.

X-Y scatter plot with multiple data series created

Once you have worked through the first part of this exercise, the multi-series X-Y scatter plot that you have created should look like this:

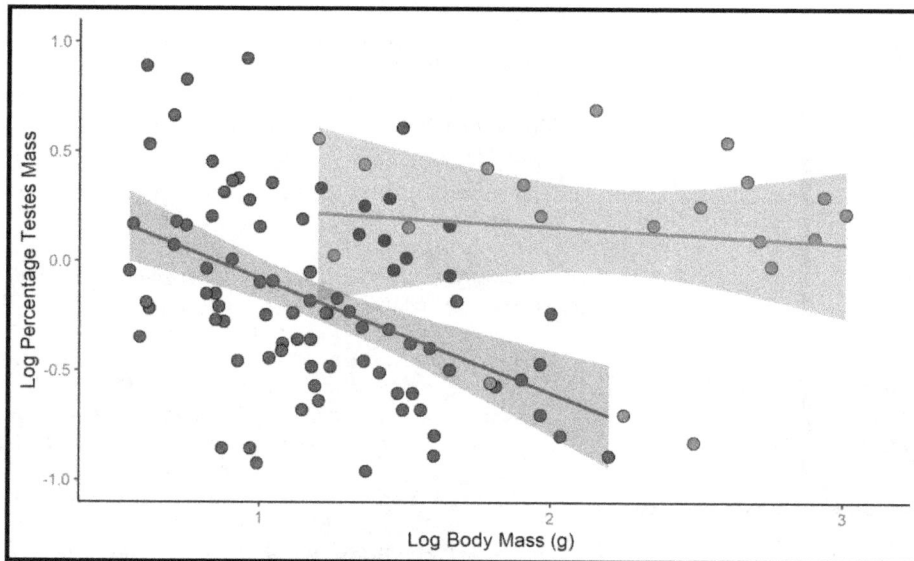

After you have created a multi-series X-Y scatter plot, you can export it from R so that you can include it in a manuscript or presentation. If you are using RGUI, you can do this by clicking on the R GRAPHICS window containing your scatter plot to select it, before clicking on FILE on the main menu bar and selecting SAVE AS. This will allow you to save it in a variety of different formats. If you are using RStudio, you can export your scatter plot by clicking on the EXPORT button at the top of the window displaying your graph and selecting SAVE AS IMAGE.

When including a multi-series scatter plot in a manuscript, it is important that you provide an appropriate figure legend for it. This legend should provide all the information required for the reader to interpret the contents of the graph. For the above multi-series scatter plot, an appropriate legend would be:

Figure 1: The relationship between log body mass and log percentage testes mass (PTM) in microchiropteran (blue) and macrochiropteran (red) bat species. For each data series, the lines represent a linear line of best fit, while the shaded areas represent a 95% confidence interval for this line.

To gain more experience in making multi-series X-Y scatter plots, you will add a third data series to the above scatter plot. This data series will be for data on body mass and testes mass from a new taxonomic group, the cetaceans. These data are held in a file called `cetacean_data.csv`. To do this, you will first need to create a new object containing these cetacean data by entering the following command into R:

```
cetacean_data <- read.table(file="cetacean_data.csv",
          sep=",",as.is=FALSE,header=TRUE)
```

This will create a new R object called `cetacean_data` containing the cetacean data set. To check that this has been done correctly, you can view this newly created R object by entering the following command into R:

```
View (cetacean_data)
```

Once you are happy that the cetacean data have been correctly imported into your analysis project, you are ready to add these data to the above multi-series X-Y plot showing the body mass and percentage testes mass data for the two groupings of bats. To do this, you need to add a third `geom_point` command to the code block from step 6 of the above flow diagram to allow you to add the cetacean data to this graph, and a third `geom_smooth` command so that a line of best fit for the cetacean data will also be added to it. In both cases, you should use a green colour to display this new information so that it can be easily separated from the existing data series that are already present on the graph. You will also need to change the values in the `xlim` and `ylim` style commands so that the axes are long enough to display the data from the newly added data series. To do this, edit the code block from step 6 of the above flow diagram (this is CODE BLOCK 83 in the document R_ CODE_DATA_VISUALISATION_WORKBOOK.DOC) so that it looks like the code at the top of the next page (the newly added `geom_point` and `geom_smooth` commands, as well as the required modifications to the existing `xlim` and `ylim` commands, are highlighted in **bold**).

NOTE: Rather than having to type in the new `geom_point` and `geom_smooth` commands from scratch, you can copy these pair of commands from the start of the existing code block and paste them into the required position as new versions before updating them with the settings required for the cetacean testes mass data series.

```
      ggplot() + geom_point(data=micro_bat_data,
aes(x=log_body_mass,y=log_ptm),shape=21,colour="black",
  fill="blue",size=3) + geom_smooth(data=micro_bat_data,
aes(x=log_body_mass,y=log_ptm),method="lm",colour="blue",
    lty="solid",size=1,se=TRUE,level=0.95,fill="blue",
      alpha=0.3) + geom_point(data=macro_bat_data,
aes(x=log_body_mass,y=log_ptm),shape=21,colour="black",
  fill="red",size=3) + geom_smooth(data=macro_bat_data,
aes(x=log_body_mass,y=log_ptm),method="lm",colour="red",
    lty="solid",size=1,se=TRUE,level=0.95,fill="red",
      alpha=0.3) + geom_point(data=cetacean_data,
aes(x=log_body_mass,y=log_ptm),shape=21,colour="black",
  fill="green",size=3) + geom_smooth(data=cetacean_data,
aes(x=log_body_mass,y=log_ptm),method="lm",colour="green",
  lty="solid",size=1,se=TRUE,level=0.95,fill="green",
    alpha=0.3) + labs(x="Log Body Mass (g)",y=
  "Log Percentage Testes Mass") + xlim(c(0.5,8.5))+
      ylim(c(-2.0,1.0)) + theme_classic()
```

When you run this modified block of code, it should produce a multi-series X-Y scatter plot that looks like this:

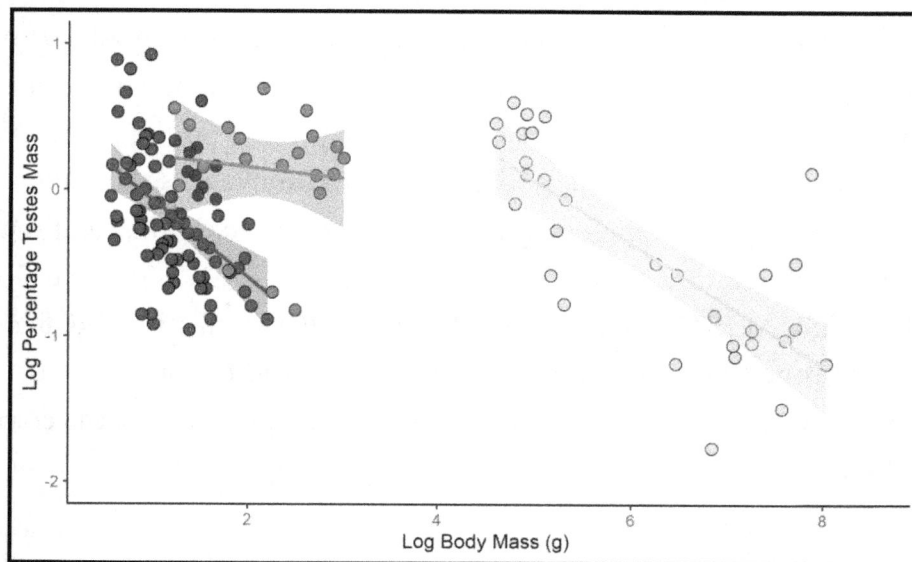

If you examine this graph, you will notice that while the data for the microchiropteran (blue) and macrochiropteran (red) bat species occupy a similar space on it, the data for the cetacean data (green) occupies a very different space. This is because while the log-transformed percentage testes masses (shown on the Y axis) are similar between the different groupings, the body masses of cetaceans are very different from those of the two groups of bats. The result of this is that the data for the two groups of bats are squashed up at the left hand end of the graph, making them harder to examine in detail. These differences in body mass ranges between the cetaceans and the bats make it difficult to display the data for all three groupings on a single multi-series scatter plot. Instead, it would be better to make a multi-panel image with each grouping represented on its own scatter plot within it. To do this, you first need to create a separate scatter plot for each data series and save it as a new R object. This can be done by editing the last block of code you entered (from page 174) to remove the `geom_point` and `geom_smooth` commands for the `macro_bat_data` and `cetacean_data` data series, changing the arguments in the `xlim` and `ylim` commands and adding a name of the object where the resulting graph will be stored to its start. This edited block of code should look like this (the required modifications are highlighted in **bold**):

```
micro_bat_scatter_plot <- ggplot() +
geom_point(data=micro_bat_data,aes(x=log_body_mass,
y=log_ptm),shape=21,colour="black",fill="blue",size=3) +
geom_smooth(data=micro_bat_data,aes(x=log_body_mass,
y=log_ptm),method="lm",colour="blue",lty= "solid",
size=1,se=TRUE,level=0.95,fill="blue",alpha=0.3) +
labs(x="Log Body Mass (g)",y="Log Percentage
Testes Mass") + xlim(c(0.5,2.5)) +
ylim(c(-1.0,1.0)) + theme_classic()
```

This block of code creates a scatter plot from the microchiropteran data and saves it as a new R object called `micro_bat_scatter_plot`. To view this graph, enter the following command into R:

```
micro_bat_scatter_plot
```

This code can then copied and edited to create individual graphs for the other two taxonomic groupings. To do this for the macrochiropteran bat data, edit the above block of code so that it looks like this (the required modifications are highlighted in **bold**):

```
macro_bat_scatter_plot <- ggplot() +
geom_point(data=macro_bat_data,aes(x=log_body_mass,
y=log_ptm),shape=21,colour="black",fill="red",size=3) +
geom_smooth(data=macro_bat_data,aes(x= log_body_mass,
y=log_ptm),method="lm",colour="red",lty="solid",
size=1,se=TRUE,level=0.95,fill="red",alpha=0.3) +
labs(x="Log Body Mass (g)",y="Log Percentage Testes Mass")
+ xlim(c(0.5,3.1)) + ylim(c(-1.0,1.0))
+ theme_classic()
macro_bat_scatter_plot
```

To do this for the cetacean data, edit the above block of code so that it looks like this R (the required modifications are highlighted in **bold**):

```
cetacean_scatter_plot <- ggplot() +
geom_point(data=cetacean_data,aes(x=log_body_mass,
y=log_ptm),shape=21,colour="black",fill="green",size=3) +
geom_smooth(data=cetacean_data,aes(x=log_body_mass,
y=log_ptm),method="lm",colour="green",lty="solid",size=1,
se=TRUE,level=0.95,fill="green",alpha=0.3) + labs(x="Log
Body Mass (g)",y="Log Percentage Testes Mass") +
xlim(c(4,8.5)) + ylim(c(-2.0,1.0)) + theme_classic()
cetacean_scatter_plot
```

Once you have created the individual graphs showing the data for each data series, you can use the ggarrange command to create a multi-panel figure containing the three individual scatter plots. To do this, enter the following commands into R (**NOTE:** This assumes that the ggpubr package has been installed in your version of R – see pages 65 and 66 for details of how to do this):

```
library(ggpubr)
ggarrange(micro_bat_scatter_plot,
macro_bat_scatter_plot,cetacean_scatter_plot,labels=
c("A","B","C"),ncol=3,nrow=1)
```

The resulting multi-panel figure should look like this:

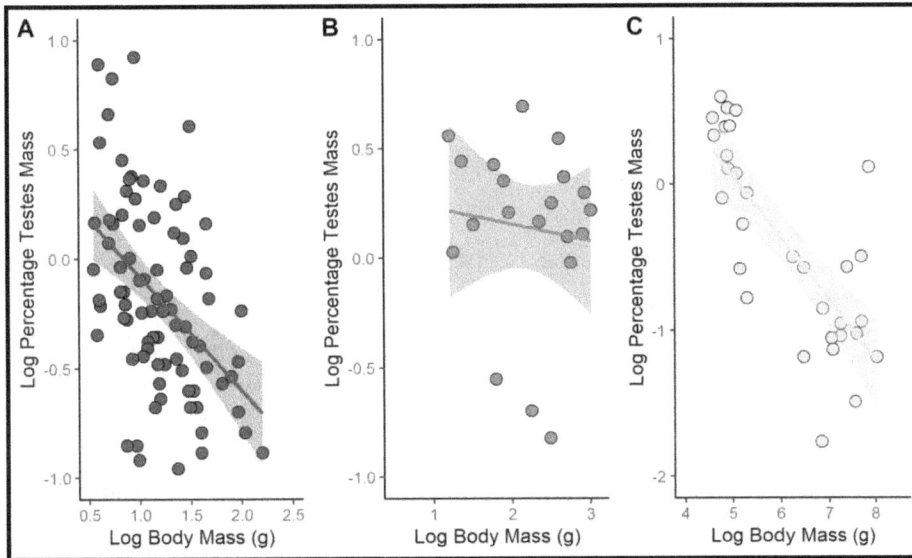

So far, the multi-series X-Y scatter plots which have been created in this exercise have assumed that the data for each series is held in their own R object in your analysis project. However, there may be occasions when you wish to create a multi-series scatter plot from different groups held in a single data set. To demonstrate how to do this, you will now make a multi-series X-Y scatter plot from the `all_bat_data` data set that you imported into R at the start of Exercise 3.1. This graph will include two data series, one for the microchiropteran bat species and a second for the macrochiropteran bat species. A code is included in the column called `type` in this data set that identifies which group each bat species in this data set belongs to (`micro` for microchiropteran species and `macro` for macrochiropteran species).

This multi-series scatter plot can be created by editing the final block of code provided in step 6 of the above flow diagram (this is CODE BLOCK 83 in the document R_CODE_DATA_VISUALISATION_WORKBOOK.DOC). The modifications required involve moving the `data` argument into the `ggplot` command and moving the `x`, `y`, `colour` and `fill` arguments into the `aes` element of the `ggplot` command. The column containing the identifier signifying which series each line of data belongs to is then specified in the `colour` and `fill` arguments in this `aes` element. As these no longer need to be included in separate `geom_point` and `geom_smooth` commands for each data series, you only need to include a single version of each command in the modified code block and the remaining version needs to be edited to remove the `data` argument and `aes` element (including the `x` and `y` arguments they contain). However, you do need to add a `scale_colour_manual` and a `scale_fill_manual` command which contains arguments to tell R what colours to use for the colour and fill for each data series that will be plotted on your final graph. The edited block of code should look like this (the required modifications are highlighted in **bold**):

```
ggplot(data=all_bat_data,aes(x=log_body_mass,y=log_ptm,
colour=type,fill=type)) + geom_point(shape=21,size=3) +
   geom_smooth(method="lm",lty="solid",size=1,se=TRUE,
level=0.95,alpha=0.3) + scale_color_manual(values=c("red",
   "blue")) + scale_fill_manual(values=c("red","blue")) +
labs(x="Log Body Mass (g)",y="Log Percentage Testes Mass")
  + xlim(c(0.5,3.1)) + ylim(c(-1.0,1.0)) + theme_classic()
```

When you run this modified block of code in R, you should get a multi-series scatter plot that looks like this:

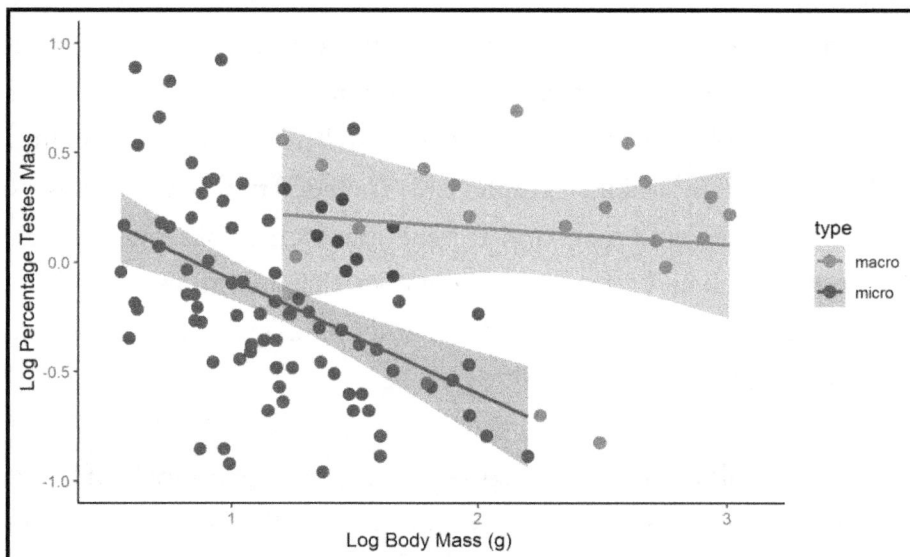

EXERCISE 3.3: HOW TO CREATE AN X-Y SCATTER PLOT WITH A MINIMUM CONVEX POLYGON (MCP) ENCLOSING ALL THE RECORDS FOR EACH DATA SERIES:

While plotting lines of best fit on a multi-series X-Y scatter plot allows you to identify the trends between two variables, and how these trends may differ between different groups of data, this is not the only way to compare the spread of data in different data series. In particular, for some types of data, and some research questions, it is not the trends within the different data series that you wish to compare, but the overall spread of data across the ecological space defined by the two variables being plotted on a scatter plot. This can be done by creating what is called a minimum convex polygon (or MCP for short) which encloses the two-dimensional area occupied by all the points belonging to each data series and plotting them on the graph itself. This allows you to identify and compare the ecological space occupied by different groupings of data in relation to the variables being plotted on the X and Y axes.

To explore how to do this, you will work through an example that uses morphometric data from anglehead lizards sampled from three locations in Malaysia. These data are held in a data set called `all_locations.csv` and were previously used in Exercises 2.3 and 2.5 of this workbook. During this exercise, you will create a multi-series scatter plot with body length plotted on the X axis and forelimb length plotted on the Y axis for lizards from locations A, B and C. You will then create minimum convex polygons which enclose the all the morphological variations in body length and forelimb length found at each location and before adding them to your scatter plot. To do this, work through the flow diagram that starts on the next page.

NOTE: These instructions assume that you have already installed the `ggplot2` package in your version of R and that you have loaded its command library into your analysis project. If you have not already done this, you will need to do so before you start this exercise. Instructions for how to install this package and load its command library into your analysis project can be found in Exercise 1.1. They also assume that you have already installed the `dplyr` package into your version of R. Instructions for how to install this package can be found in Exercise 2.1. You will load the command library from this package into your analysis project as part of the workflow detailed in the flow diagram.

Data set held in a comma separated values (.CSV) file

For this example, the data set you will use is stored in a file called `all_locations.csv` that is located in the WORKING DIRECTORY folder you created during the introduction to this chapter.

1. Set the WORKING DIRECTORY for your analysis project

Before you start any analysis in R, you first need to set the WORKING DIRECTORY. To do this, enter the text `setwd("` and then type the address of your WORKING DIRECTORY, using slashes (/) as the folder separators, before entering a second quotation mark followed by a closing bracket, like this `")`. For example, if your WORKING DIRECTORY has the address C:\STATS_FOR_BIOLOGISTS_TWO, your `setwd` command should look like this:

```
setwd("C:/STATS_FOR_BIOLOGISTS_TWO")
```

If you are using RGUI, enter your `setwd` command in the R CONSOLE window (remembering to use the address of your own WORKING DIRECTORY folder in it) and then press the ENTER key on your keyboard. If you are using RStudio, enter your `setwd` command into the SCRIPT EDITOR window. To run it, select it and then click on the RUN button at the top of this window. You will enter all the remaining commands for this exercise in a similar manner, depending on the user interface you are using.

To check that your WORKING DIRECTORY has been set properly, enter the command `getwd()` and carefully check that the address it returns is the same as the one for the STATS_FOR_BIOLOGISTS_TWO folder you created at the start of this chapter.

Before you move on to step 2, make sure that all the data you wish to use in your analysis project are located in this WORKING DIRECTORY folder. In this case, this is a file called `all_locations.csv`. **NOTE:** If the data you are going to import into R in step 2 are not located in the WORKING DIRECTORY you set in this step, the import code provided in the next step will not work.

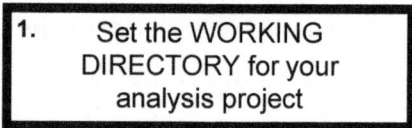

```
2.    Load your data into
R using the read.table
         command
```

The `read.table` command provides the easiest way to load data held in a .CSV file (and stored in the WORKING DIRECTORY you set in step 1) into R so you can analyse it. To do this for the data set being used in this example, enter the following command into R:

```
all_locations <-
read.table(file="all_locations.csv",
    sep=",",as.is=FALSE,header=TRUE)
```

This code has to be entered exactly as it is written here or it will not work. If you wish to use the copy-and-paste approach for entering this command, copy the text directly below CODE BLOCK 84 in the document R_CODE_DATA_VISUALISATION_WORKBOOK.DOC and paste it into R.

This command will create a new object in R called `all_locations` which will contain the data from the specified .CSV file. To import a different .CSV file into R, all you need to do is change the file name in the `file` argument to the name of the one you wish to import. You can also use whatever name you wish for the R object which will be created by this command. To do this, simply replace `all_locations` at the start of the first line of the above code with the name you wish to use for it. **NOTE:** If your .CSV data set uses a semicolon as the column separator, you would need to replace the `sep=","` argument with `sep=";"`.

```
3.    Check the data have
loaded into R correctly by
checking the names of the
columns and by viewing it
```

Whenever you import any data into R you need to check that they have loaded correctly. First, you need to check that all the required columns are present in the R object you just created. To do this, enter the following command into R:

```
names(all_locations)
```

This is CODE BLOCK 85 in the document R_CODE_DATA_VISUALISATION_WORKBOOK.DOC. This command will return the names used for each column in the R object you just created. For this example, the names should be: `id`, `body_length`, `forelimb_length`, `location` and `sex`.

Next, you should view the contents of the whole table using the `View` command. This is done by entering following code into R:

```
View(all_locations)
```

This is CODE BLOCK 86 in the document R_CODE_DATA_VISUALISATION_WORKBOOK.DOC. This command will open a DATA VIEWER window where you can examine your data set and check that the correct data have been loaded into R.

181

4. Create your initial scatter plot based on the data set imported in step 2

After you have successfully imported your data set into R, you are ready to use it to create your initial scatter plot. In this case, it will be based on the data on the morphometrics of anglehead lizards from three locations in Malaysia that are held in the R object called `all_locations` created in step 2. To do this, enter the following block of code into R:

```
ggplot(data=all_locations,
aes(x=body_length,y=forelimb_length,
colour=location)) + geom_point() +
scale_colour_manual(values=c("red",
       "green","blue"))
```

This is CODE BLOCK 87 in the document R_CODE_DATA_ VISUALISATION_WORKBOOK.DOC, and it contains three commands separated by + symbols. These are the `ggplot` command, the `geom_point` command and the `scale_ colour_manual` command. The `ggplot` command sets the data set which will be used for the graph. This is done using the `data` argument and, in this case, it will be the R object called `all_locations` created in step 2 of this exercise. The data that will be plotted on the X axis of the resulting graph is set using the `x` argument of the `aes` element of this `ggplot` command. In this case, it is the column called `body_length`. The data that will be plotted on the Y axis of the resulting graph is set using the `y` argument of the `aes` element of this `ggplot` command. In this case, it is the column called `forelimb_length`. Finally, the `colour` argument is also included in the `aes` element of the `ggplot` command to identify the column that contains information on which of the three locations each lizard in the data set was sampled from. This information is held in the column called `location`.

The second command in this code block, `geom_point`, sets the type of graph that will be created from the data specified in the `ggplot` command. In this case, it will be a scatter plot consisting of one point for each row of data in the data set specified in the `data` argument of the `ggplot` command. As this information has been specified in the `ggplot` command, no arguments need to be included in the `geom_point` command for this particular data set

The third command in this code block, `scale_colour_ manual`, sets the colours that will be used to represent the data from the different groupings found in the column set by the `colour` argument in the `aes` element of the `ggplot` command. In this instance, this is the column called `location`, and it contains three different categories (A, B and C). As a result, three colours need to be specified in the `scale_colour_manual` command. For this example, these are `red`, `green` and `blue`.

5. Create a minimum convex polygon for each data series on your initial scatter plot

In order to create a minimum convex polygon for each of your data series, you first need to create new subsets that contain just the data for each series using the `subset` command. To do this for the lizard data from the three locations listed in the `location` column of the `all_locations` dataset, enter the following three commands (each of which needs to be entered on a separate line) into R:

```
location_a <- subset(all_locations,
          location=="A")
location_b <- subset(all_locations,
          location=="B")
location_c <- subset(all_locations,
          location=="C")
```

This is CODE BLOCK 88 in the document R_CODE_DATA_ VISUALISATION_WORKBOOK.DOC.

Next, you need to load the command library for the `dplyr` package into your analysis project. To do this, enter the following command into R:

```
library(dplyr)
```

This is CODE BLOCK 89 in the document R_CODE_DATA_ VISUALISATION_WORKBOOK.DOC.

You can now use the `slice` command from the `dplyr` command library to create a minimum convex polygon that will enclose all the data on the scatter plot from the lizards sampled at each of the three locations. To do this, enter the following three commands (each of which needs to be entered on a separate line) into R:

```
polygon_location_a <- location_a %>%
slice(chull(body_length,forelimb_length))
polygon_location_b <- location_b %>%
slice(chull(body_length,forelimb_length))
polygon_location_c <- location_c %>%
slice(chull(body_length,forelimb_length))
```

This is CODE BLOCK 90 in the document R_CODE_DATA_ VISUALISATION_WORKBOOK.DOC. In this code, there are three separate commands (each starting on a new line), one for each sampling location. In these commands, the data set containing the data for each series (e.g. `location_a` in the first of these commands) is defined before the `slice` command (which will create the polygon based on that data set), and is linked to it by the code `%>%`. In the `slice` command itself, the `chull` argument sets the command to create a minimum convex polygon based on the data from the two variables listed after it (in this case, `body_length` and `forelimb_length`). These variables need to be provided in the order X variable followed by Y variable. The polygon created by each individual `slice` command in this code block will be held in a new R object with the name provided at the very start of each command.

6. Add the minimum convex polygons for each data series to your initial scatter plot

Once you have created a minimum convex polygon for each data series, you can add them to your initial X-Y scatter plot by adding three new graphing commands to the block of code from step 4. To do this, edit the code from that step so that it looks like this (the newly added graphing commands are highlighted in **bold**):

```
ggplot(data=all_locations,
aes(x=body_length,y=forelimb_length,
colour=location)) + geom_point() +
scale_colour_manual(values=c("red","green",
"blue")) + geom_polygon(data=
polygon_location_a,fill="red",alpha=0.3) +
geom_polygon(data=polygon_location_b,fill=
"green",alpha=0.3) + geom_polygon(data=
polygon_location_c,fill="blue",alpha=0.3)
```

This is CODE BLOCK 91 in the document R_CODE_DATA_VISUALISATION_WORKBOOK.DOC. It contains three new `geom_polygon` commands separated by + symbols, one for each of the minimum convex polygons you wish to add to your scatter plot. In these newly added `geom_polygon` commands, the `data` argument is used to identify the R object which contains the minimum convex polygon that you have just created for each data series (such as `polygon_location_a` for the first data series). The `fill` argument sets the fill colour that will be used for the polygon (such as `red` for the minimum convex polygon for the first data series), while the `alpha` argument sets the level of transparency for this polygon (in this case, it is set to `0.3`, which means that it will be 30% opaque). Once you have finished editing this code block, you can run it again to create an updated version of your graph.

184

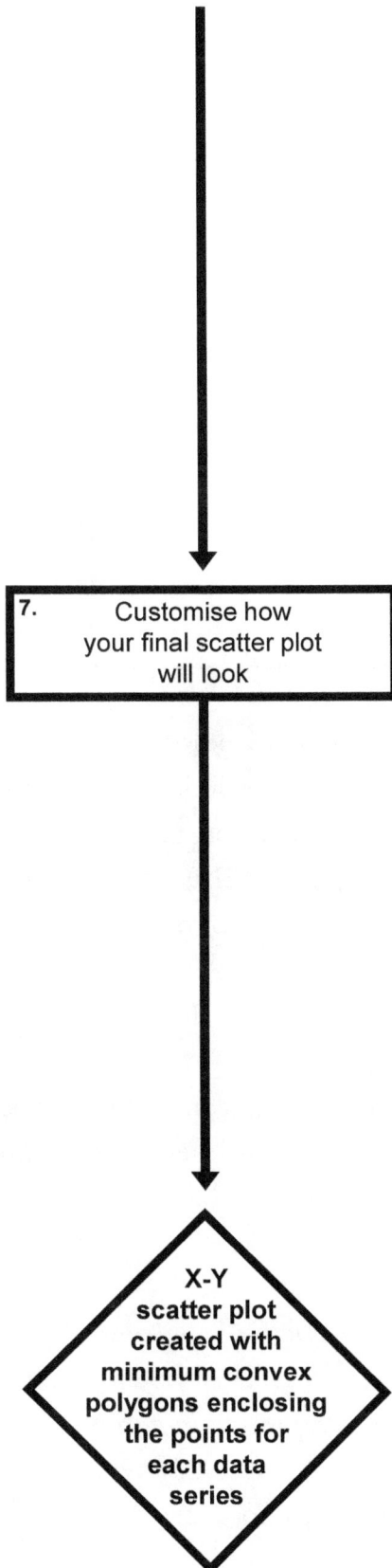

After you have created your initial scatter plot, and added the required minimum convex polygons enclosing the data for each data series, you can customise how the final graph will look. This can be done by adding a number of new style commands to the block of code used to produce it. To do this, edit the code from step 6 so that it looks like this (the newly added style commands are highlighted in **bold**):

```
ggplot(data=all_locations,aes(x=
body_length,y=forelimb_length,colour=
location)) + geom_point() +
scale_colour_manual(values=c("red","green",
"blue")) + geom_polygon(data=
polygon_location_a,fill="red",alpha=0.3) +
geom_polygon(data=polygon_location_b,
fill="green",alpha=0.3) + geom_polygon(data=
polygon_location_c,fill="blue",alpha=0.3) +
labs(x="Body Length (cm)",y="Forelimb Length
(cm)") + xlim(c(15,50))+ ylim(c(3.5,10)) +
theme_classic()
```

This is CODE BLOCK 92 in the document R_CODE_DATA_ VISUALISATION_WORKBOOK.DOC, and it adds four new style commands separated by + symbols to the code block used to create your scatter plot with the minimum convex polygons on it in step 6. These are the `labs` command, the `xlim` command, the `ylim` command, and the `theme_ classic` command. The `labs` command sets the labels that will be used for the X axis (using the `x` argument) and the Y axis (using the `y` argument). The `xlim` command sets the minimum and maximum values for the X axis of the graph. In this case, these will be `15` for the minimum value and `50` for the maximum value. The `ylim` command sets the minimum and maximum values for the Y axis of the graph. In this case, these will be `3.5` for the minimum value and `10` for the maximum value. Finally, the `theme_ classic` command sets the remaining style elements of the final graph to those of the pre-existing classic theme. Once you have finished editing this code block, you can run it again to create the final version of your graph.

7. Customise how your final scatter plot will look

X-Y scatter plot created with minimum convex polygons enclosing the points for each data series

The final X-Y scatter plot with minimum convex polygons enclosing the data from each data series created using the code block provided in step 7 should look like this:

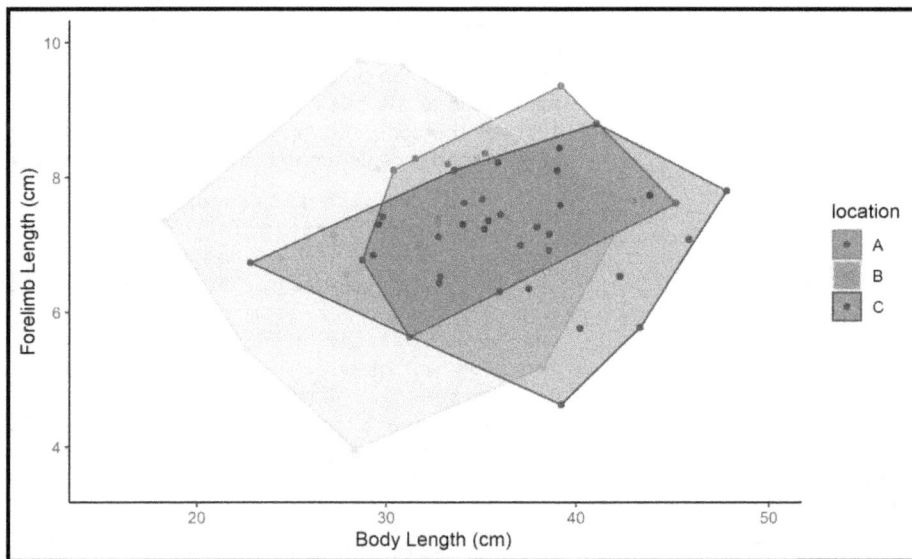

Once you have created an X-Y scatter plot with minimum convex polygons for each data series on it, you can export it from R so that you can include it in a manuscript or presentation. If you are using RGUI, you can do this by clicking on the R GRAPHICS window containing your scatter plot to select it, before clicking on FILE on the main menu bar and selecting SAVE AS. This will allow you to save it in a variety of different formats. If you are using RStudio, you can export your graph by clicking on the EXPORT button at the top of the window displaying your scatter plot and selecting SAVE AS IMAGE.

When including a scatter plot with minimum convex polygons on it in a manuscript, it is important that you provide an appropriate figure legend for it. This legend should provide all the information required for the reader to interpret the contents of the graph. For the above X-Y scatter plot with minimum convex polygons, an appropriate legend would be:

Figure 1: A comparison of the body length and forelimb length of anglehead lizards sampled from three locations in Malaysia. The data from Location A is shown in red, from Location B in green and Location C in blue. The polygons represent minimum convex polygons enclosing all the records from each location and show the combined variation for these variables for all individuals sampled at each location.

In the first part of this exercise, the minimum convex polygons for each data series were created separately. However, it is also possible to create them all at once within a single `slice` command. To do this, you would remove the `subset` commands from step 5 in the above flow diagram and replace the three separate `slice` commands in it with a new `slice` command which not only specifies the data set that will be used for it, but also the column containing the groups you wish to create separate polygons for using the `group_by` argument. For example, to do this for the data in the `all_locations` data set, you would need to replace the three separate `slice` commands in step 5 with the following new, single `slice` command (the required modifications are highlighted in **bold**):

```
polygons <- all_locations %>% group_by(location) %>%
        slice(chull(body_length,forelimb_length))
```

NOTE: This code assumes that you have already downloaded and installed the `dplyr` package, and loaded its command library into your analysis project.

This new `slice` command will create a new R object called `polygons` that will contain minimum convex polygons for all the different data series identified by the contents of the `location` column in the `all_locations` data set. After you have run the above `slice` command, you can add the minimum convex polygons generated by it to your scatter   plot by editing the block of code provided in step 7 of the above flow diagram (this is CODE BLOCK 92 from the document R_CODE_DATA_VISUALISATION_ WORKBOOK.DOC) so that it looks like this (the required modifications are highlighted in **bold**):

```
ggplot(data=all_locations,aes(x=body_length,
   y=forelimb_length,colour=location,fill=location)) +
geom_point() + scale_colour_manual(values=c("red","green",
   "blue")) + geom_polygon(data=polygons,alpha=0.3) +
   labs(x="Body Length (cm)",y="Forelimb Length (cm)") +
      xlim(c(15,50)) + ylim(c(3.5,10)) + theme_classic()
```

When you run this code block, it should create a new scatter plot containing minimum convex polygons for each data series that looks like the image at the top of the next page.

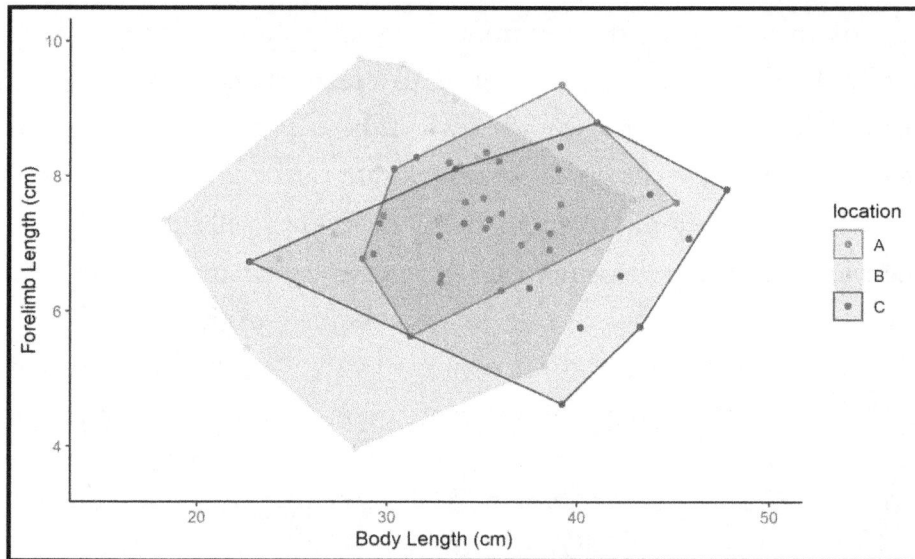

EXERCISE 3.4: HOW TO CREATE A LINE GRAPH TO DISPLAY TIME SERIES DATA:

So far, the graphs created in the exercises in this chapter have been scatter plots that plot the value of one variable against the value of another. However, these are only useful for displaying certain types of data and for certain purposes. In particular, they are not very useful for showing time series data. This is because time series data need to be plotted in a specific temporal order. In addition, for such graphs, it is often the change in the values of a particular variable over time that is of most interest rather than the trend between the values. As a result, rather than being displayed on a scatter plot, time series data are usually plotted as a graph with the line linking the values from successive time periods together to show how they have changed over time. Such time series line graphs are typically created with the GGPlot graphing package using the `geom_line` command. In order to create such graphs, you need to have your data arranged in a spreadsheet or table with one row per time period (such as year or month), and with one column containing the time period each record represents and a second column containing its value.

To explore how to create a line graph displaying time series data using the `geom_line` command from GGPlot, you will work through an example that uses a time series of beaked whale strandings from the coasts of the UK and Ireland collected between 1913 and

2002 (see page 322 for sources of these data). This data set contains separate columns with the number of recorded strandings for three species of beaked whales (Cuvier's beaked whale, the northern bottlenose whale and Sowerby's beaked whale) in each year, along with the year itself and the combined total number of strandings of these three species recorded in that year. In this exercise, you will start by creating a line graph of the time series of the recorded number of Cuvier's beaked whale strandings in the UK and Ireland between 1913 and 2002. To do this, work through the following flow diagram:

Data set held in a comma separated values (.CSV) file

For this example, the data set you will use is stored in a file called `beaked_whales.csv` that is located in the WORKING DIRECTORY folder you created during the introduction to this chapter.

1. Set the WORKING DIRECTORY for your analysis project

Before you start any analysis in R, you first need to set the WORKING DIRECTORY. To do this, enter the text `setwd("` and then type the address of your WORKING DIRECTORY, using slashes (/) as the folder separators, before entering a second quotation mark followed by a closing bracket, like this `")`. For example, if your WORKING DIRECTORY has the address C:\STATS_FOR_BIOLOGISTS_TWO, your `setwd` command should look like this:

```
setwd("C:/STATS_FOR_BIOLOGISTS_TWO")
```

If you are using RGUI, enter your `setwd` command in the R CONSOLE window (remembering to use the address of your own WORKING DIRECTORY folder in it) and then press the ENTER key on your keyboard. If you are using RStudio, enter your `setwd` command into the SCRIPT EDITOR window. To run it, select it and then click on the RUN button at the top of this window. You will enter all the remaining commands for this exercise in a similar manner, depending on the user interface you are using.

To check that your WORKING DIRECTORY has been set properly, enter the command `getwd()` and carefully check that the address it returns is the same as the one for the STATS_FOR_BIOLOGISTS_TWO folder you created at the start of this chapter.

Before you move on to step 2, make sure that all the data you wish to use in your analysis project are located in this WORKING DIRECTORY folder. In this case, this is a file called `beaked_whales.csv`. **NOTE:** If the data you are going to import into R in step 2 are not located in the WORKING DIRECTORY you set in this step, the import code provided in the next step will not work.

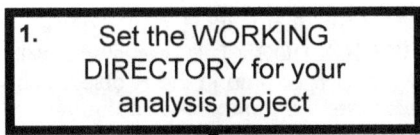

The `read.table` command provides the easiest way to load data held in a .CSV file (and stored in the WORKING DIRECTORY you set in step 1) into R so you can analyse it. To do this for the data set being used in this example, enter the following command into R:

```
beaked_whales <-
read.table(file="beaked_whales.csv",
sep=",",as.is=FALSE,header=TRUE)
```

This code has to be entered exactly as it is written here or it will not work. If you wish to use the copy-and-paste approach for entering this command, copy the text directly below CODE BLOCK 93 in the document R_CODE_DATA_ VISUALISATION_WORKBOOK.DOC and paste it into R.

This command will create a new object in R called `beaked_whales` which will contain the data from the specified .CSV file. To import a different .CSV file into R, all you need to do is change the file name in the `file` argument to the name of the one you wish to import. You can also use whatever name you wish for the R object which will be created by this command. To do this, simply replace `beaked_whales` at the start of the first line of the above code with the name you wish to use for it. **NOTE:** If your .CSV data set uses a semicolon as the column separator, you would need to replace the `sep=","` argument with `sep=";"`.

---

2. Load your data into R using the `read.table` command

Whenever you import any data into R you need to check that they have loaded correctly. First, you need to check that all the required columns are present in the R object you just created. To do this, enter the following command into R:

```
names(beaked_whales)
```

This is CODE BLOCK 94 in the document R_CODE_DATA_ VISUALISATION_WORKBOOK.DOC. This command will return the names used for each column in the R object you just created. For this example, the names should be: `year`, `cuviers_beaked_whale`, `northern_bottlenose_ whale`, `sowerbys_beaked_whale` and `total`.

Next, you should view the contents of the whole table using the `View` command. This is done by entering following code into R:

```
View(beaked_whales)
```

This is CODE BLOCK 95 in the document R_CODE_DATA_ VISUALISATION_WORKBOOK.DOC. This command will open a DATA VIEWER window where you can examine your data set and check that the correct data have been loaded into R.

---

3. Check the data have loaded into R correctly by checking the names of the columns and by viewing it

---

Once you have imported the required data into R, you are ready to create your initial line graph showing how the number of records in this data set change over time. In this example, you will create a time series showing the number of Cuvier's beaked whales stranding on UK and Irish coasts in each year between 1913 and 2002. This means the first step is to create an initial line graph showing how the number of strandings of this species changes from one year to the next. To do this, enter the following block of code into R (these instructions assume that you have already installed the `ggplot2` package in your version of R and that you have loaded its command library into your analsys project - see Exercise 1.1 for details of how to do this):

```
ggplot(data=beaked_whales,
aes(x=year,y=cuviers_beaked_whale)) +
geom_line(lty="solid",size=1,
colour="blacK")
```

This is CODE BLOCK 96 in the document R_CODE_DATA_VISUALISATION_WORKBOOK.DOC and it contains two commands separated by a + symbol. These are the `ggplot` command and the `geom_line` command. The `ggplot` command sets the data set which will be used for the graph. This is done using the `data` argument and, in this case, it will be the R object called `beaked_whales` created in step 2 of this exercise. The data that will be plotted on the X axis of the resulting graph is set using the `x` argument of the `aes` element of this `ggplot` command. In this case, it is the column called `year` in the `beaked_whales` data set. The data that will be plotted on the Y axis of the resulting graph is set using the `y` argument of the `aes` element of this `ggplot` command. In this case, it is the column called `cuviers_beaked_whale`, which contains the number of strandings of Cuvier's beaked whales recorded on UK and Irish coasts in each year.

The second command in this code block, `geom_line`, sets the type of graph that will be created from the data specified in the `ggplot` command. In this case, it will be a line graph showing how the number of records in the column specified in the `y` argument varies according to the temporal variable being plotted on the X axis. In this command, the type of line is set by the `lty` argument (in this case, it will be a solid line), while the thickness of the line is set by the `size` argument and its colour is set by the `colour` argument. In this case, the `size` argument will be set to 1 and the `colour` to black.

4. **Create an initial line graph showing how the number of records changes across the time period being examined**

While the initial line graph will show how the values of the variable you are plotting change over time, it can sometimes be hard to identify the individual values for each time period. As a result, it is often useful to add markers to show the individual values for each point in the time series to your initial line graph. This is done by adding a `geom_point` graphing command to the code block used to create your initial line graph. To do this, edit the code block from step 4 so that it looks like this (the newly added graphing command is highlighted in **bold**):

```
ggplot(data=beaked_whales,
aes(x=year,y=cuviers_beaked_whale)) +
geom_line(lty="solid",size=1,
colour="black") + geom_point(shape=21,
size=2,colour="black",fill="white")
```

5. Add markers to show the individual values for each point in the time series

This is CODE BLOCK 97 in the document R_CODE_DATA_ VISUALISATION_WORKBOOK.DOC and it adds a new graphing command to the code used to create the initial line graph in step 4. This is a `geom_point` command which will add the markers for each data point in the time series to the graph based on the X and Y variables specified in the `aes` element of the `ggplot` command. In this newly added `geom_point` command, the symbol used to mark each data point is set by the `shape` argument, while their size is set by the `size` argument, their outline colour is set by the `colour` argument and their fill colour by the `fill` argument. **NOTE:** The `fill` argument is only needed for markers specified by the pch codes 21 to 25 in the `shape` argument. Once you have finished editing this block of code, you can run it again to create an updated version of your initial time series line graph.

As this `geom_point` command is entered into the code block after the `geom_line` command, the marker points will be plotted on top of the line rather than underneath it. If you wanted your marker points to be plotted underneath the line, you would need to include the `geom_point` command before the `geom_line` command in the above code block.

Once you have created your initial time series line graph, and you have added markers for the individual data points to it, you can customise how the final graph will look. This can be done by adding a number of new style commands to the block of code used to produce it. To do this, edit the code block from step 5 so that it looks like this (the newly added style commands are highlighted in **bold**):

```
ggplot(data=beaked_whales,aes(x=year,
y=cuviers_beaked_whale)) + geom_line(lty=
    "solid",size=1,colour="black") +
geom_point(shape=21,size=2,colour="black",
fill="white") + labs(x="Year",y="Number of
    Strandings") + ylim(c(0,3)) +
        theme_classic()
```

This is CODE BLOCK 98 in the document R_CODE_DATA_ VISUALISATION_WORKBOOK.DOC, and it adds three new style commands separated by + symbols to the code block used to create your updated graph in step 5. These are the `labs` command, the `ylim` command, and the `theme_classic` command. The `labs` command sets the labels that will be used tor the X axis (using the `x` argument) and the Y axis (using the `y` argument). The `ylim` command sets the minimum and maximum values for the Y axis of the graph. In this case, these will be 0 for the minimum value and 3 for the maximum value. Finally, the `theme_classic` command sets the remaining style elements of the final graph to those of the pre-existing classic theme. Once you have finished editing this code block, you can run it again to create the final version of your time series line graph.

6. **Customise how your final time series line graph will look**

Time series line graph created from your data set

The time series line graph of the number of strandings of Cuvier's beaked whales recorded in each year between 1913 and 2002 in the UK and Ireland created using the final block of code provided in step 6 of the above flow diagram should look like this:

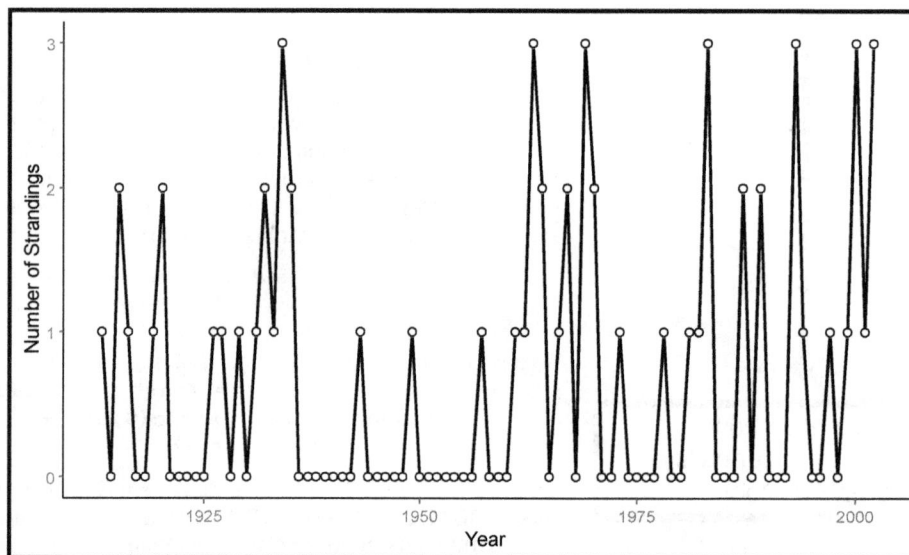

Once you have created a time series line graph, you can export it from R so that you can include it in a manuscript or presentation. If you are using RGUI, you can do this by clicking on the R GRAPHICS window containing your graph to select it, before clicking on FILE on the main menu bar and selecting SAVE AS. This will allow you to save it in a variety of different formats. If you are using RStudio, you can export your graph by clicking on the EXPORT button at the top of the window displaying your time series graph and selecting SAVE AS IMAGE.

When including a time series line graph in a manuscript, it is important that you provide an appropriate figure legend for it. This legend should provide all the information required for the reader to interpret the contents of the graph. For the above graph, an appropriate legend would be:

Figure 1: *The number of strandings of Cuvier's beaked whales recorded on the coasts of the UK and Ireland in each year between 1913 and 2002.*

In the first part of this exercise, the values you plotted in your time series were the individual data points. However, the levels of fluctuations between successive time periods in such individual data points can sometimes make it difficult to see any temporal trends within the data set. The way round this is to not only plot the individual data points, but to also plot a running average value (also known as a rolling mean, a rolling average or moving average). For a specific time period, a running average is calculated as an average of one or more of the preceding values, the value for that time period itself and one or more succeeding values. For example a five-year running average would be calculated as the average of the values for the two years before, the value for the year itself and the values for the two year after it. Such running averages can provide a smoothed line which reduces the influence of inter-time period fluctuations on the overall trend over time.

In order to plot a running average on a time series line graph, you first need to calculate the running average values for your data set. This can be done in R using the `rollmean` command from the `tidyquant` package. As a result, before you can use this command to calculate a running average, you first need to ensure that you have this package installed in your version of R. To do this, first enter the command `library()` into R. This will allow you to check whether or not you already have the `tidyquant` package installed. If it isn't, you can download and install it by using the following command:

```
install.packages("tidyquant")
```

Once you have ensured that the `tidyquant` package is installed in your version of R, you can load its command library into your analysis project using the following command:

```
library(tidyquant)
```

You can now use the `rollmean` command from this package to calculate the running average for the time frame you wish to calculate it from. To explore how you can do this, you will now calculate a five-year running average for the number of strandings of Cuvier's beaked whale from the UK and Ireland between 1913 and 2002. To do this, enter the following command into R:

```
beaked_whales$cbw_ra_5=rollmean(beaked_whales$
cuviers_beaked_whale,5,fill=list(NA,NULL,NA))
```

This command will create a new column in the `beaked_whales` data set called `cbw_ra_5` which will contain the 5 year running average of the number of Cuvier's beaked whale strandings. Within the `rollmean` command itself, the data set and column of data to be used to calculate the running average are specified first (in this case, it is the column called `cuviers_beaked_whale` in the data set called `beaked_whales`), before the size of the time frame that will be used to calculate the running average is specified (in this case, it will be calculated based on a time frame of five successive temporal records centred on each year in the data set). Finally, the `fill` argument is used to specify what should happen to any time periods where missing values mean that there are not the full number of records to calculate the running average. In this case, the `list(NA,NULL,NA)` term is used to specify that any rows of data that do not have values for every time period needed to calculate the specified running average will be filled with an `na` instead of an actual value.

After you have run the command to calculate the running average, you need to view the data set to ensure that it has been calculated successfully. To do this, enter the following command into R:

```
View(beaked_whales)
```

Once the five-year running average for Cuvier's beaked whale has been successfully calculated and added to the new column called `cbw_ra_5` in the R object called `beaked_whales`, it can be added as a new line to the line graph created in step 6 of the above flow diagram. This is done by editing the code from step 6 of the above flow diagram (this is CODE BLOCK 98 in the document R_CODE_DATA_VISUALISATION_WORKBOOK.DOC) so that it looks like this (the newly added command is highlighted in **bold**):

```
ggplot(data=beaked_whales,aes(x=year,
  y=cuviers_beaked_whale)) + geom_line(lty="solid",
size=1,colour="black") + geom_point(shape=21,size=2,
  colour="black",fill="white") + geom_line(data=
beaked_whales,aes(x=year,y=cbw_ra_5),lty="solid",size=2
colour="red") + labs(x="Year",y="Number of Strandings") +
          ylim(c(0,3)) + theme_classic()
```

When you run this block of code, you should get a new time series line graph which displays not only the underlying data of the number of strandings from each year (in black), but also the five-year running average that you have calculated (in red), and it should look like this:

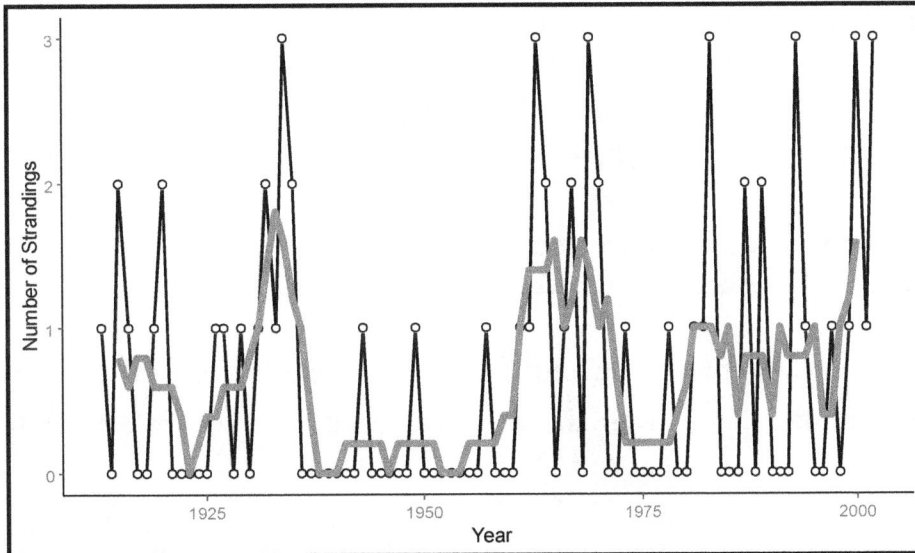

NOTE: When you run this block of code, you will get a warning message in your R CONSOLE window that says *Removed 4 row(s) containing missing values (geom_path).* This is okay and it refers to the first two and last two values in the running average column that have been filled in with `na` as there are not enough prior and successive records to calculate the specified running average.

Within the beaked whale data set, there are data on strandings for another two species, the northern bottlenose whale and Sowerby's beaked whale. If you wish to compare the time series for these three species, you could plot data for the individual years on the same graph by adding new `geom_line and geom_point` commands to the code block from step 6 of the above flow diagram. However, given the high levels of inter-annual fluctuations in these data, this is unlikely to show any clear patterns. Instead, in this exercise, you will create a multi-series graph based on the five-year running average for each species. The first step in doing this will be to calculate a five-year running average for northern bottlenose whales and for Sowerby's beaked whales using the `rollmean` command. This can be done by modifying the command you used to calculate the five-year running average for Cuvier's beaked whale.

The command to calculate the five-year running average for northern bottlenose whales should look like this (the required modifications are highlighted in **bold**):

```
beaked_whales$nbw_ra_5=rollmean(beaked_whales$
northern_bottlenose_whale,5,fill=list(NA,NULL,NA))
```

While the command for calculating the five-year running average for Sowerby's beaked whales should look like this (the required modifications are highlighted in **bold**):

```
beaked_whales$sbw_ra_5=rollmean(beaked_whales$
sowerbys_beaked_whale,5,fill=list(NA,NULL,NA))
```

Once you have run these two commands, you need to check the `beaked_whales` data set to ensure that they have given the required results. To do this, enter the following command into R:

```
View(beaked_whales)
```

Two new columns should have been added to this data set, one called `nbw_ra_5` which contains the five-year running average for the number of northern bottlenose whale strandings, and one called `sbw_ra_5` which contains the five-year running average for Sowerby's beaked whale. Once you are sure that these new columns have been added correctly, and do indeed contain the five-year running averages for each species, you are ready to create your multi-series line graph showing the five-year running average for each of the three species. This can be done by first creating a time series line graph with just the Cuvier's beaked whale five-year running average on it. To do this, edit the code from step 6 of the above flow diagram so that it looks like this (the required modifications are highlighted in **bold**):

```
ggplot() + geom_line(data=beaked_whales,aes(x=year,
    y=cbw_ra_5),lty="solid",size=2,colour="red") +
labs(x="Year",y="Number of Strandings") + ylim(c(0,3)) +
                theme_classic()
```

NOTE: If you examine this code, you will notice that the `data`, `x` and `y` arguments have been removed from the `ggplot` command, and instead are specified in the `geom_line` command. This is needed because each line on the final graph will be based on a different series of data. You will also see that the `geom_point` command has been remove. This is because the final graph will not include a marker for each individual data point.

Once you have created this new block of code which will create a time series line graph displaying just the Cuvier's beaked whale five-year running averages, you can then add a new `geom_line` command to it so that the five-year running averages for the northern bottlenose whale will be added to the graph. To do this edit the above block of code so that it look like this (the newly added `geom_line` command is highlighted in **bold**):

```
ggplot() + geom_line(data=beaked_whales,aes(x=
year,y=nbw_ra_5),lty="solid",size=2,colour="blue")+
geom_line(data=beaked_whales,aes(x=year,y=cbw_ra_5),
lty="solid",size=2,colour="red") + labs(x="Year",y="Number
of Strandings") + ylim(c(0,3)) + theme_classic()
```

This code block can then be further edited to add a third `geom_line` command which will add the five-year running average data for Sowerby's beaked whale to the graph. To do this edit the above block of code so that it look like this (the newly added `geom_line` command is highlighted in **bold**):

```
ggplot() + geom_line(data=beaked_whales,aes(x=
year,y=sbw_ra_5),lty="solid",size=2,colour="green") +
geom_line(data=beaked_whales,aes(x=year,y=nbw_ra_5),
lty="solid",size=2,colour="blue) + geom_line(data=
beaked_whales,aes(x=year,y=cbw_ra_5),lty="solid",size=2,
colour="red") + labs(x="Year",y="Number of Strandings") +
ylim(c(0,3)) + theme_classic()
```

When you run this block of code, you should get a multi-series line graph showing the five-year running averages for strandings of the three beaked whale species in the UK and Ireland that looks like the image at the top of the next page. **NOTE:** Since the `geom_line` commands for the Sowerby's beaked whale data and the northern bottlenose whale data have been added to the code block before the `geom_line` command for the Cuvier's beaked whale data, they will be displayed behind the line for the Cuvier's beaked whale data and not the front of it.

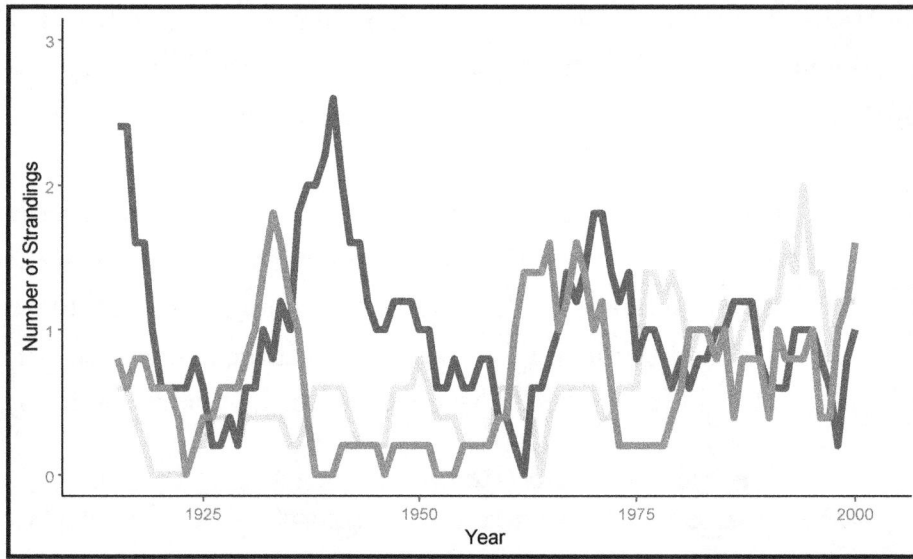

At the moment, this graph is quite confusing and it is difficult to compare the trends in the five-year running averages between the three species. This is because the lines for the different species overlap with each other, especially towards the end of the time series. In situations like this, you can make the trends for each species clearer by creating a stacked time series graph instead. In such graphs, the data for each species are stacked on top of each other (you can see an example of this type of graph at the bottom of page 202), making it easier to see how the trends in the different lines compare. In order to be able to create a stacked time series graph, you first need to process the data which you are going to use for the different series. In this example, you will create a stacked time series graph with the data for the Cuvier's beaked whales on the bottom, the data for northern bottlenose whales in the middle and the data for Sowerby's beaked whales on the top.

In order to do this, you need to create two new columns of data, one which contains the values you will use to plot the stacked northern bottlenose whale data, and one which contains the values you will use to plot the stacked Sowerby's beaked whale data. The values for the stacked northern bottlenose whale data line are calculated by adding the five-year running average value for northern bottlenose whales for each year to the value for the five-year running average for Cuvier's beaked whale for the same year. This can be done by entering the following command into R:

```
beaked_whales$nbw_ra_5_stacked=beaked_whales$cbw_ra_5+
                beaked_whales$nbw_ra_5
```

The values for the stacked Sowerby's beaked whale data line are calculated by adding the five-year running average for Sowerby's beaked whale for each year to the values for the five-year running averages for both Cuvier's beaked whale and the northern bottlenose whale for the same year. This can be done by entering the following command into R:

```
beaked_whales$sbw_ra_5_stacked=beaked_whales$cbw_ra_5+
         beaked_whales$nbw_ra_5+beaked_whales$sbw_ra_5
```

NOTE: As it will be plotted as the bottom series on the graph, you can just use the raw five-year running average data for Cuvier's beaked whale, and so you do not need to calculate a new stacked value for it.

Once you have run these two commands, you need to check the `beaked_whales` data set to ensure that they have given the required results. To do this, enter the following command into R:

```
View(beaked_whales)
```

Two new columns should have been added to this data set, one called `nbw_ra_5_stacked` which contains the stacked five-year running average for the number of northern bottlenose whale strandings, and one called `sbw_ra_5_stacked` which contains the stacked five-year running average for Sowerby's beaked whale. Once you are sure that these new columns have been added correctly, and do indeed contain the stacked five-year running averages for these two species, you are ready to create your stacked time series graph showing the five-year running average for each of the three species. This can be done by editing the code used to create the unstacked time series graph from page 199 so that it looks like the block of code at the top of the next page (the required modifications are highlighted in **bold**).

NOTE: The order of the `geom_area` commands in this code block is critical. The `geom_area` command for the data series that is at the top of the stack (in this case, Sowerby's beaked whale) needs to appear first in the code block followed by the `geom_area` command next data series down the stack (in this case, the northern bottlenose whale), and finally by the `geom_area` command for data series that is at the bottom of the stack (in this case, Cuvier's beaked whale). If these commands are not included in this order, not all the data series will be visible on the resulting graph.

```
Ggplot() + geom_area(data=beaked_whales,aes(x=year,
y=sbw_ra_5_stacked),fill="green",lty="solid",size=0.2,colour
="black") + geom_area(data=beaked_whales,aes(x=year,
y=nbw_ra_5_stacked),fill="blue",lty="solid",size=0.2,
colour="black") + geom_area(data=beaked_whales,aes(x=
year,y=cbw_ra_5),fill="red",lty="solid",
size=0.2,colour="black") + labs(x="Year",y="Number of
Strandings") + ylim(c(0,4)) + theme_classic()
```

If you examine this code, you will see that you have made a number of changes to it. Firstly, the geom_line commands have been replaced by geom_area commands. This means that the data for each species will be displayed as a filled polygon on the stacked time series graph rather than a line. Secondly, in these geom_area commands, the name of the columns for the unstacked data for northern bottlenose whales and Sowerby's beaked whales have been replaced with the names of the columns for the stacked versions of these data in the y argument of their aes elements. Finally, the maximum value in the ylim command has been changed from 3 to 4 to ensure the Y axis is high enough to display the new stacked values. When you run this block of code, you should get a stacked multi-series line graph that looks like this:

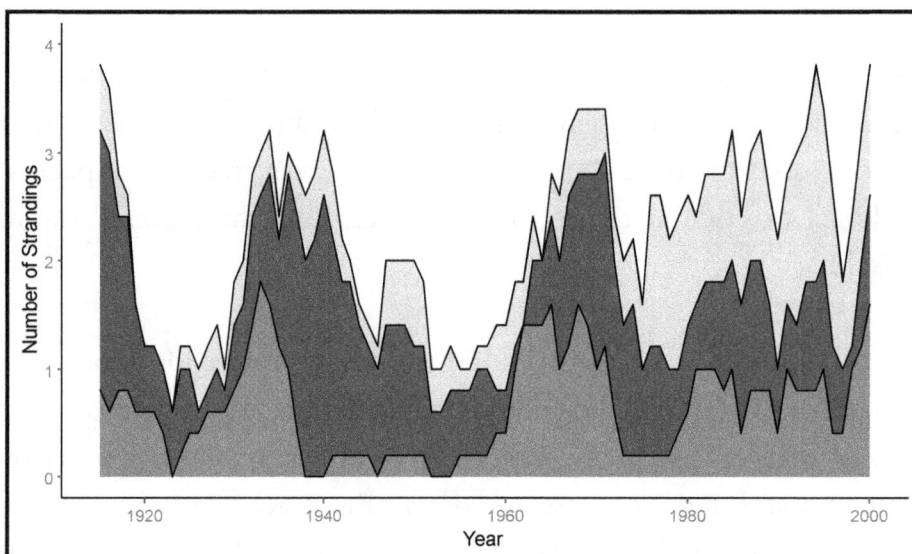

EXERCISE 3.5: HOW TO CREATE A MATRIX OF PAIR PLOTS TO CHECK FOR COVARIANCE BETWEEN VARIABLES IN A DATA SET:

As well as creating graphs of individual data points to examine the relationships between individual pairs of variables, you can also create a matrix which compares all the possible pairs of relationships in a set for three of more variables. This is done using something called a pair plot matrix. These are also known as pairs plots, scatter plot matrices, correlation matrices and correlograms. In a pair plot matrix, individual scatter plots are created for each possible pair of variables in the data set being used to make it, along with information about the relationship between each pair of variables. While they can serve a number of possible functions, pair plot matrices are most commonly used as part of regression analyses to explore the levels of covariance between pairs of independent or explanatory variables prior to conducting the main analysis. This is because it allows you to identify any pairs of variables which have a level of covariance that is too high for them to be considered separate variables in such analyses.

The basis for creating a pair plot matrix in R is either the `pairs` command or the `corrgram` command. Both of these commands identify all the possible pair-wise comparisons for a set of variables and then creates scatter plots for each one which are presented in the form of a pair plot matrix. To explore how to create a pair plot matrix using these commands, you will work through a series of examples using a data set called `porpoise_modelling_data.csv` that has been created to model the relationship between harbour porpoise presence (held in a column called `hp_pa`) and local environmental variables for a study area in the northern North Sea. Specifically, you will create pair plot matrices to measure the covariance between the four environmental variables in this data set that would be used as independent or explanatory variables in the modelling process. These are water depth (held in a column called `depth`), slope of the seabed (held in a column called `slope`), the variation in the slope of the sea bed (held in a column called `sd_slope`), and the distance to the nearest section of coast (held in a column called `dist_coast`). To create the first of these pair plot matrices, work through the flow diagram that starts at the top of the next page.

Data
set held
in a comma
separated values
(.CSV) file

For this example, the data set you will use is stored in a file called `porpoise_modelling_data.csv` that is located in the WORKING DIRECTORY folder you created during the introduction to this chapter.

Before you start any analysis in R, you first need to set the WORKING DIRECTORY. To do this, enter the text `setwd("` and then type the address of your WORKING DIRECTORY, using slashes (/) as the folder separators, before entering a second quotation mark followed by a closing bracket, like this `")`. For example, if your WORKING DIRECTORY has the address C:\STATS_FOR_BIOLOGISTS_TWO, your `setwd` command should look like this:

```
setwd("C:/STATS_FOR_BIOLOGISTS_TWO")
```

If you are using RGUI, enter your `setwd` command in the R CONSOLE window (remembering to use the address of your own WORKING DIRECTORY folder in it) and then press the ENTER key on your keyboard. If you are using RStudio, enter your `setwd` command into the SCRIPT EDITOR window. To run it, select it and then click on the RUN button at the top of this window. You will enter all the remaining commands for this exercise in a similar manner, depending on the user interface you are using.

1. Set the WORKING DIRECTORY for your analysis project

To check that your WORKING DIRECTORY has been set properly, enter the command `getwd()` and carefully check that the address it returns is the same as the one for the STATS_FOR_BIOLOGISTS_TWO folder you created at the start of this chapter.

Before you move on to step 2, make sure that all the data you wish to use in your analysis project are located in this WORKING DIRECTORY folder. In this case, this is a file called `porpoise_modelling_data.csv`. **NOTE:** If the data you are going to import into R in step 2 are not located in the WORKING DIRECTORY you set in this step, the import code provided in the next step will not work.

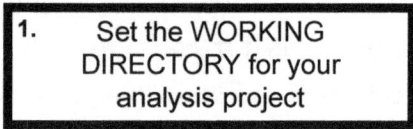

The `read.table` command provides the easiest way to load data held in a .CSV file (and stored in the WORKING DIRECTORY you set in step 1) into R so you can analyse it. To do this for the data set being used in this example, enter the following command into R:

```
porpoise_modelling_data <- read.table(
  file="porpoise_modelling_data.csv",
   sep=",",as.is=FALSE,header=TRUE)
```

This code has to be entered exactly as it is written here or it will not work. If you wish to use the copy-and-paste approach for entering this command, copy the text directly below CODE BLOCK 99 in the document R_CODE_DATA _ VISUALISATION_WORKBOOK.DOC and paste it into R.

This command will create a new object in R called `porpoise_modelling_data` which will contain the data from the specified .CSV file. To import a different .CSV file into R, all you need to do is change the file name in the `file` argument to the name of the one you wish to import. You can also use whatever name you wish for the R object which will be created by this command. To do this, simply replace `porpoise_modelling_data` at the start of the first line of the above code with the name you wish to use for it. **NOTE**: If your .CSV data set uses a semicolon as the column separator, you would need to replace the `sep=","` argument with `sep=";"`.

Whenever you import any data into R you need to check that they have loaded correctly. First, you need to check that all the required columns are present in the R object you just created. To do this, enter the following command into R:

```
names(porpoise_modelling_data)
```

This is CODE BLOCK 100 in the document R_CODE_DATA _VISUALISATION_WORKBOOK.DOC. This command will return the names used for each column in the R object you just created. For this example, the names should be: `hp_ pa`, `sd_slope`, `dist_coast`, `depth`, `slope`, `X` and `Y`.

Next, you should view the contents of the whole table using the `View` command. This is done by entering following code into R:

```
View(porpoise_modelling_data)
```

This is CODE BLOCK 101 in the document R_CODE_DATA _VISUALISATION_WORKBOOK.DOC. This command will open a DATA VIEWER window where you can examine your data set and check that the correct data have been loaded into R.

---

2. Load your data into R using the `read.table` command

3. Check the data have loaded into R correctly by checking the names of the columns and by viewing it

4. Create your pair plot matrix for the desired pairs of variables from your data set

Pair plot matrix created for a given set of variables in a data set

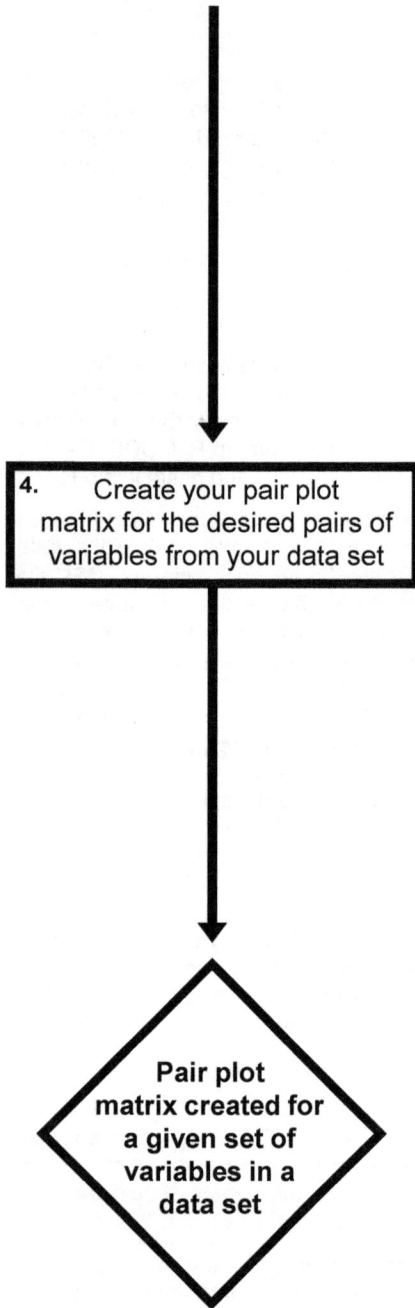

Once you have imported the required data into R, you are ready to create a pair plot matrix for a set of variables in your data set. In this example, you will make a pair plot matrix for the variables called `depth`, `slope`, `sd_slope` and `dist_coast` using the `pairs` command. To do this, enter the following command into R:

```
pairs(~depth+slope+sd_slope+dist_coast,
    data=porpoise_modelling_data,pch=20,
    main="Environmental Variables Pair Plot
                    Matrix")
```

This is CODE BLOCK 102 in the document R_CODE_ DATA _VISUALISATION_WORKBOOK.DOC. In this command, the columns containing the variables to be included in the pair plot matrix are specified as a list after a tilde symbol (which looks like this: ~). In this list, the name of each column is separated by a + symbol. For this example, the variables that will be plotted on the resulting pair plot matrix are `depth`, `slope`, `sd_slope` and `dist_coast`. The `data` argument is used to specify the data set which contains these variables. In this case, it is the data set called `porpoise_modelling_data` created in step 2 of this flow diagram. The `pch` argument is used to set the symbols that will be used on the individual scatter plots in the pair plot matrix (this means that it is serving the same role as the `shape` argument used in other graphing commands for making scatter plots). In this case, it is filled circles, which are defined by the pch code `20`. Finally, the `main` argument is used to set the title of resulting pair plot matrix. In this case, it is the text `Environmental Variables Pair Plot Matrix`.

The pair plot matrix created by working through this flow diagram should look like the image at the top of the next page.

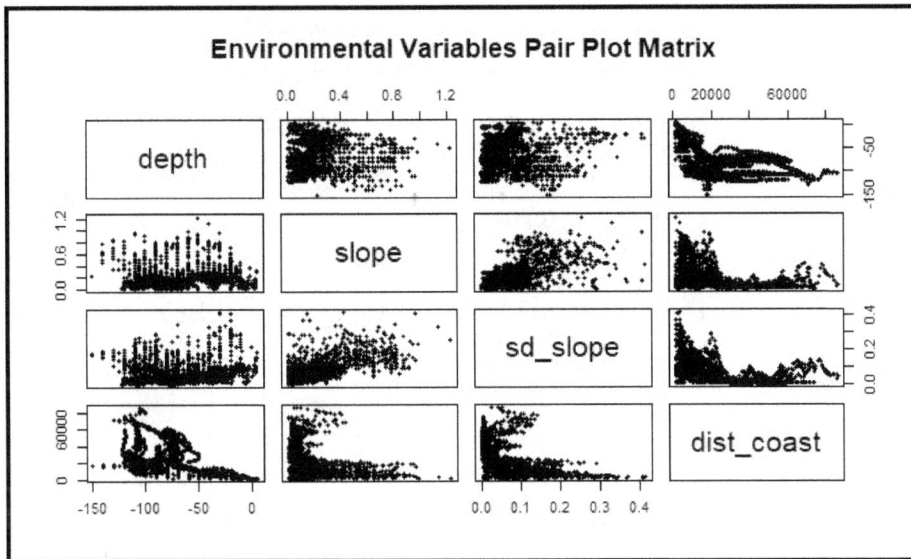

Once you have created a pair plot matrix, you can export it from R so that you can include it in a manuscript or presentation. If you are using RGUI, you can do this by clicking on the R GRAPHICS window containing your pair plot matrix to select it, before clicking on FILE on the main menu bar and selecting SAVE AS. This will allow you to save it in a variety of different formats. If you are using RStudio, you can export your graph by clicking on the EXPORT button at the top of the window displaying your pair plot matrix and selecting SAVE AS IMAGE.

When including a pair plot matrix in a manuscript, it is important that you provide an appropriate figure legend for it. This legend should provide all the information required for the reader to interpret its contents. For the above pair plot matrix, an appropriate legend would be:

Figure 1: *The relationships between four environmental variables used to model the habitat preferences of harbour porpoises in the northern North Sea. These variables are water depth, slope, standard deviation of slope and distance from the nearest section of coastline.*

In the first part of this exercise, the pair plot matrix you created had the same information plotted on the graphs in the top right and bottom left sections of the matrix. However, by including an `upper.panel` and/or a `lower.panel` argument in the `pairs` command, you can change the information that is presented in one or both of these sections of the resulting matrix. To give you experience in doing this, you will now customise the above pair plot matrix so that different types of information are presented in the upper right and lower left sections. Firstly, you will add a line of best fit using a smoother function to the scatter plots in the lower left section. This is done by adding a `lower.panel` argument with the `panel.smooth` term to the `pairs` command. To do this, edit the `pairs` command from step 4 of the above flow diagram (this is CODE BLOCK 102 in the document R_CODE_DATA_VISUALISATION_WORKBOOK.DOC) so that it looks like this (the required modifications are highlighted in **bold**):

```
pairs(~depth+slope+sd_slope+dist_coast,data=
porpoise_modelling_data,lower.panel=panel.smooth,pch=20,
    main="Environmental Variables Pair Plot Matrix")
```

When you run this command, you should get a pair plot matrix that looks like this (with a red line of best fit smoother added to the scatter plots on its lower left section):

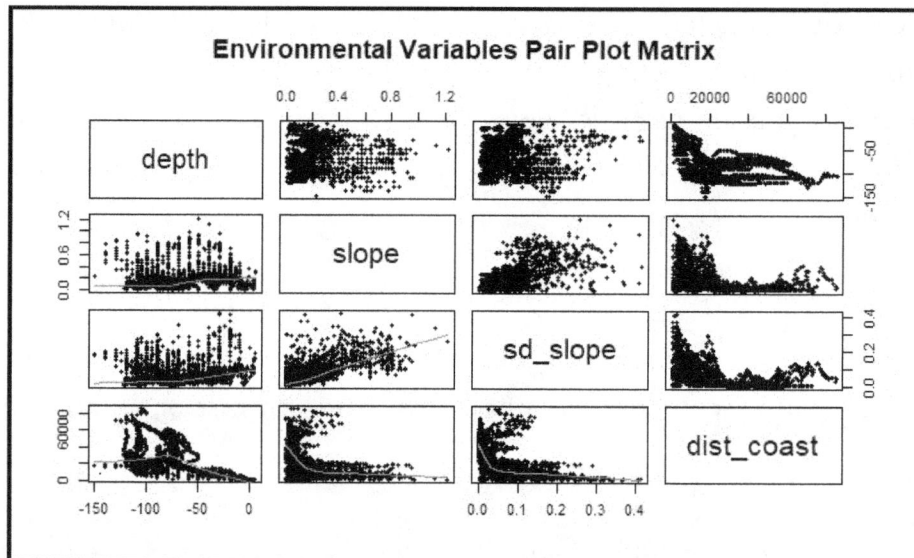

You can also create pair plot matrices which show the scatter plots in one section of the matrix while information on the strength and direction of the relationships between the pairs of variables is shown on the other. The easiest way to do this is to use the `corrgram` command from the `corrgram` package. This command provides an alternative way to make pair plot matrices, and to make ones that display the information in different ways to those available in the `pairs` command. To explore the how you can use the `corrgram` command to do this, you first need to ensure that the you have the `corrgram` package installed in your version of R. This can be done by entering the command `library()` into R. If the `corrgram` package is not listed in the R PACKAGES AVAILABLE window which opens, you will need to download and install it by entering the following command into R.

```
install.packages("corrgram")
```

Once you have ensured that you have the `corrgram` package installed in your version of R, you need to load the associated command library into your analysis project by entering the following command into R:

```
library(corrgram)
```

You can then use the `corrgram` command from this package to create pair plot matrices with information about the strength of the relationships in one of the sections and the individual scatter plots in the other. To do this for the `porpoise_modelling_data` data set, you first need to create a new R object that only contains the variables you wish to include in your pair plots. In this case, it is the four environmental variables (`depth`, `slope`, `sd_slope` and `dist_coast`). This can be done by entering the following `subset` command into R:

```
porpoise_egvs=subset(porpoise_modelling_data,
     select=c(depth,slope,sd_slope,dist_coast))
```

NOTE: Rather that specifying individual rows of data that should be included in the new R object created by this command, the `select=c(depth,slope,sd_slope, dist_coast)` argument in this version of the `subset` command specifies the names of the columns that should be included in the new R object (in this case, one called `porpoise_egvs`).

Once you have created a new object in R containing just the variables you wish to include on your pair plot matrix, you can then use the `corrgram` command to create your desired matrix. In this case, it will be one with scatter plots representing the relationships between the pairs of variables in the lower left section, and correlation coefficients which tell you the strength and direction of the relationships between the two variables in the upper right section.

To do this, enter the following command into R:

```
corrgram(porpoise_egvs,order=FALSE,
cor.method="pearson",lower.panel=panel.pts,
upper.panel=panel.cor,text.panel=panel.txt)
```

When you run this command in R, it should produce a pair plot matrix that looks like this:

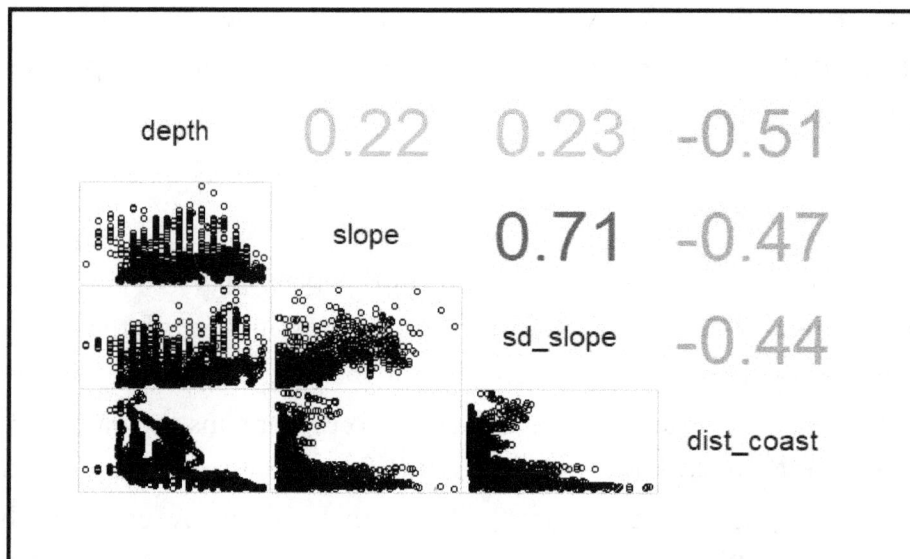

Other options for the information you can display in the different sections of your pair plot matrix using the `corrgram` command (and other methods of correlation you can apply to your data using it) can be found in the documentation for this command at *www.rdocumentation.org/packages/corrgram/versions/1.13/topics/corrgram.*

Chapter Six

How To Create Other Types Of Graphs

In Chapter Three, you learned how to make histograms in R using the commands from the GGPlot graphing package, while in Chapter Four you learned how to make a variety of graphs showing data from different groups, and in Chapter Five, you learned how to make graphs showing individual data points. While this means that the exercises in these chapters show you how to make many of the types of graphs that biologists need to be able to make on a regular basis, there are other types of graphs that you may wish to use which are not covered in these exercises as they do not neatly fall into the types of graphs included in Chapters Three to Five. These are pie charts, bubble graphs, multi-type graphs (such as graphs that include both lines and bars) and X-Y graphs showing the tracks of the movements of individual animals. In this chapter, you will learn how to make these additional types of graphs.

If you have not already done so, before you start the exercises in this chapter, you first need to create a WORKING DIRECTORY folder on your computer and load the necessary data into it (**NOTE**: If you have already created this folder and downloaded data for a previous chapter in this workbook, you do not need to do this again). To do this on a computer with a Windows operating system, open Windows Explorer and navigate to the location where you would like to create the folder (such as your C:\ drive or your DOCUMENTS folder). Next, right click anywhere in this location and select NEW> FOLDER. Now call this folder STATS_FOR_BIOLOGISTS_TWO by typing this into the folder name section to replace what it is currently called (which will most likely be NEW FOLDER). To create a WORKING DIRECTORY folder on a computer running a Mac operating system, open Finder and navigate to the location where you would like to create the folder (such as your DOCUMENTS folder or your DESKTOP). Next, click on FILE> NEW FOLDER, and then type the name STATS_FOR_BIOLOGISTS_TWO before pressing the ENTER key on your keyboard.

Once you have created your WORKING DIRECTORY folder, you are ready to download the data sets you will use for the exercises in this workbook from *www.gisinecology.com/ stats-for-biologists-2*. After you have downloaded the compressed folder containing the required data by following the instructions provided on that page, you need to extract all the data files from it and copy them into the folder called STATS_FOR_BIOLOGISTS_TWO that you have just created.

Next, you need to check that the required data have been extracted to the correct folder. If you are using a computer with a Windows operating system, you can use Windows Explorer to open your newly created WORKING DIRECTORY folder and examine its contents. If all the files from the compressed folder are present in it (there should be a total of 90 of them), you can click on the folder icon at the left hand end of the ADDRESS BAR at the top of the WINDOWS EXPLORER window to reveal its full address. Write this address down as you will need it to set this folder as your WORKING DIRECTORY during the exercises provided in this workbook (see pages 12 and 13 for details of how to modify folder addresses so they will be recognised by R).

If you are using a computer with a Mac operating system, you can use Finder to open your newly created WORKING DIRECTORY folder and examine its contents. If all the required data files are present in it (there should be a total of 90 of them), press the CMD and I keys on your keyboard at the same time. This will open the GET INFO window where you will find its address (which is also called the pathway). Write this address down somewhere as you will need it to set this folder as your WORKING DIRECTORY during the exercises provided in this workbook (see pages 12 and 13 for details of how to modify folder addresses so they will be recognised by R).

After you have loaded the required data into your WORKING DIRECTORY folder, you can open RGUI or RStudio, depending on which option you wish to use (see Chapter 2 for more details). Once you have opened your preferred R user interface, you need to create a file called CHAPTER_SIX_EXERCISES where you will save the results of your analyses from your R CONSOLE window as you work through this chapter. To do this using RGUI, click on the FILE menu and select SAVE WORKSPACE. To do this in RStudio, click on SESSION and select SAVE WORKSPACE AS. In both cases, save it as a WORKSPACE file with the name CHAPTER_SIX_EXERCISES.RDATA in your WORKING

DIRECTORY folder (this will be the one called STATS_FOR_BIOLOGISTS_TWO that you have just created). If you are using RStudio, you will also want to save the contents of your SCRIPT EDITOR window (where you will enter and edit the R code you will use to carry out specific commands). To do this, click on the FILE menu and select SAVE AS. Save your file as an R SCRIPT file with the name CHAPTER_SIX_EXERCISES.R in your WORKING DIRECTORY folder. As you work through the exercises in this chapter, remember to regularly save the contents of your R CONSOLE window (which will contain the R objects you have created up to that point) to your WORKSPACE file and, if you are using RStudio, the contents of your SCRIPT EDITOR window to your R SCRIPT file.

Finally, you need to remove any data that are currently held in R's temporary memory. To do this, enter the following command into R:

```
rm(list=ls())
```

If you are using RGUI, you can simply type this code after the command prompt at the bottom of the R CONSOLE window (it looks like this: >) and then press the ENTER key on your keyboard to run it. If you are using RStudio, you can type this command into the SCRIPT EDITOR window (the upper left hand window). To run this command, select it and then click on the RUN button at the top of this window. This will run it in the R CONSOLE window (the lower left hand one in the main RStudio user interface). You are now ready to start the exercises in this chapter.

EXERCISE 4.1: HOW TO MAKE A PIE CHART USING GGPLOT:

While not widely used in biology, pie charts can be useful for showing the relative contributions of different groups of data in a data set to an overall total. In particular, pie charts can be very useful for showing the relative diversity of different taxonomic groups within a data set. In order to make a pie chart from such data, they need to be arranged in a spreadsheet or table with one row for each taxonomic group, and individual columns which contain the names of the taxonomic groups and the number of species recorded for each one within the study.

In this exercise, you will learn how to make a pie chart by making one which shows the relative biodiversity of different groups in a data set of tropical trees. These data come from Cardosa *et al.* (2017), and consist of information on the number of species recorded in the ten most speciose families of trees in the Amazon rainforest[1]. To create a pie chart showing the relative diversity of tree species in each family from this data set, work through the flow diagram which starts on the next page.

NOTE: These instructions assume that you have already installed the `ggplot2` package in your version of R. To find out if it is already installed in your version of R, enter the command `library()`. If it is listed in the R PACKAGES AVAILABLE window which opens, it is already installed. If it is not listed here, you will need to installed it before you start this exercise. Instructions for how to do this can be found in Exercise 1.1. Once you have ensured that this packages is installed in your version of R, you need to load its command library into your analysis project. This can be done by entering the following command into R:

```
library(ggplot2)
```

---

[1] Cardoso et al. (2017). Amazon plant diversity revealed by a taxonomically verified species list. PNAS October, 3 2017 114 (40): 10695 – 10700.

Data set held in a comma separated values (.CSV) file

For this example, the data set you will use is stored in a file called `tropical_trees.csv` that is located in the WORKING DIRECTORY folder you created during the introduction to this chapter.

1. Set the WORKING DIRECTORY for your analysis project

Before you start any analysis in R, you first need to set the WORKING DIRECTORY. To do this, enter the text `setwd("` and then type the address of your WORKING DIRECTORY, using slashes (/) as the folder separators, before entering a second quotation mark followed by a closing bracket, like this `")`. For example, if your WORKING DIRECTORY has the address C:\STATS_FOR_BIOLOGISTS_TWO, your `setwd` command should look like this:

```
setwd("C:/STATS_FOR_BIOLOGISTS_TWO")
```

If you are using RGUI, enter your `setwd` command in the R CONSOLE window (remembering to use the address of your own WORKING DIRECTORY folder in it) and then press the ENTER key on your keyboard. If you are using RStudio, enter your `setwd` command into the SCRIPT EDITOR window. To run it, select it and then click on the RUN button at the top of this window. You will enter all the remaining commands for this exercise in a similar manner, depending on the user interface you are using.

To check that your WORKING DIRECTORY has been set properly, enter the command `getwd()` and carefully check that the address it returns is the same as the one for the STATS_FOR_BIOLOGISTS_TWO folder you created at the start of this chapter.

Before you move on to step 2, make sure that all the data you wish to use in your analysis project are located in this WORKING DIRECTORY folder. In this case, this is a file called `tropical_trees.csv`. **NOTE:** If the data you are going to import into R in step 2 are not located in the WORKING DIRECTORY you set in this step, the import code provided in the next step will not work.

2. Load your data into R using the `read.table` command

3. Check the data have loaded into R correctly by checking the names of the columns and by viewing it

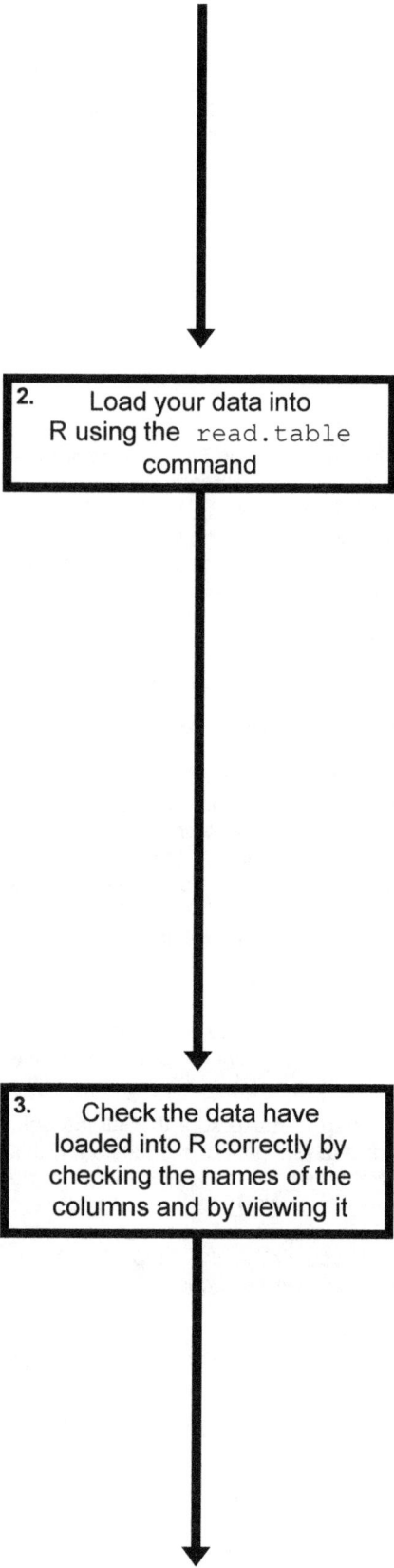

The `read.table` command provides the easiest way to load data held in a .CSV file (and stored in the WORKING DIRECTORY you set in step 1) into R so you can analyse it. To do this for the data set being used in this example, enter the following command into R:

```
tropical_trees <-
read.table(file="tropical_trees.csv",
    sep=",",as.is=FALSE,header=TRUE)
```

This code has to be entered exactly as it is written here or it will not work. If you wish to use the copy-and-paste approach for entering this command, copy the text directly below CODE BLOCK 103 in the document R_CODE_DATA _VISUALISATION_WORKBOOK.DOC and paste it into R.

This command will create a new object in R called `tropical_trees` which will contain the data from the specified .CSV file. To import a different .CSV file into R, all you need to do is change the file name in the `file` argument to the name of the one you wish to import. You can also use whatever name you wish for the R object which will be created by this command. To do this, simply replace `tropical_trees` at the start of the first line of the above code with the name you wish to use for it. **NOTE:** If your .CSV data set uses a semicolon as the column separator, you would need to replace the `sep=","` argument with `sep=";"`.

Whenever you import any data into R you need to check that they have loaded correctly. First, you need to check that all the required columns are present in the R object you just created. To do this, enter the following command into R:

```
names(tropical_trees)
```

This is CODE BLOCK 104 in the document R_CODE_DATA _VISUALISATION_WORKBOOK.DOC. This command will return the names used for each column in the R object you just created. For this example, the names should be: `family`, `no_of_species` and `rank`.

Next, you should view the contents of the whole table using the `View` command. This is done by entering following code into R:

```
View(tropical_trees)
```

This is CODE BLOCK 105 in the document R_CODE_DATA VISUALISATION_WORKBOOK.DOC. This command will open a DATA VIEWER window where you can examine your data set and check that the correct data have been loaded into R.

216

Once you have imported the required data into R, you are ready to create your initial pie chart based on the species diversity data in the data set on tropical trees created in step 2. To do this, enter the following block of code into R:

```
ggplot(data=tropical_trees,
aes(x="",y=no_of_species,fill=family)) +
geom_bar(stat="identity") + coord_polar("y",
                    start=0)
```

This is CODE BLOCK 106 in the document R_CODE_DATA VISUALISATION_WORKBOOK.DOC, and it contains three commands separated by + symbols. These are the `ggplot` command, the `geom_bar` command and the `coord_polar` command. The `ggplot` command sets the data set which will be used for the graph. This is done using the `data` argument and, in this case, it will be the R object called `tropical_trees` created in step 2 of this exercise. The data which will be plotted on the X axis of the resulting graph is set using the `x` argument of the `aes` element of this `ggplot` command. In this case, it is set to "" as there is effectively no X axis on a pie chart. The data which will be plotted on the Y axis of the resulting graph is set using the `y` argument of the `aes` element of this `ggplot` command (this will determine the width of each wedge on the resulting pie chart). In this case, it is the column called `no_of_species` in the `tropical_trees` data set, which contains the number of species recorded in each family of tropical trees included in it. Finally, the column which contains the groupings that will determine the number of wedges on the final pie chart is set by the `fill` argument, In this case, it is the column called `family`.

The second command in this code block, `geom_bar`, sets the type of graph that will be created from the data specified in the `ggplot` command. In this case, it will be a bar graph. This may seem odd as you are trying to make a pie chart, but in the `ggplot2` package, pie charts are created as circular bar graphs. Within the `geom_bar` command, the argument `stat="identity"` is included to ensure that it is the actual number provided in the column determined by the `y` argument in the `aes` element of the `ggplot` command which sets the size of each wedge on your final pie chart rather than any other potential statistic. In this case, it is the number of species in each family of tropical trees provided in the column called `no_of_species`.

Finally, the `coord_polar` command is included to turn a standard bar graph created by the `geom_bar` command into a circular pie chart. In this command, the coordinate which will be used to create the size of the wedges is set by the argument `"y"`, and then the starting point for the graph (in degrees from the top of the graph) is set using the `start=0` argument.

4. Create an initial pie chart based on your data set

5. Customise how your final pie chart will look

Pie chart created from your data set

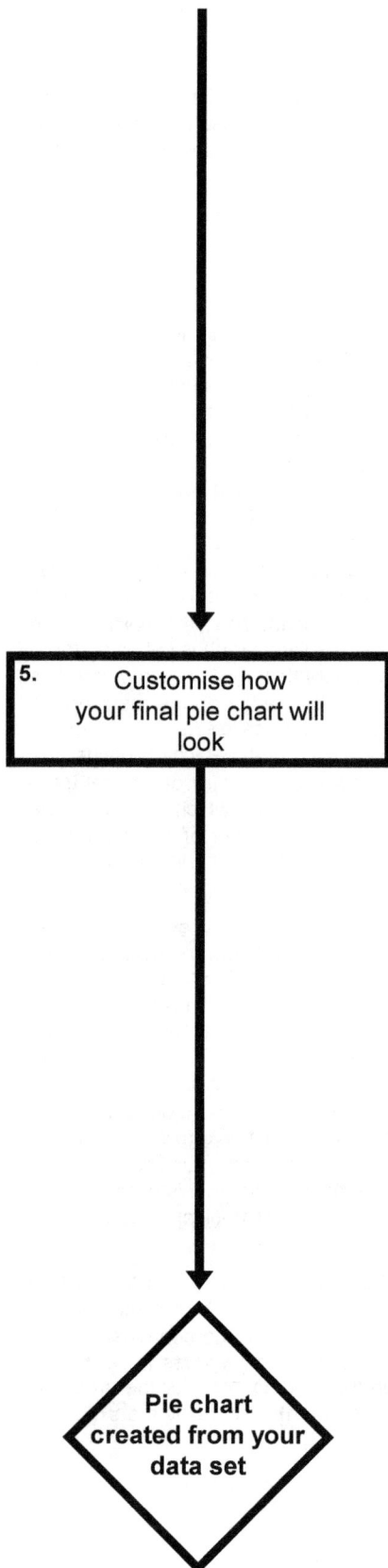

Once you have created your initial pie chart, and you are sure that it is displaying the intended data for the relative sizes of the wedges on it, you can customise how the final chart will look. This can be done by adding a number of new style commands to the block of code used to produce it. To do this, edit the code block from step 4 so that it looks like this (the newly added style commands are highlighted in **bold**):

```
ggplot(data=tropical_trees,aes(x="",
   y=no_of_species,fill=family)) +
geom_bar(stat="identity") + coord_polar("y",
   start=0) + scale_fill_manual(values=
  c("red","blue","green","orange","pink",
 "yellow","brown","tan","grey","black")) +
    theme(axis.text.x=element_blank(),
 axis.title.y=element_blank(),axis.title.x=
        element_blank(),legend.title=
 element_blank()) + theme(panel.background=
        (element_rect(fill="white")))
```

This is CODE BLOCK 107 in the document R_CODE_DATA _VISUALISATION_WORKBOOK.DOC, and it adds three new style commands to the code block used to create your initial pie chart in step 4. These are a `scale_fill_ manual` command and two `theme` commands.

The `scale_fill_manual` command is used to set the colours for each wedge on the pie chart. This means that in this command you need to include one colour for every group from the column you have specified in the `fill` argument of the `aes` element in the `ggplot` command. In this case, this is the column called `family` and it contains ten groups, so ten colours need to be specified in this command.

The first of the new `theme` commands contains four arguments which remove unwanted labels from your final pie chart. In this `theme` command, the `axis.text= element_blank()` argument removes the labels from the X axis, the `axis.title.y=element_blank()` argument removes the title from the Y axis, the `axis.title. x=element_blank()` argument removes the title from the X axis and the `legend.title=element_blank()` argument removes the title from the legend.

The second new `theme` command contains a single argument. This is the `panel.background=(element_ rect(fill="white"))` argument, and it sets the background colour for the chart to white. Once you have finished editing this code block, you can run it again to create the final version of your pie chart.

At the end of the first part of this exercise, the pie chart showing the relative diversity of trees in different families in the Amazon rain forest should look like this:

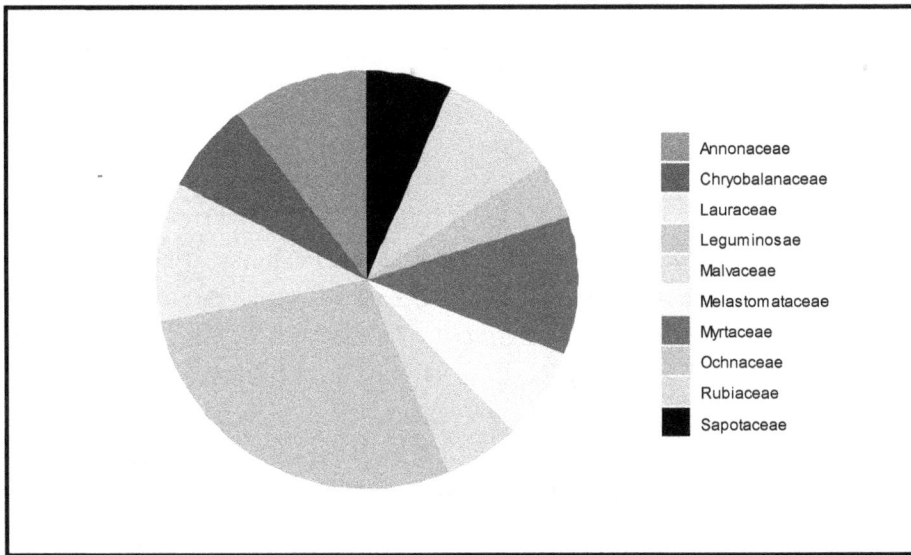

Once you have created a pie chart, you can export it from R so that you can include it in a manuscript or presentation. If you are using RGUI, you can do this by clicking on the R GRAPHICS window containing your pie chart to select it, before clicking on FILE on the main menu bar and selecting SAVE AS. This will allow you to save it in a variety of different formats. If you are using RStudio, you can export your pie chart by clicking on the EXPORT button at the top of the window displaying the chart and selecting SAVE AS IMAGE.

When including a pie chart in a manuscript, it is important that you provide an appropriate figure legend for it. This legend should provide all the information required for the reader to interpret its contents. For the above pie chart, an appropriate legend would be:

Figure 1: The relative diversity of tropical tree species in different families in the Amazon rainforest based on data from Cardoso et al. 2017.

In the first part of this exercise, the order of the different wedges and their colours was set automatically based on default values used by the `geom_bar` command. However, there will be times when you wish change the order of the wedges. To explore how to do this, you will change the order of the wedges so that they arranged in a clockwise order from the 12 o'clock position based on the number of species each family contains. This is done by including a `reorder` term in the `fill` argument of the `aes` element of the `ggplot` command. To do this, edit the final block of code from step 5 in the above flow diagram (this is CODE BLOCK 107 in the document R_CODE_DATA_VISUALISATION_ WORKBOOK.DOC) so that it looks like this (the required modifications are highlighted in **bold**):

```
ggplot(data=tropical_trees,aes(x="",y=no_of_species,
  fill=reorder(family,no_of_species))) + geom_bar(stat=
"identity") + coord_polar("y",start=0) + scale_fill_manual(
  values=c("red","blue","green","orange","pink","yellow",
    "brown","tan","grey","black")) + theme(axis.text.x=
element_blank(),axis.title.y=element_blank(),axis.title.x=
    element_blank(),legend.title= element_blank())+
  theme(panel.background=(element_rect(fill="white")))
```

When you run this block of code, you should get a new pie chart where the wedges are ordered based on their width from largest to smallest in a clockwise order and which looks like this:

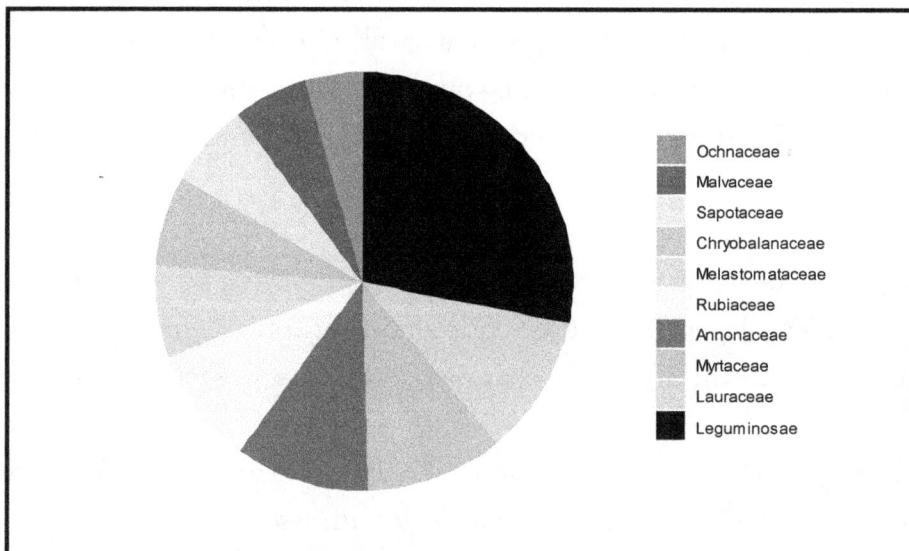

When you examine this pie chart, you will see that while the relative diversity for each family can be identified from the relative position and widths of the wedges, there is no way to identify the number of species which each family contains. This information can be added to a standard pie chart by adding labels to each wedge. However, a more interesting way to display this information is to change it from a standard pie chart to one where the length of each wedge is directly related to its value. This is done by changing the variables specified in the x and y arguments of the aes element of the ggplot command. Specifically, the values to be plotted on the pie chart (in this case, the data from the no_of_species column) is moved from the y argument to the x argument, and the setting in the y argument is replaced with a copy of the term from the fill argument which appears immediately after it (in this case, it is the term reorder(family,no_of_species)). As well as doing this, you need to add a new theme command containing a panel.grid.major style argument. This will add a grey grid to the background of the graph so that you can see a scale against which the length of each wedge can be measured. This creates a modified block of code that should look like this (the required modifications are highlighted in **bold**):

```
ggplot(data=tropical_trees,aes(x=no_of_species,y=
reorder(family,no_of_species),fill=reorder(family,
no_of_species))) + geom_bar(stat="identity") +
coord_polar("y", start=0) + scale_fill_manual(values=
c("red","blue","green","orange","pink","yellow","brown",
"tan","grey","black")) + theme(axis.text.x=element_blank(),
axis.title.y = element_blank(), axis.title.x =
element_blank(),legend.title= element_blank()) +
theme(panel.background=(element_rect(fill="white"))) +
theme(panel.grid.major=element_line(colour="grey"))
```

When you run this block of code, you should get a new pie chart that looks like the image at the top of the next page.

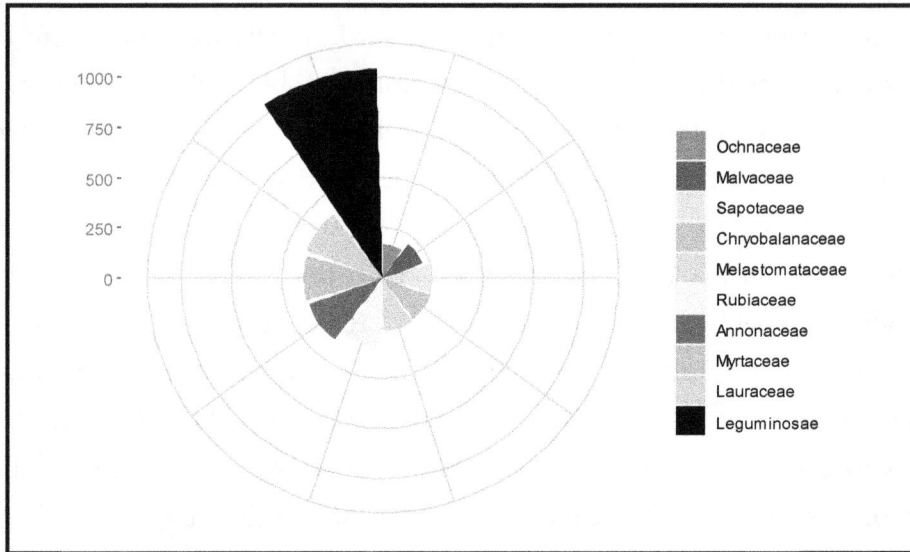

When you examine this pie chart, you will see that the number of species represented by each wedge is now indicated by its length rather than by its relative width. In addition, the length of each wedge can now be estimated from the grey-coloured circular grid that has now been added to the background of the graph.

EXERCISE 4.2: HOW TO CREATE A BUBBLE GRAPH:

On an X-Y scatter plot of the type you made in Exercise 3.1, the values of two variables are represented, one on the X axis and one on the Y axis. A bubble graph is an extension of this type of graph, but it uses the size of each symbol (the "bubbles") to represent the values of the third variable. For example, on a graph of morphological measurements, the X axis might represent body length, the Y axis body mass and the "bubble" size the relative brain size. This would allow you to easily see if there is a relationship between these three variables.

In this exercise, you will learn how to make a bubble graph by creating one which shows the percentage of the total land area of thirty different countries that constitutes legally protected areas (these values will be represented by the sizes of the "bubbles" on the final graph) in relation to the log of their population densities (plotted on the Y axis) and the log of their total land areas (plotted on the X axis). To do this, work through the flow diagram below. **NOTE:** These instructions assume that you have already installed the `ggplot2` package in your version of R and that you have loaded its command library into your analysis project. If you have not already done this, you will need to do so before you start this exercise. Instructions for how to ensure you have this package installed and how to load its command library into your analysis project can be found in Exercise 4.1.

Data set held in a comma separated values (.CSV) file

For this example, the data set you will use is stored in a file called `protected_areas_data.csv` that is located in the WORKING DIRECTORY folder you created during the introduction to this chapter.

1. Set the WORKING DIRECTORY for your analysis project

Before you start any analysis in R, you first need to set the WORKING DIRECTORY. To do this, enter the text `setwd("` and then type the address of your WORKING DIRECTORY, using slashes (/) as the folder separators, before entering a second quotation mark followed by a closing bracket, like this `")`. For example, if your WORKING DIRECTORY has the address C:\STATS_FOR_BIOLOGISTS_TWO, your `setwd` command should look like this:

```
setwd("C:/STATS_FOR_BIOLOGISTS_TWO")
```

If you are using RGUI, enter your `setwd` command in the R CONSOLE window (remembering to use the address of your own WORKING DIRECTORY folder in it) and then press the ENTER key on your keyboard. If you are using RStudio, enter your `setwd` command into the SCRIPT EDITOR window. To run it, select it and then click on the RUN button at the top of this window. You will enter all the remaining commands for this exercise in a similar manner, depending on the user interface you are using.

To check that your WORKING DIRECTORY has been set properly, enter the command `getwd()` and carefully check that the address it returns is the same as the one for the STATS_FOR_BIOLOGISTS_TWO folder you created at the start of this chapter.

Before you move on to step 2, make sure that all the data you wish to use in your analysis project are located in this WORKING DIRECTORY folder. In this case, this is the file called `protected_areas_data.csv`. **NOTE:** If the data you are going to import into R in step 2 are not located in the WORKING DIRECTORY you set in this step, the import code provided in the next step will not work.

223

The `read.table` command provides the easiest way to load data held in a .CSV file (and stored in the WORKING DIRECTORY you set in step 1) into R so you can analyse it. To do this for the data set being used in this example, enter the following command into R:

```
protected_areas_data <-
read.table(file="protected_areas_data.csv",
    sep=",",as.is=FALSE,header=TRUE)
```

This code has to be entered exactly as it is written here or it will not work. If you wish to use the copy-and-paste approach for entering this command, copy the text directly below CODE BLOCK 108 in the document R_CODE_DATA _VISUALISATION_WORKBOOK.DOC and paste it into R.

This command will create a new object in R called `protected_areas_data` which will contain the data from the specified .CSV file. To import a different .CSV file into R, all you need to do is change the file name in the `file` argument to the name of the one you wish to import. You can also use whatever name you wish for the R object which will be created by this command. To do this, simply replace `protected_areas_data` at the start of the first line of the above code with the name you wish to use for it. **NOTE:** If your .CSV data set uses a semicolon as the column separator, you would need to replace the `sep=","` argument with `sep=";"`.

---

2. **Load your data into R using the `read.table` command**

Whenever you import any data into R you need to check that they have loaded correctly. First, you need to check that all the required columns are present in the R object you just created. To do this, enter the following command into R:

```
names(protected_areas_data)
```

This is CODE BLOCK 109 in the document R_CODE_DATA _VISUALISATION_WORKBOOK.DOC. This command will return the names used for each column in the R object you just created. For this example, the names should be: `country`, `region`, `log_population`, `log_area`, `percentage_protected` and `status`.

Next, you should view the contents of the whole table using the `View` command. This is done by entering following code into R:

```
View(protected_areas_data)
```

3. **Check the data have loaded into R correctly by checking the names of the columns and by viewing it**

This is CODE BLOCK 110 in the document R_CODE_DATA VISUALISATION_WORKBOOK.DOC. This command will open a DATA VIEWER window where you can examine your data set and check that the correct data have been loaded into R.

4. Create your initial bubble graph based the data set imported in step 2

Once you have imported the required data into R, you are ready to create your initial bubble graph. To do this, enter the following block of code into R:

```
ggplot(data=protected_areas_data,
aes(x=log_area,y=log_population,size=
percentage_protected)) + geom_point(shape=
21,colour="black",fill=NA,alpha=1)
```

This is CODE BLOCK 111 in the document R_CODE_DATA _VISUALISATION_WORKBOOK.DOC, and it contains two commands separated by a + symbol. These are the `ggplot` command and the `geom_point` command. The `ggplot` command sets the data set which will be used for the graph. This is done using the `data` argument and, in this case, it will be the R object called `protected_areas_data` created in step 2 of this exercise. The data which will be plotted on the X axis of the resulting graph is set using the `x` argument of the `aes` element of this `ggplot` command. In this case, it is the column called `log_area` in the `protected_areas_data` data set. The data which will be plotted on the Y axis of the resulting graph is set using the `Y` argument of the `aes` element of this `ggplot` command. In this case, it is the column called `log_population`. Finally, the `size` argument is included in the `aes` element of the `ggplot` command to identify the column which contains the variable that will determine the size of the bubble representing each data point. In this case, it is the column called `percentage_protected`, which contains information about the proportion of the total area of each country that is covered by legally protected areas.

The second command in this code block, `geom_point`, sets the type of graph that will be created from the data specified in the `ggplot` command. In this case, it will be an X-Y scatterplot. Within this command, four arguments are included. These are `shape`, `colour`, `fill` and `alpha`. The `shape` command is used to set the shape of the symbol representing the bubbles. In this case, it will open circles, which are set using the code `21`. The `colour` argument is used to set the colour to be used for each point (in this case, `black`), while the `fill` argument is set to `NA`. This means that the points will have no fill, making it easier to see the sizes of any overlapping bubbles. Finally, the `alpha` argument is used to set the level of transparency for the points. In this case, it is set to `1`, making the points opaque.

```
5.  Customise how
    your final bubble graph
    will look
```

```
Bubble
graph created
from your data
set
```

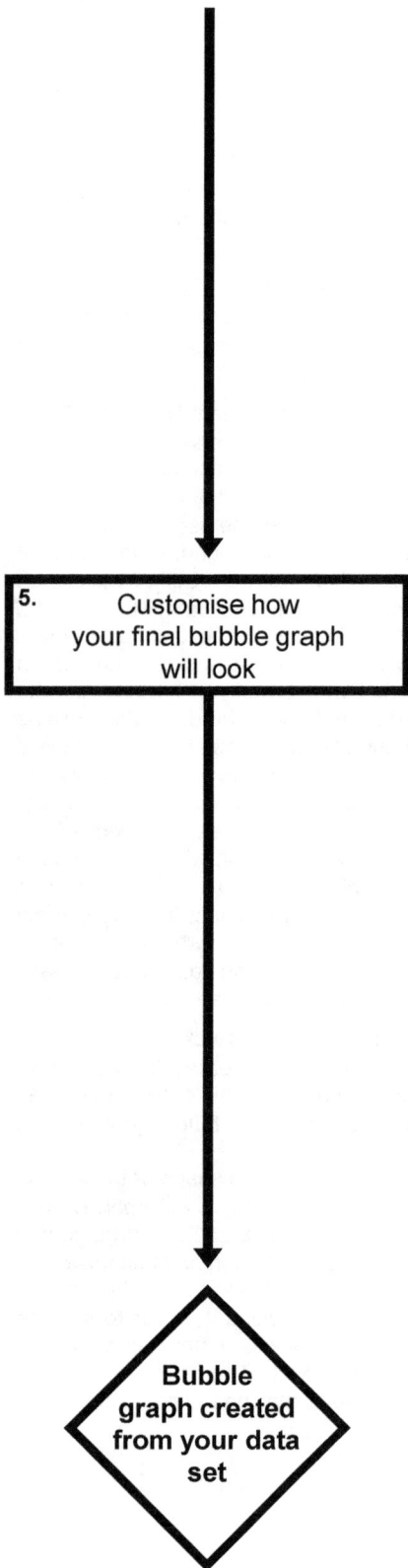

Once you have created your initial bubble graph, and you are sure that it is displaying the intended data on the X and Y axes as well as with the size of the bubbles, you can customise how the final graph will look. This can be done by adding a number of new style commands to the block of code used to produce it. To do this, edit the block of code from step 4 so that it looks like this (the newly added style commands are highlighted in **bold**):

```
ggplot(data=protected_areas_data,
  aes(x=log_area,y=log_population,size=
percentage_protected)) + geom_point(shape=
  21,colour="black",fill=NA,alpha=1) +
scale_size(range=c(0.1,10)) + labs(x="Total
  Area",y="Population Size",size="% Area
    Protected") + theme_classic()
```

This is CODE BLOCK 112 in the document R_CODE_DATA _VISUALISATION_WORKBOOK.DOC, and it adds a three new style command to the code block used to create your initial bubble graph. These are the `scale_size` command, the `labs` command and the `theme_classic` command. The `scale_size` command is used to set the relative sizes of the bubbles representing different values. This is done using the `range=c(0.1,10)` argument. To use a different size range of symbols on a bubble graph, you simply need to change the values in this argument, which represent the minimum bubble size (the first number) and maximum bubble size (the second number). The `labs` command is used to set the labels for X and Y axes using the `x` and `y` arguments. In this case, they are `Log Total Area` and `Log Population Size` respectively. The `size` argument is used in the `labs` command to provide the label for the legend which tells the reader what values each bubble size represents. In this case, it is set to `% Area Protected`. Finally, the `theme_classic` command sets all other elements of the final graph to those used by the pre-set classic theme. Once you have finished editing this code block, you can run it again to create the final version of your bubble graph.

At the end of the first part of this exercise, you should have a bubble graph that looks like this:

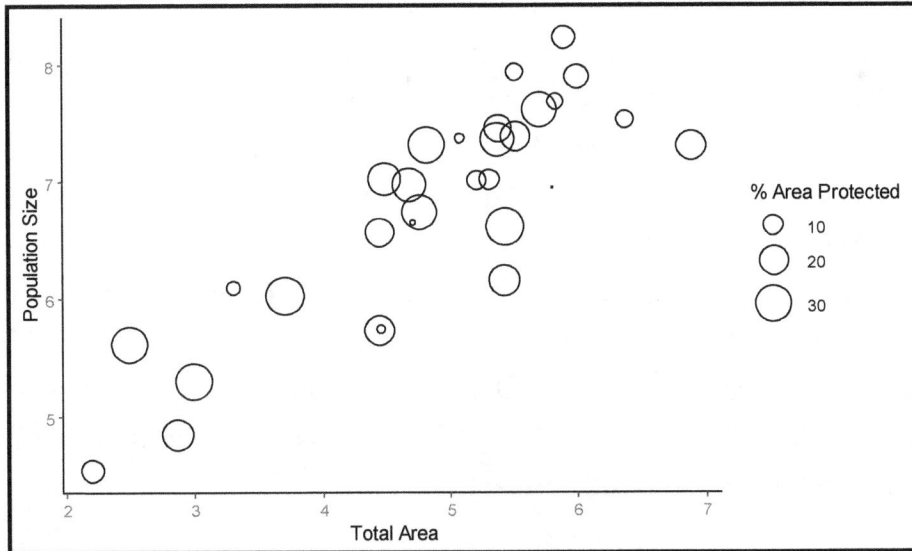

Once you have created a bubble graph, you can export it from R so that you can include it in a manuscript or presentation. If you are using RGUI, you can do this by clicking on the R GRAPHICS window containing your graph to select it, before clicking on FILE on the main menu bar and selecting SAVE AS. This will allow you to save it in a variety of different formats. If you are using RStudio, you can export your graph by clicking on the EXPORT button at the top of the window displaying it and selecting SAVE AS IMAGE.

When including a bubble graph in a manuscript, it is important that you provide an appropriate figure legend for it. This legend should provide all the information required for the reader to interpret its contents. For the above bubble graph, an appropriate legend would be:

Figure 1: *The relationship between the area of land in a set of 30 countries (log-transformed and plotted on the X axis), their population size (log-transformed and plotted on the Y axis) and the percentage of their total area that is legally protected (the size of the symbols).*

The bubble graph you have just created only displayed three variables, one for each axis and a third using the size of the bubbles. However, you can also display a fourth variable on this type of graph by using the values from another column in the data set to set the colours used for the different data points. This information can be added to the graph by moving the `colour` argument from the `geom_point` command to the `aes` element of the `ggplot` command and using it to specify which column contains the groupings that will be displayed using different colours. You can then set which colours are used for each group by adding a `scale_colour_manual` command to the code block. To demonstrate how you can do this, you will add a new variable to the above bubble graph. This will be whether each country represented on it is classified as developed or developing. This information is contained in a column called `status`. To create a version of the above bubble graph where the bubbles representing developed countries are shown in red, while those for developing countries are shown in blue, you will need to edit the block of code from step 5 of the above flow diagram (this is CODE BLOCK 112 in the document R_CODE_DATA_VISUALISATION_WORKBOOK.DOC) so that it looks like this (the required modifications are highlighted in **bold**):

```
ggplot(data=protected_areas_data,aes(x=log_area,
  y=log_population,size=percentage_protected,colour=
  status)) + geom_point(shape=21,alpha=1) + scale_size(
range=c(.1,10))+ labs(x="Log Total Area",y="Log Population
  Size",size="% Area Protected",colour="Status") +
    theme_classic() + scale_colour_manual(values=
                    c("red","blue"))
```

The bubble graph created from this modified block of code should look like this:

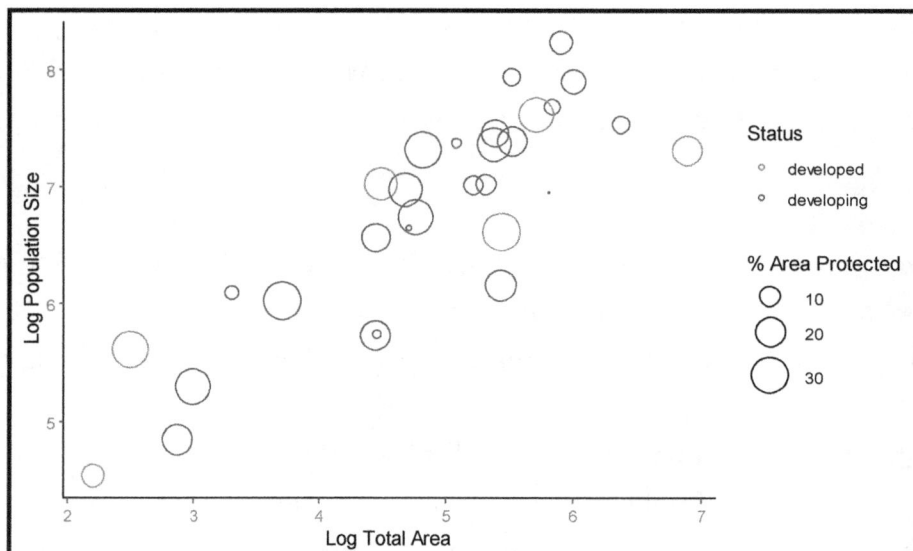

EXERCISE 4.3: HOW TO CREATE A MIXED TYPE GRAPH WHICH SHOWS DIFFERENT DATA SERIES IN DIFFERENT WAYS:

There will be times when you wish to not only display more than one data series on the same graph, but also to display different data series in different ways. For example, you may wish to plot one data series on a graph with bars (like a standard bar graph) and a second using a line. These are called mixed type graphs, and they are relatively easy to create using the GGPlot package simply by using different graphing commands for each of the data series. However, if the two data series have very different ranges of values for the variable which is plotted on the Y axis, this simple approach can result in a mis-match between the two data series meaning that they do not plot on top of each other in an appropriate way. As a result, when creating mixed type graphs, you will usually need to add a second Y axis to the right hand side of your graph which displays the scale used for the second data series.

To demonstrate how this can be done, in this exercise, you will create a mixed type graph showing how the body mass of a small passerine bird, the great tit, varies across the day and how this compares to the temperature variations across the same time period. The data on great tit body mass will be displayed as bars showing the mean values for each hourly period of the day and error bars which show the variations in body mass in these time periods. In contrast, the data on the mean temperature in each hour will be plotted as a line. The starting point for creating this mixed type graph will be a data set with a single row per individual great tit measured, and columns containing information about its body mass, the time of day it was measured and the temperature at that time, as well as a number of other variables which will not be used in this exercise. These data are held in a data set called `great_tit_daily_mass.csv`. To create this mixed type graph, work through the flow diagram that starts on the next page.

NOTE: These instructions assume that you have already installed the packages called `ggplot2`, `Rcpp`, `plyr` and `dplyr` into your version of R and loaded their command libraries into your analysis project (these command libraries need to be loaded in this exact order for all the commands used in this exercise to work properly). If you have not already done this, you will need to do it before working through this flow diagram (see Exercises 2.1 and 2.3 for instructions on how to do this).

Data set held in a comma separated values (.CSV) file

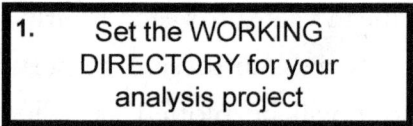

For this example, the data set you will use is stored in a file called `great_tit_daily_mass.csv` that is located in the WORKING DIRECTORY folder you created during the introduction to this chapter.

Before you start any analysis in R, you first need to set the WORKING DIRECTORY. To do this, enter the text `setwd("` and then type the address of your WORKING DIRECTORY, using slashes (/) as the folder separators, before entering a second quotation mark followed by a closing bracket, like this `")`. For example, if your WORKING DIRECTORY has the address C:\STATS_FOR_BIOLOGISTS_TWO, your `setwd` command should look like this:

```
setwd("C:/STATS_FOR_BIOLOGISTS_TWO")
```

If you are using RGUI, enter your `setwd` command in the R CONSOLE window (remembering to use the address of your own WORKING DIRECTORY folder in it) and then press the ENTER key on your keyboard. If you are using RStudio, enter your `setwd` command into the SCRIPT EDITOR window. To run it, select it and then click on the RUN button at the top of this window. You will enter all the remaining commands for this exercise in a similar manner, depending on the user interface you are using.

1. Set the WORKING DIRECTORY for your analysis project

To check that your WORKING DIRECTORY has been set properly, enter the command `getwd()` and carefully check that the address it returns is the same as the one for the STATS_FOR_BIOLOGISTS_TWO folder you created at the start of this chapter.

Before you move on to step 2, make sure that all the data you wish to use in your analysis project are located in this WORKING DIRECTORY folder. In this case, this is a file called `great_tit_daily_mass.csv`. **NOTE:** If the data you are going to import into R in step 2 are not located in the WORKING DIRECTORY you set in this step, the import code provided in the next step will not work.

2. Load your data into R using the `read.table` command

3. Check the data have loaded into R correctly by checking the names of the columns and by viewing it

The `read.table` command provides the easiest way to load data held in a .CSV file (and stored in the WORKING DIRECTORY you set in step 1) into R so you can analyse it. To do this for the data set being used in this example, enter the following command into R:

```
great_tit_daily_mass <-
read.table(file="great_tit_daily_mass.csv",
    sep=",",as.is=FALSE,header=TRUE)
```

This code has to be entered exactly as it is written here or it will not work. If you wish to use the copy-and-paste approach for entering this command, copy the text directly below CODE BLOCK 113 in the document R_CODE_DATA _VISUALISATION_WORKBOOK.DOC and paste it into R.

This command will create a new object in R called `great_tit_daily_mass` which will contain the data from the specified .CSV file. To import a different .CSV file into R, all you need to do is change the file name in the `file` argument to the name of the one you wish to import. You can also use whatever name you wish for the R object which will be created by this command. To do this, simply replace `great_tit_daily_mass` at the start of the first line of the above code with the name you wish to use for it. **NOTE:** If your .CSV data set uses a semicolon as the column separator, you would need to replace the `sep=","` argument with `sep=";"`.

Whenever you import any data into R you need to check that they have loaded correctly. First, you need to check that all the required columns are present in the R object you just created. To do this, enter the following command into R:

```
names(great_tit_daily_mass)
```

This is CODE BLOCK 114 in the document R_CODE_DATA _VISUALISATION_WORKBOOK.DOC. This command will return the names used for each column in the R object you just created. For this example, the names should be: `X`, `mass`, `wing`, `sex`, `age`, `julday`, `hourday`, `meantemp`, `location`, `dayprop` and `daypart`.

Next, you should view the contents of the whole table using the `View` command. This is done by entering following code into R:

```
View(great_tit_daily_mass)
```

This is CODE BLOCK 115 in the document R_CODE_DATA _VISUALISATION_WORKBOOK.DOC. This command will open a DATA VIEWER window where you can examine your data set and check that the correct data have been loaded into R.

The first step in creating a mixed type graph is to create a summary table containing the values you wish to plot on it. This can be done using the `ddply` command. For this example, you wish to plot the mean mass value for different hours of the day as a bar graph with error bars and then the mean temperature in each hour as a line graph. This means that your summary table needs to contain the mean mass, the standard deviation of mass and the mean temperature for each hour of the day. To do this for the data being used for this example, enter the following code into R:

```
great_tit_mixed_graph_data <-
ddply(great_tit_daily_mass,c("hourday"),
  summarise,n=length(mass),mean_mass=
     mean(mass),sd_mass=sd(mass),
  mean_temperature=mean(meantemp))
```

This is CODE BLOCK 116 in the document R_CODE_DATA _VISUALISATION_WORKBOOK.DOC. This command will create a summary table from the data held in the data set called `great_tit_daily_mass`. In this summary table, the individual rows are defined by data groupings provided in the `hourday` column. This is set by the `c("hourday")` argument. Other arguments are then used to provide a name and a calculation for each summary statistic that will be added to the table based on these groupings. For example, the `n=length(mass)` argument will create a column in the summary table called `n` which will give the count of the data in each group (which is calculated using the `length` argument). Similarly, the `mean_mass=mean (mass)` argument creates a column in the summary table called `mean_mass` which will contain the mean of the `mass` data for each group of data, while the `sd_mass=sd(mass)` argument creates a column in the summary table called `sd_mass` which will contain the standard deviation of the `mass` data for each group. Finally, the `mean_ temperature=mean(meantemp)` argument creates a column in the summary table called `mean_temperature` which will contain the mean of the `meantemp` data for each group.

Once your summary table has been created, you should view it to check that it contains the information you need it to contain by entering following code into R:

```
View(great_tit_mixed_graph_data)
```

This is CODE BLOCK 117 in the document R_CODE_DATA _VISUALISATION_WORKBOOK.DOC. This command will open a DATA VIEWER window where you can review the summary table that you just created to ensure it contains the information required to make your intended mixed type graph.

4. Create a summary table containing the values you wish to plot on your mixed type graph

Once you have successfully created your summary table and checked that it contains the required information, you are ready to use it to create an initial graph with the first data series on it. In this example, this will be a bar graph data series showing the mean body mass of great tits for each hour of the day, with error bars representing standard deviations. To do this, enter the following code into R:

```
ggplot() + geom_bar(data=
great_tit_mixed_graph_data,aes(x=hourday,y=
    mean_mass),stat="identity",colour=
"black",fill="white") + geom_errorbar(data=
    great_tit_mixed_graph_data,aes(x=
    hourday,ymin=mean_mass-sd_mass,ymax=
        mean_mass+sd_mass),width=0.2)
```

This is CODE BLOCK 118 in the document R_CODE_DATA_VISUALISATION_WORKBOOK.DOC, and it contains three commands separated by + symbols. These are the `ggplot` command, the `geom_bar` command, and the `geom_errorbar` command. In this case, the `ggplot` command is used to create a blank graph onto which all the elements specified by the other commands will be plotted.

The second command in this code block, `geom_bar`, sets the type of graph which will be created for the first data series. In this case, it will be a bar graph. In this command, the `data` argument is used to specify the data set on which the bar graph will be based. In this case, it will be the R object called `great_tit_mixed_graph_data` created in step 4. The column of data which will be plotted on the X axis of the resulting graph is set using the `x` argument of the `aes` element of this `geom_bar` command. In this case, it is the column called `hourday` in the `great_tit_mixed_graph_data` data set. The column of data which will be plotted on the Y axis is set using the `y` argument of the `aes` element of this `geom_bar` command. In this case, it is the column called `mean_mass`. The exact values to be plotted on the Y axis are set by the `stat` argument. In this case, the argument used is `stat="identity"`, meaning that the height of the bars will be set by the values provided in the column defined by the `y` argument. Finally, the `colour` and `fill` arguments are used to set the outline colour and the fill colour that will be used for the bars on the graph. In this case, these will be `black` and `white` respectively.

The third command, `geom_errorbar`, creates the error bars that will be added to the resulting graph. The `data` argument in this command is used to set the data which will be used to plot the error bars. In this case, it is `great_tit_mixed_graph_data`. The `ymin` and `ymax` arguments in the `aes` element of this command set the upper and lower limits of the error bars. In this case, they are set to be the value from the `mean_mass` column in the summary table data set minus the value from the `sd_mass` column (which contains the standard deviation for body mass for each group of data) for the `ymin` argument and `mean_mass` plus the value from the `sd_mass` column to the `ymax` argument. The `width=0.2` argument sets the width of the horizontal line at the end of each error bar to `0.2`.

5. Create a graph displaying the first data series based on the data in your summary table

Once you have created a graph with your first data series plotted on it using one graph type, you can then add your second data series using a second graph type. In this case, this will be a line data series showing the mean temperature for each hour of the day where the great tit body masses were also recorded. This is done by editing the code block from step 5 so that it looks like this (the newly added graphing command is highlighted in **bold**):

```
ggplot() + geom_bar(data=
great_tit_mixed_graph_data,aes(x=hourday,y=
    mean_mass),stat= "identity",colour=
"black",fill="white") + geom_errorbar(data=
    great_tit_mixed_graph_data,aes(x=
    hourday,ymin=mean_mass-sd_mass,ymax=
    mean_mass+sd_mass),width=0.2) +
geom_line(data=great_tit_mixed_graph_data,
aes(x=hourday,y=mean_temperature*2.5),stat="
    identity",colour="blue",size=2)
```

This is CODE BLOCK 119 in the document R_CODE_DATA _VISUALISATION_WORKBOOK.DOC. The newly added command is a `geom_line` command. This command sets how the second data series will be displayed on the graph. In this case, it will be displayed as a line data series. Within this command, the `data` argument is used to specify the data set on which the line data series will be based. In this case, it will be the R object called `great_tit_mixed_ graph_data` created in step 4 of this exercise. The column of data which will be plotted on the X axis of the resulting graph is set using the `x` argument of the `aes` element of this `geom_line` command. In this case, it is the column called `hourday` in the `great_tit_mixed_graph_data` data set. The column of data which will be plotted on the Y axis is set using the `y` argument in the `aes` element. In this case, it is the column called `mean_temperature`. However, as the temperature scale is very different from the mass scale, you need to add a scaling factor to this `y` argument so that the line data series will plot in the right place on the Y axis relative to the bars from the first data series. In this case, the scaling factor is `*2.5`. This scaling factor is included in the `y` argument directly after the name of the column containing the data that will be plotted on the Y axis for the second data series. **NOTE:** The best way to work out what scaling factor you should use for the second data series is to use trial-and-error until you find a value which results in it plotting in the right location on the graph relative to the first data series. The `stat` argument is used to set the values which will be plotted on the Y axis for the second data series. In this case, the argument is `stat="identity"` meaning the values from the table will be used to plot these data. The `colour` argument is used to set the colour that will be used for the line. In this case, it will be `blue`. Finally, the `size` argument is used to set the thickness of the line. In this case, it will be `2`. Once you have finished editing this code block, you can run it again to create an updated version of your graph.

6. Add the second data series to your graph using a second graph type

234

If you examine the graph created by the code from step 6, you will see that it does indeed display each data series in a different way (one as bars with error bars and the second as a line). However, you will also see that there is only a single Y axis on the left hand side of the graph, meaning that it is difficult to work out what values your second data series represent. To deal with this, you need to add a second Y axis to the right hand side of your graph using the `scale_y_continuous` command. To do this, edit the code from step 6 so that it looks like this (the newly added command is highlighted in **bold**):

```
        ggplot() + geom_bar(data=
great_tit_mixed_graph_data,aes(x=hourday,y=
    mean_mass),stat= "identity",colour=
"black",fill="white") + geom_errorbar(data=
    great_tit_mixed_graph_data,aes(x=
    hourday,ymin=mean_mass-sd_mass,ymax=
    mean_mass+sd_mass),width=0.2) +
geom_line(data=great_tit_mixed_graph_data,
aes(x=hourday,y=mean_temperature*2.5),stat="
    identity",colour="blue",size=2) +
scale_y_continuous(name="Mean Body Mass",
    sec.axis=sec_axis(~./2.5,name=
                "Temperature"))
```

This is CODE BLOCK 120 in the document R_CODE_DATA _VISUALISATION_WORKBOOK.DOC. The newly added `scale_y_continuous` command contains two arguments. The first of these is `name`, and this provides a label for the existing left hand Y axis. In this case, it is `Mean Body Mass`. The second argument in this command is `sec.axis`, and it creates the second axis on the right hand side of the graph. In this argument, the `~./` term is the inverse of the scaling factor included in the `y` argument in the `geom_line` command. In this case, the scaling factors is `2.5`, so the inverse term included in the `sec.axis` argument of the `scale_y_continuous` command is `~./2.5`. This ensures that the scale displayed on the second axis matches the position where the second data series is plotted on the graph. The label to be used for the second Y axis is specified by the `name` term in `sec.axis` argument. In this case, it is `Temperature`. Once you have finished editing this code block, you can run it again to create an updated version of your graph.

7. Add a second Y axis to the right hand side of your graph displaying the scale for your second data series

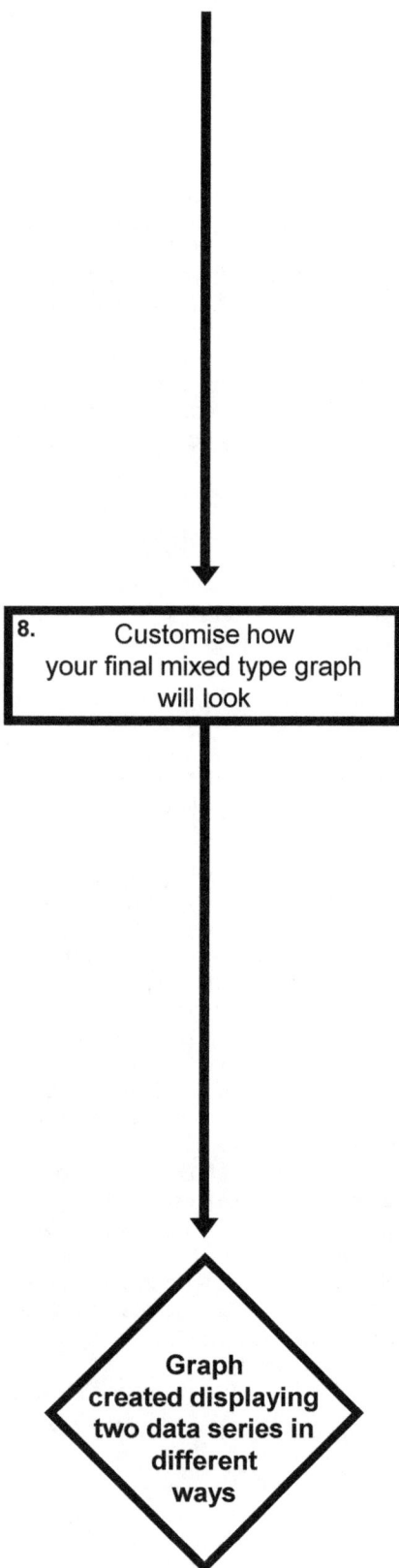

Once you have your two data series plotted on your graph in the required ways, and you have added a second Y axis displaying the scale for the second data series, you can finalise how it looks. This can be done by adding a variety of style commands to the code block used to create it. For this example, you will add three new style commands. These are `labs`, `coord_carteasian` and `theme_classic`. To do this, edit the block of code from step 7 so that it looks like this (the newly added style commands are highlighted in **bold**):

```
ggplot() + geom_bar(data=
great_tit_mixed_graph_data,aes(x=hourday,y=
    mean_mass),stat= "identity",colour=
"black",fill="white") + geom_errorbar(data=
    great_tit_mixed_graph_data,aes(x=
    hourday,ymin=mean_mass-sd_mass,ymax=
    mean_mass+sd_mass),width=0.2) +
geom_line(data=great_tit_mixed_graph_data,
aes(x=hourday,y=mean_temperature*2.5),stat="
    identity",colour="blue",size=2) +
scale_y_continuous(name="Mean Body Mass",
    sec.axis=sec_axis(~./2.5,name=
"Temperature")) + labs(x="Hour of Day") +
    coord_cartesian(ylim=c(15,23)) +
              theme_classic()
```

This is CODE BLOCK 121 in the document R_CODE_DATA _VISUALISATION_WORKBOOK.DOC. The newly added `labs` command contains a single argument. This is x, and it sets the label to be used for the X axis of the final graph (the labels for the two Y axes have already been set in the `scale_y_continuous` command added to the code block in step 7). In this case, this is `Hour of Day`. The `coord_cartesian` command allows you to zoom in on a specific portion of your graph, meaning that the axes of your graph do not need to start at zero. In this case, this command contains a single argument. This is `ylim`, and it determines that the area displayed on the graph should start at a Y value of `15` and end at a Y value of `23` (**NOTE:** These values are determined by the scale used for the left hand Y axis of the graph). Finally, the `theme_classic` command sets the remaining style elements of the final graph to those of the pre-existing classic theme. Once you have finished editing this code block, you can run it again to create the final version of your graph.

8. Customise how your final mixed type graph will look

Graph created displaying two data series in different ways

The mixed type graph created by the block of code from the final step in this flow diagram should look like this:

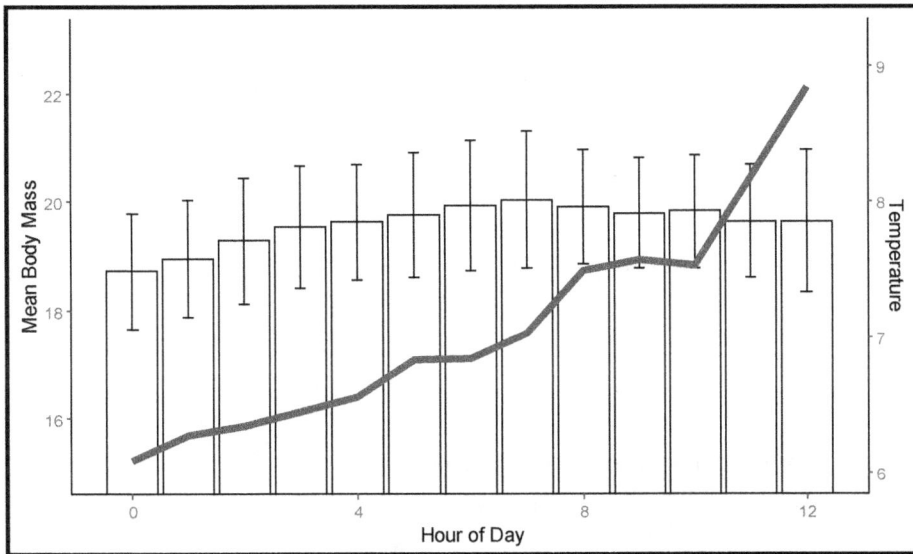

Once you have created a mixed type graph, you can export it from R so that you can include it in a manuscript or presentation. If you are using RGUI, you can do this by clicking on the R GRAPHICS window containing your mixed type graph to select it, before clicking on FILE on the main menu bar and selecting SAVE AS. This will allow you to save it in a variety of different formats. If you are using RStudio, you can export your graph by clicking on the EXPORT button at the top of the window displaying it and selecting SAVE AS IMAGE.

When including a mixed type graph in a manuscript, it is important that you provide an appropriate figure legend for it. This legend should provide all the information required for the reader to interpret its contents. For the above mixed type graph an appropriate legend would be:

Figure 1: *Variations in body mass of great tits at different hours of the day (bars represent mean values ± standard deviations) and in mean temperature at that time of day (line).*

You can make mixed graphs of almost any combination of graph types that you like. To do this, it is simply a matter of changing the graphing commands associated with each data series and altering the scaling factor, the setting for the second Y axis and the `coord_cartesian` command to ensure that both data series plot in appropriate positions relative to each other on the final graph. However, in order to be able to see the complete details of each data series, it is important that you include the different graphing commands in the right order in your code block. This is because the order these commands appear in a code block determines which data series are plotted at the back of the graph (and so appear underneath the other data series), and which are plotted at the front (and so appear on top). When creating graphs with the GGPlot package, the data series or the graphing commands that are included at the start of a code block are plotted at the back of the final graph, while those that appear later in the code block are plotted on top of the earlier data series.

To explore how to change the types of graphs included a mixed graph type and how to change the order in which the different data series appear, you will now make a new version of the above mixed graph. However, rather than being displayed as bars, the data for the mean mass of great tits in each hourly time period will be plotted as points. This is done by replacing the `geom_bar` command for this data series with a `geom_point` command, and changing the arguments contained in it to those required by this new command. To do this, edit the block of code from step 8 of the above flow diagram (this is CODE BLOCK 121 in the document R_CODE_DATA_VISUALISATION_ WORKBOOK.DOC) so that it looks like this (the required modifications are highlighted in **bold**):

```
ggplot() + geom_point(data=great_tit_mixed_graph_data,
  aes(x=hourday,y=mean_mass),stat="identity",shape=21,
 size=4,colour="black",fill="white") + geom_errorbar(data=
  great_tit_mixed_graph_data,aes(x=hourday,ymin=mean_mass-
     sd_mass,ymax=mean_mass+sd_mass),width=0.2) +
 geom_line(data= great_tit_mixed_graph_data,aes(x=hourday,
y=mean_temperature*2.5),stat="identity",colour="blue",size=
        2) + scale_y_continuous(name="Mean Body
  Mass",sec.axis=sec_axis(~./2.5,name="Temperature")) +
 labs(x="Hour of Day") + coord_cartesian(ylim=c(15,23)) +
                theme_classic()
```

The mixed type graph created by this modified block of code should look like this:

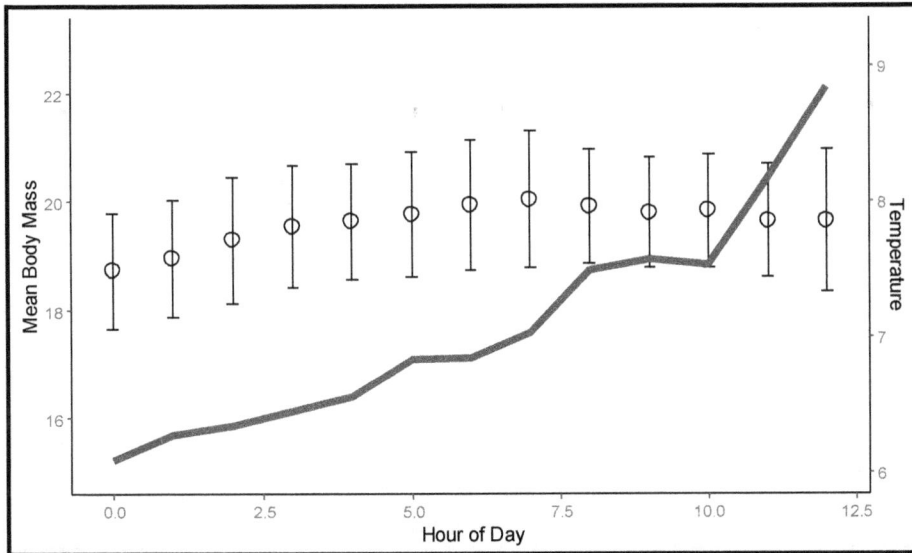

If you examine this graph, you will see that while it displays the intended data, there are a few things that look a bit odd on it. Firstly, the error bars for the mean body mass data have been plotted on top of the points, and this makes the points themselves less clear. As a result, it would be better if the error bars were plotted behind the points. Similarly, the line representing the temperature data is plotted on top of the mean body mass data, and this partly obscures the ends of some of the error bars. Again, it would be better if this line data series was plotted behind the error bars. These issues can be sorted by changing the order of the graphing commands in the above code block. To see the effect of doing this, re-arrange the above code so that it looks like this (with the graphing commands provided in the order geom_line, geom_errorbar, and then geom_point):

```
ggplot() + geom_line(data=great_tit_mixed_graph_data,
aes(x=hourday,y=mean_temperature*2.5),stat="identity",
    colour="blue",size=2) + geom_errorbar(data=
great_tit_mixed_graph_data,aes(x=hourday,ymin=mean_mass-
    sd_mass,ymax=mean_mass+sd_mass),width=0.2) +
    geom_point(data=great_tit_mixed_graph_data,aes(x=
hourday,y=mean_mass),stat="identity",shape=21,size=4,
colour="black",fill="white") + scale_y_continuous(name=
    "Mean Body Mass",sec.axis=sec_axis(~./2.5,name=
        "Temperature")) + labs(x="Hour of Day") +
coord_cartesian(ylim=c(15,23)) + theme_classic()
```

The mixed type graph created by this modified block of code should look like this:

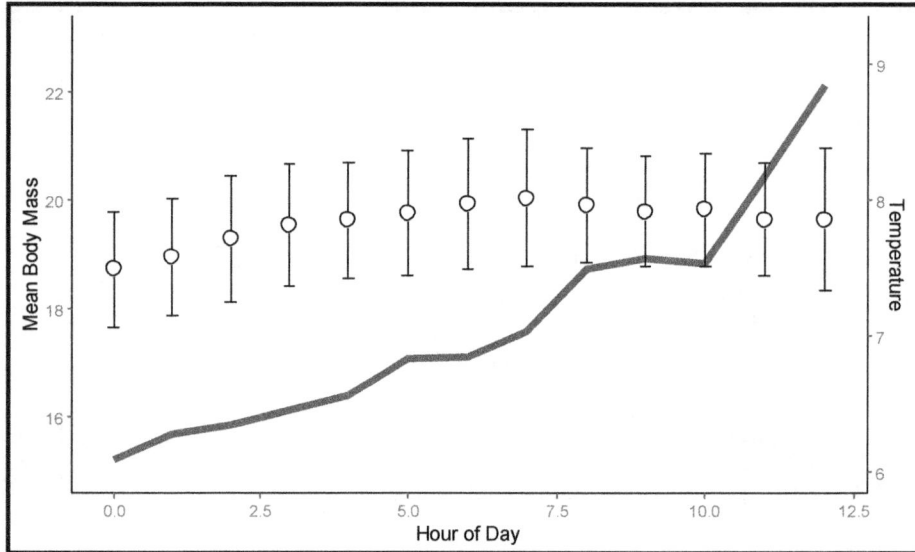

EXERCISE 4.4: HOW TO CREATE A SIMPLE X-Y GRAPH OF THE MOVEMENTS OF A TAGGED ANIMAL:

In Exercises 3.4 and 3.5, you learned how to make a line graph of time series data to show how the values of a specific variable changed over time. However, there is another type of time series data that are common in biological research. These are data sets of information about the movements of individual animals. While this can include information from focal follows, such data are most commonly obtained from radio or satellite tags which have been attached to animals to record a list of positions where they went while the tag was active. Data collected from radio or satellite tags can be visualised in a number of ways, including through the use of a Geographic Information System (GIS). However, they can also be visualised in R using a simple X-Y scatter plot. On such graphs, the individual locations recorded by the tag can be linked together by a line to show the route the animal took through space and time. Plotting the movements of tagged animals in this way is often a useful first step in analysing them as it allows you to start visualising, exploring and error-checking them.

In order to be able to create a simple X-Y graph of the movements of tagged animals, your data need to be arranged in a spreadsheet or table with one row per location you wish to plot. In this table, you also need to have columns which provide information on the temporal order that the positions were recorded in, and information on the X and Y coordinates for each location. These coordinates can either be in latitude and longitude (although when plotting such data remember than longitude is the X coordinate and latitude is the Y coordinate), or in easting and northing of a projected coordinate system (see Chapter Seven for more information). If you have data from more than one track in your data set, such as tracks from several days from the same individual, or tracks from more than one individual, you will also need a column which identifies which positions belong to the same track, and which belong to different ones.

In R, a simple X-Y graph of the movements of tagged animals can be created using a combination of the geom_point and the geom_path graphing commands from the ggplot2 package. By combining these commands, you can not only plot the location of each point, but also add a line that joins each point together based on the order it was recorded. To explore how this can be done, in this exercise, you will create a number of X-Y graphs showing the movements of a tagged African white-backed vulture around its nest site in an arcacia tree in the Maasai Mara National Nature Reserve in southern Kenya over a five day period. The data set that you will be using is a simulated one, but it is based on the movements of real vultures. The positions in this data set have been recorded using eastings and northings in the UTM Zone 36 S projected coordinate system. You will start by creating a simple X-Y graph of the movements on this vulture on the first day of data collection. To do this, work through the flow diagram that starts on the next page.

NOTE: This work flow assumes that you have the ggplot2 package installed in your version of R and that its command library has been loaded into your analysis project. If you have not already done this, you will need to do it before working through this flow diagram (see Exercise 1.1 for instructions on how to do this).

Data sets held in individual comma separated values (.CSV) files

For this example, the data sets you will use are stored in two file, one called `vulture_data.csv` and one called `vulture_nest_site.csv` that are located in the WORKING DIRECTORY folder you created during the introduction to this chapter.

Before you start any analysis in R, you first need to set the WORKING DIRECTORY. To do this, enter the text `setwd("` and then type the address of your WORKING DIRECTORY, using slashes (/) as the folder separators, before entering a second quotation mark followed by a closing bracket, like this `")`. For example, if your WORKING DIRECTORY has the address C:\STATS_FOR_BIOLOGISTS_TWO, your `setwd` command should look like this:

```
setwd("C:/STATS_FOR_BIOLOGISTS_TWO")
```

If you are using RGUI, enter your `setwd` command in the R CONSOLE window (remembering to use the address of your own WORKING DIRECTORY folder in it) and then press the ENTER key on your keyboard. If you are using RStudio, enter your `setwd` command into the SCRIPT EDITOR window. To run it, select it and then click on the RUN button at the top of this window. You will enter all the remaining commands for this exercise in a similar manner, depending on the user interface you are using.

1. Set the WORKING DIRECTORY for your analysis project

To check that your WORKING DIRECTORY has been set properly, enter the command `getwd()` and carefully check that the address it returns is the same as the one for the STATS_FOR_BIOLOGISTS_TWO folder you created at the start of this chapter.

Before you move on to step 2, make sure that all the data you wish to use in your analysis project are located in this WORKING DIRECTORY folder. In this case, these are the files called `vulture_data.csv` and `vulture_nest_site.csv`. **NOTE:** If the data you are going to import into R in step 2 are not located in the WORKING DIRECTORY you set in this step, the import code provided in the next step will not work.

```
2.    Load your data into
      R using the read.table
      command
```

The `read.table` command provides the easiest way to load data held in a .CSV file (and stored in the WORKING DIRECTORY you set in step 1) into R so you can analyse it. To do this for the first data set being used in this example, enter the following command into R:

```
vulture_data <-
read.table(file="vulture_data.csv",sep=",",
        as.is=FALSE,header=TRUE)
```

This code has to be entered exactly as it is written here or it will not work. If you wish to use the copy-and-paste approach for entering this command, copy the text directly below CODE BLOCK 122 in the document R_CODE_DATA _VISUALISATION_WORKBOOK.DOC and paste it into R. This command will create a new object in R called `vulture_data` which will contain the data from the specified .CSV file.

To import the second data set being used in this example, enter the following command into R:

```
vulture_nest_site <-
read.table(file="vulture_nest_site.csv",
    sep=",",as.is=FALSE,header=TRUE)
```

This code has to be entered exactly as it is written here or it will not work. If you wish to use the copy-and-paste approach for entering this command, copy the text directly below CODE BLOCK 123 in the document R_CODE_DATA _VISUALISATION_WORKBOOK.DOC and paste it into R. This command will create a new object in R called `vulture_nest_site` which will contain the data from the specified .CSV file.

```
3.   Check the data have
     loaded into R correctly by
     checking the names of the
     columns and by viewing it
```

Whenever you import any data into R you need to check that they have loaded correctly. First, you need to check that all the required columns are present in the R object you just created. To do this for the `vulture_data` data set, enter the following command into R:

```
names(vulture_data)
```

This is CODE BLOCK 124 in the document R_CODE_DATA _VISUALISATION_WORKBOOK.DOC. This command will return the names used for each column in the R object called `vulture_data`. In this case, these should be `id`, `day`, `position`, `easting` and `northing`.

Next, you should view the contents of the whole table using the `View` command. This is done by entering following code into R:

```
View(vulture_data)
```

This is CODE BLOCK 125 in the document R_CODE_DATA _VISUALISATION_WORKBOOK.DOC. This command will open a DATA VIEWER window where you can examine your data set and check that the correct data have been loaded into R.

Repeat these checks for the second data set (called `vulture_nest_site`) using the following two commands:

```
names(vulture_nest_site)
View(vulture_nest_site)
```

This is CODE BLOCK 126 in the document R_CODE_DATA _VISUALISATION_WORKBOOK.DOC. The column names for this data set should be `nest_location`, `easting` and `northing`. **NOTE:** There is only one line of data in the `vulture_nest_site` data set as it only contains the location of the nest site itself and no other data.

```
4.    Create a new R object
with just the data from the time
period you wish to plot
```

If your data set contains data from multiple tracks (e.g. data from the same animal recorded on different days or data from different individuals of the same species), it is likely that you will want to extract the data from a single time period or individual before you create your X-Y graph showing the movements of your tagged animal. This can be done with the `subset` command, as long as you have a column which contains a variable that allows you to identify the time period or individual each data point belongs to. In the data set called `vulture_data` that you will use in this example, this is a column called `day`. This column contains a code that allows you to identify which day each data point was recorded on. To create a new R object with just the data from the first day, enter the following command into R:

```
vulture_day_1 <- subset(vulture_data,
                day=="Day 1")
```

This is CODE BLOCK 127 in the document R_CODE_DATA _VISUALISATION_WORKBOOK.DOC. This command will create a new subset of data from the data set called `vulture_data` and save it as a new R object called `vulture_day_1`. It will contain just the positions from the first day of tracking (which are can be identified by the code `Day 1` in the column called `day`).

Next, you should view the contents of your new tavle using the View command. This is done using by entering the following command into R:

```
View(vulture_day_1)
```

This is CODE BLOCK 128 in the document R_CODE_DATA _VISUALISATION_WORKBOOK.DOC. This command will open a DATA VIEWER window where you can examine your data set and check that it contains the required subset of data. In this example, it should only contain the tracking data from Day 1 and not from any of the other days.

```
5.    Create your initial X-Y
      graph from the locations in
      your tracking data set
```

After you have successfully created the required subset of data, you are ready to use it to make your initial X-Y graph based these data. To do this for the subset of data created in step 4, enter the following block of code into R:

```
ggplot() + geom_path(data=vulture_day_1,
aes(x=easting,y=northing),colour="black",
size=0.3) + geom_point(data=vulture_day_1,
aes(x=easting,y=northing),shape=21,size=3,
                fill="black")
```

This is CODE BLOCK 129 in the document R_CODE_DATA _VISUALISATION_WORKBOOK.DOC, and it contains three commands separated by + symbols. These are the `ggplot` command, the `geom_path` command and the `geom_point` command. In this case, the `ggplot` command is used to create a blank graph on to which the elements specified by the other commands will be plotted.

The `geom_path` command creates a line that joins all the locations in a data set together in the order they are listed in it (this should be the temporal order in which they were recorded). In this command, the `data` argument identifies the data set that will be used to create this line. In this case, it will be the data set called `vulture_day_1`. The `x` and `y` arguments in the `aes` element of this command are used to identify the columns in this data set which contain the X and Y coordinates for each location. In this case, these are the columns called `easting` and `northing` respectively. Finally, the `colour` and `size` arguments are used to set how the line will be drawn. In this case, these arguments will be set to `black` and `0.3` respectively, to plot these data as a relatively thin black line.

The `geom_point` command adds a series of points to the X-Y graph which mark the exact locations for each data point. In this command, the `data` argument identifies the data set that will be used to create these points. In this case, it will be the data set called `vulture_day_1`. The `x` and `y` arguments in the `aes` element of this command are used to identify the columns in this data set which contain the X and Y coordinates for each location. In this case, these are the columns called `easting` and `northing` respectively. Finally, the `shape`, `fill` and `size` arguments are used to set how the points will be drawn. In this case, these arguments will be set to `21`, `black` and `3` respectively, to mark each location with a solid black circle of size 3.

246

Once you have created your initial graph showing the desired track from the locations in your tracking data set, you can add additional locational information to it. In this example, you will add the location of the nest site being used by the vulture that was being tracked. This information is contained in the data set called vulture_nest_site. To do this, edit the block of code from step 5 so that it looks like this (the newly added commands are highlighted in **bold**):

```
ggplot() + geom_path(data=vulture_day_1,
aes(x=easting,y=northing),colour="black",
size=0.3) + geom_point(data=vulture_day_1,
aes(x=easting,y=northing),shape=21,size=3,
    fill="black") + geom_point(data=
    vulture_nest_site,aes(x=easting,y=
    northing),shape=22,colour="black",
    fill="red",size=4) + geom_text(data=
    vulture_nest_site,aes(x=easting,
 y=northing,label="Nest Site"),size=4,
        vjust=1.25,hjust=-0.2)
```

This is CODE BLOCK 130 in the document R_CODE_DATA _VISUALISATION_WORKBOOK.DOC. The two newly added commands are a second geom_point command and a geom_text command. The new geom_point command adds a point to the graph which marks the location of the nest site used by the tagged vulture. In this command, the data argument identifies the data set that will be used to create these points. In this case, it will be the data set called vulture_nest_site. The x and y arguments in the aes element of this command are used to identify the columns in this data set which contain the X and Y coordinates for each location. In this case, these are the columns called easting and northing respectively. Finally, the shape, fill and size arguments are used to set how the point will be drawn. In this case, these arguments will be set to 22, red and 4 respectively, to mark this location as a solid red square.

The geom_text command adds a text label to indicate what the red square created by the second geom_point command represents. In this command, the data argument identifies the data series which will be labelled by this command. In this case, it will be one called vulture_nest_site. The x and y arguments in the aes element of this command are used to identify the columns in this data set which contain the X and Y coordinates where the label will be placed on the graph. In this case, these are the columns called easting and northing respectively. The label argument is used to provide the text for the label that will be created by this command. In this case, this will be the text Nest Site. The size argument is used to set the size of the text used for the label (in this case, this will be 4), while the vjust and hjust arguments are used to offset the position of the text so that it is beside, rather than on top of, the marker being used to mark the nest site. Once you have finished editing this code block, you can run it again to create an updated version of your graph.

6. **Add any required additional information to your initial X-Y graph**

247

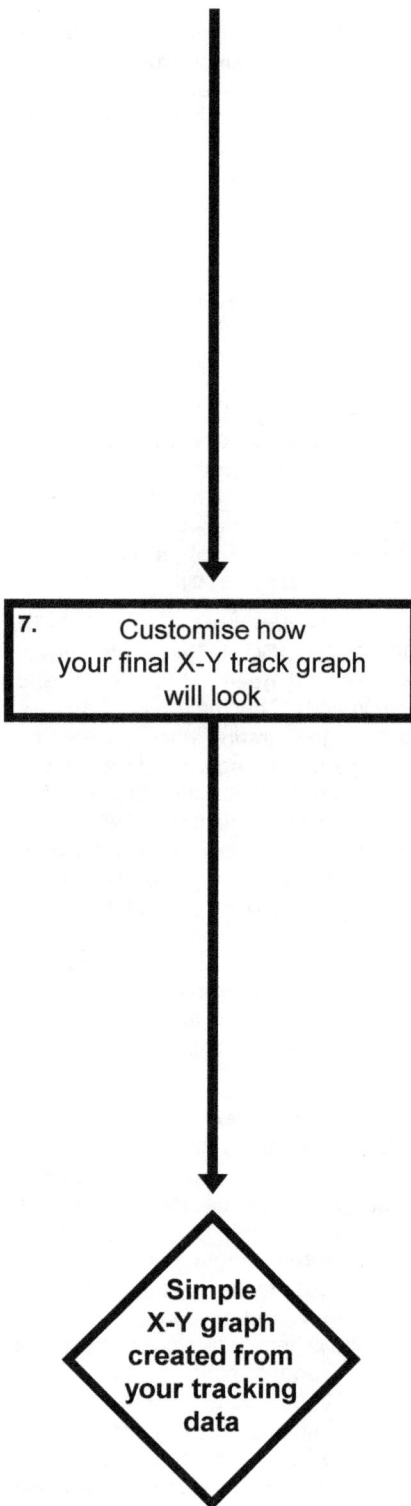

Once you have all the required data added to the initial X-Y graph of your tracking data, you can customise how the final graph will look. This can be done by adding a number of new style commands to the block of code used to produce it. To do this, edit the block of code from step 6 so that it looks like this (the newly added style commands are highlighted in **bold**):

```
ggplot() + geom_path(data=vulture_day_1,
aes(x=easting,y=northing),colour="black",
size=0.3) + geom_point(data=vulture_day_1,
aes(x=easting,y=northing),shape=21,size=3,
   fill="black") + geom_point(data=
   vulture_nest_site,aes(x=easting,y=
northing),shape=22,colour="black",fill=
   "red",size=4) + geom_text(data=
   vulture_nest_site,aes(x=easting,
y=northing,label="Nest Site"),size=4,
vjust=1.25,hjust=-0.2) + xlim(702500,
   730000) + ylim(9840000,9870000) +
   labs(x="Easting",y="Northing") +
         theme_classic()
```

This is CODE BLOCK 131 in the document R_CODE_DATA _VISUALISATION_WORKBOOK.DOC, and it adds four new style commands to the code block used to create your initial X-Y graph in step 6. These are the `xlim` command, the `ylim` command, the `labs` command and the `theme_ classic` command. The `xlim` command is used to set the minimum and maximum limits of the X axis. In this case, these are `702500` and `730000` respectively. The `ylim` command is used to set the minimum and maximum limits of the Y axis. In this case, these are `9840000` and `9870000` respectively. The `labs` command is used to set the labels which will used for the X and Y axes. In this case, these are `Easting` for `x` and `Northing` for `y`. Finally, the `theme_ classic` command sets all other elements of the final graph to those used by the pre-set classic theme. Once you have finished editing this code block, you can run it again to create the final version of your graph.

7. Customise how your final X-Y track graph will look

Simple X-Y graph created from your tracking data

The X-Y graph of your tracking data created by the block of code from the final step in this flow diagram should look like the image at the top of the next page.

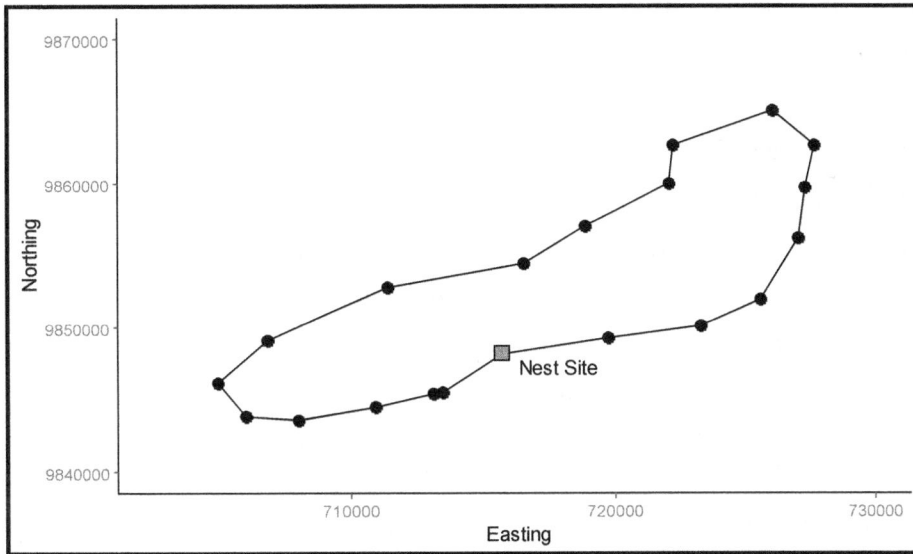

Once you have created a simple X-Y graph from a tracking data set, you can export it from R so that you can include it in a manuscript or presentation. If you are using RGUI, you can do this by clicking on the R GRAPHICS window containing your X-Y graph to select it, before clicking on FILE on the main menu bar and selecting SAVE AS. This will allow you to save it in a variety of different formats. If you are using RStudio, you can export your graph by clicking on the EXPORT button at the top of the window displaying it and selecting SAVE AS IMAGE.

When including an X-Y graph made from tracking data in a manuscript, it is important that you provide an appropriate figure legend for it. This legend should provide all the information required for the reader to interpret its contents. For the above graph, an appropriate legend would be:

Figure 1: The track followed by a male African white-backed vulture during a day-long foraging trip from its nest site in the Maasai Mara National Nature Reserve in southern Kenya. The coordinates are in the UTM Zone 36 S projected coordinate system.

If you examine the above graph, you will see that there is no information on it that you can use to tell which order the individual points along the track were recorded in, and so which direction the animal was moving in. This information can be added to the graph by labelling the individual points marking each location plotted by the first `geom_point` command.

This can be done by adding another `geom_text` command (like the one used to add the label to the red square marking the location of the nest site). However, if you use this option, you will find that some or all of the labels will either overlap with each other, or with the lines connecting the points, making them difficult to read. As a result, it is better to use a command called `geom_text_repel` to add such labels to the points along your track to show the order they were recorded in. This command automatically varies the exact location of the labels to minimise the chance that they will overlap with other features on your graph. To demonstrate how you can use this command to add labels to a graph, you will now add a number indicating the order the data points were recorded along the track the vulture followed on Day 1. However, before you can do this you first need to ensure that the `Rcpp` and `ggrepel` packages are installed in your version of R. To do this, enter the command `library()` into R. If the names of these two packages are not listed in the R PACKAGES AVAILABLE, enter the following command into R:

```
install.packages(c("Rcpp","ggrepel")
```

If only one of these packages is already installed in your version of R, remove its name from the above command before you run it, and if both of these packages are already installed, you do not need to run this command at all.

Once you have ensured that these packages are installed in your version of R, you need to load their associated command libraries into your analysis project. To do this, enter the following pair of command into R:

```
library(Rcpp)
library(ggrepel)
```

You are now ready to use the `geom_text_repel` command to add labels to the X-Y graph created by working through in the above flow diagram. To do this, edited the block of code from step 7 (this is CODE BLOCK 131 in the document R_CODE_DATA_ VISUALISATION_WORKBOOK.DOC) so that it looks like the code at the top of the next page (the newly added `geom_text_repel` command is highlighted in **bold**).

```
Ggplot() + geom_path(data=vulture_day_1,aes(x=easting,y=
   northing),colour="black",size=0.3) + geom_point(data=
vulture_day_1,aes(x=easting,y=northing),shape=21,size=3,
   fill="black") + geom_point(data=vulture_nest_site,
 aes(x=easting,y=northing),shape=22,colour="black",fill=
 "red",size=4) + geom_text(data=vulture_nest_site, aes(x=
 easting,y=northing,label="Nest Site"),size=4,vjust=1.25,
   hjust=-0.2) + geom_text_repel(data=vulture_day_1,
aes(x=easting,y=northing,label=position),size=4,vjust=-0.5,
hjust=0.5) + xlim(702500,730000) + ylim(9840000,9870000) +
   labs(x="Easting",y="Northing") + theme_classic()
```

In the newly added `geom_text_repel` command, the `data` argument is used to identify the data set which contains the locations you wish to label. In this case, it is the data set called `vulture_day_1`. The columns containing the coordinates for these locations are set by the `x` and `y` arguments in the `aes` element of this command. The `label` argument is used to set the column which contains the labels that will be used for each point, and it is also included in the `aes` element of the command. In this case, it is the column called `position` in the `vulture_day_1` data set. Finally, the `size`, `vjust` and `hjust` arguments are used to set how big the labels will be and fine-tune exactly where they will be plotted on the graph. In this case, the labels will be plotted as size 4, while they will be offset vertically (using the `vjust` argument) by a value of -0.5 and horizontally (using the `hjust` argument) by a value of 0.5. This means that they will be plotted above and to the left of the data point each label is attached to. When you run this modified code block, you should get a new X-Y graph of tracking data that looks like this:

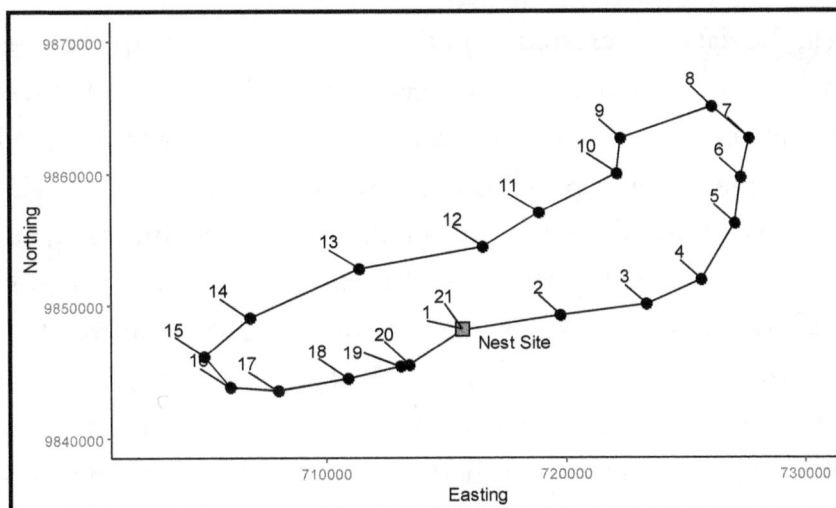

You can also add multiple tracks to the same X-Y graphs. These can be the tracks for different individuals, or for the same individual from different time periods. This is done by adding additional arguments to the `geom_path` and the first `geom_point` command, and by adding new `scale_colour_manual` and `scale_fill_manual` style commands to allow you to determine which colours will be used to plot the data from different tracks. To demonstrate how you can do this, you will now create a new X-Y graph that will show the tracks of the African white-backed vulture from the `vulture_data` data set recorded on five different days (Days 1 to 5). To do this, you will need to edit the block of code from step 7 of the above flow diagram (this is CODE BLOCK 131 in the document R_CODE_DATA_VISUALISATION_WORKBOOK.DOC) so that it looks like this (the required modifications are highlighted in **bold**):

```
ggplot()+geom_path(data=vulture_data,aes(x=easting,
        y=northing,group=day,colour=day),size=0.3) +
  geom_point(data=vulture_data,aes(x=easting, y=northing,
            group=day,fill=day),shape=21,size=3) +
scale_colour_manual(values=c("blue","green","red","brown",
    "yellow")) + scale_fill_manual(values=c("blue","green",
        "red","brown","yellow")) + geom_point(data=
  vulture_nest_site, aes(x=easting,y=northing),shape=22,
    colour="black",fill="red",size=4) + geom_text(data=
      vulture_nest_site,aes(x=easting,y=northing,label=
        "Nest Site"),size=4,vjust=-1.1,hjust=0.75) +
  labs(x="Easting",y="Northing",colour=NULL,fill=NULL) +
                    theme_classic()
```

In this code block, the data set referred to in the `data` arguments of the `geom_path` and the first `geom_point` command has been changed from `vulture_day_1` to `vulture_data` to reflect the fact that this multi-track graph will be based on the whole data set and not just the subset representing the tracking data from a single day created in step 4 of the above flow diagram. In addition, the arguments `group=day` and `colour=day` have been added to the `aes` element of the `geom_path` command. This tells R to use the groupings provided in the column called `day` in the `vulture_data` data set to create different line data series on the final graph (one representing the data from each group listed in this column). Similarly, the arguments `group=day` and `fill=day` have been added to the `aes` element of the first `geom_point` command to do the same for the points representing each individual location that will be plotted on the final graph.

The exact colours which will be used for each of these data series are set using the `scale_colour_manual` command (for the colours of the lines for each data series created by the modified `geom_path` command) and the `scale_fill_manual` command (for the colours of the points for each data series created by the modified `geom_point` command). The arguments `colour=NULL` and `fill=NULL` have been added to the `labs` command so that no label is added to the legend that will be created to tell the reader which colour represents the data for each data series. Finally, the values included in the `vjust` and `hjust` arguments in the `geom_text` command which adds a label to the marker representing the location of the nest site have been altered so that this label is now drawn above the marker, rather than below it, so that it doesn't overlap with any of the lines representing the additional data series. **NOTE:** The `geom_text_repel` command is not included in this modified code block. This is because adding labels for every point on each of tracks would create a graph which was too cluttered. The `xlim` and `ylim` commands have also been removed to ensure that the graph will show the data from all the data series and not just the first one.

When you run the above code block in R, you should get a new X-Y graph of the tracking data that looks like this:

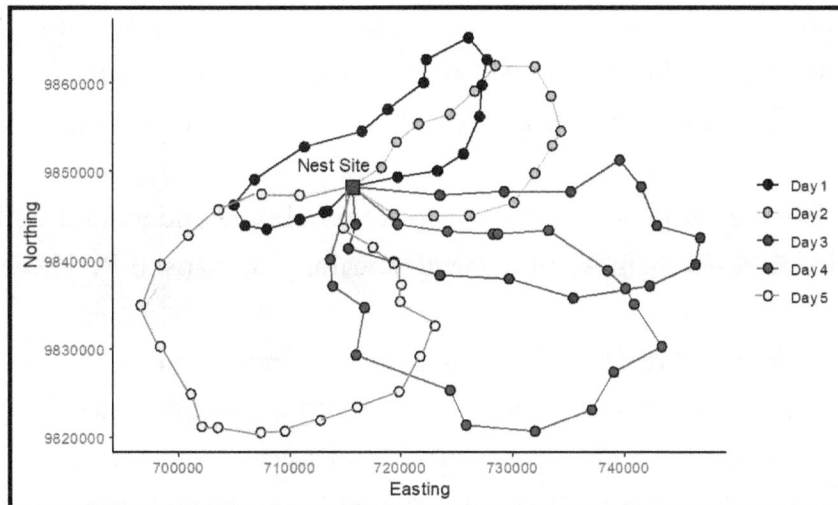

As you can see, this X-Y graph now displays the simulated tracks for a tagged African white-backed vulture over five successive days as it made foraging trips from its nest site and back again throughout the Maasai Mara National Nature Reserve in Kenya, with the data for each day being displayed in a separate colour.

--- Chapter Seven ---

How To Create Maps From Biological Data Using R

In biological research, maps are usually created using dedicated geographic information system (GIS) software packages such as QGIS, and indeed, this is usually by far the best option when you need to make a high quality map for inclusion in a presentation or publication. However, there will be times when it will be both useful and necessary to be able to make maps in R itself. Reasons for doing this can include using maps to explore your data, checking that data processing tasks, such as subsetting or joining data sets together, have been carried out successfully, creating maps to display the results of statistical analyses, and to make types of maps that cannot easily be made in GIS software (such as maps that integrate other forms of data visualisations, like pie charts).

Creating maps within R is relatively straight-forward and can be thought of as an extension of making scatter plots and other types of X-Y graphs. For example, in Exercise 4.4, you learned how to make a simple X-Y graph to show the relative positions of the locations where a tagged animal had been recorded over a specific period of time. This means that this graph was, in effect, a very simple map displaying these tracking data.

There are three key elements you will need to know about, understand and learn how to handle in order to successfully make biologically meaningful maps in R. These are:

1. **Map Projection:** A Map Projection is a way of transforming the three-dimensional surface of the Earth so that it can be displayed on a two-dimensional map. This is done using a mathematical formula. All Map Projections create some distortion, and this means it is important when you are making maps (whether in R or in another software package) that you select a Map Projection which does not distort the features that are important to your analysis (such as distances, the relationships between points and/or the shape of objects displayed on it). If you wish to display

data from different data sets on the same map in R, they all need to have been made using the same Map Projection.

2. **Shapefile:** A Shapefile is a file format for saving certain types of spatial data. Spatial data are commonly displayed in one of two ways. These are as features or vectors (such as points, lines or polygons) and as grids. The most common file format for saving spatial data that are being represented as features or vectors is the Shapefile format. All Shapefiles need to have an appropriate Map Projection associated with them in order for them to be plotted correctly.

3. **Raster Data Set:** A raster data set is a spatial data set which contains data that are arranged and saved as a grid. Each grid cell in a raster data set represents a specific area in space, and all the grid cells in a raster data set represent an area of the same size. Raster data sets can be saved in a variety of formats, but the most common is the geotiff format. Like Shapefiles, Raster Data Sets need to have an appropriate Map Projection associated with them in order for them to be plotted correctly.

In this chapter, you will learn how to make three basic types of maps in R. These are a map showing the distribution of the data in a data set in relation to other features in a specific study area; a map which includes pie charts to provide additional biological information for specific locations displayed on it; and a map which uses a raster data set to show the results of a simple spatial analysis.

If you have not already done so, before you start the exercises in this chapter, you first need to create a WORKING DIRECTORY folder on your computer and load the necessary data into it (**NOTE:** If you have already created this folder and downloaded data for a previous chapter in this workbook, you do not need to do this again). To do this on a computer with a Windows operating system, open Windows Explorer and navigate to the location where you would like to create the folder (such as your C:\ drive or your DOCUMENTS folder). Next, right click anywhere in this location and select NEW> FOLDER. Now call this folder STATS_FOR_BIOLOGISTS_TWO by typing this into the folder name section to replace what it is currently called (which will most likely be NEW FOLDER). To create a WORKING DIRECTORY folder on a computer running a Mac operating system, open Finder and navigate to the location where you would like to create the folder (such as your

DOCUMENTS folder or your DESKTOP). Next, click on FILE> NEW FOLDER, and then type the name STATS_FOR_BIOLOGISTS_TWO before pressing the ENTER key on your keyboard.

Once you have created your WORKING DIRECTORY folder, you are ready to download the data sets you will use for the exercises in this workbook from *www.gisinecology.com/ stats-for-biologists-2*. After you have downloaded the compressed folder containing the required data by following the instructions provided on that page, you need to extract all the data files from it and copy them into the folder called STATS_FOR_BIOLOGISTS_TWO that you have just created.

Next, you need to check that the required data have been extracted to the correct folder. If you are using a computer with a Windows operating system, you can use Windows Explorer to open your newly created WORKING DIRECTORY folder and examine its contents. If all the files from the compressed folder are present in it (there should be a total of 90 of them), you can click on the folder icon at the left hand end of the ADDRESS BAR at the top of the WINDOWS EXPLORER window to reveal its full address. Write this address down as you will need it to set this folder as your WORKING DIRECTORY during the exercises provided in this workbook (see pages 12 and 13 for details of how to modify folder addresses so they will be recognised by R).

If you are using a computer with a Mac operating system, you can use Finder to open your newly created WORKING DIRECTORY folder and examine its contents. If all the required data files are present in it (there should be a total of 90 of them), press the CMD and I keys on your keyboard at the same time. This will open the GET INFO window where you will find its address (which is also called the pathway). Write this address down somewhere as you will need it to set this folder as your WORKING DIRECTORY during the exercises provided in this workbook (see pages 12 and 13 for details of how to modify folder addresses so they will be recognised by R).

After you have loaded the required data into your WORKING DIRECTORY folder, you can open RGUI or RStudio, depending on which option you wish to use (see Chapter 2 for more details). Once you have opened your preferred R user interface, you need to create a file called CHAPTER_SEVEN_EXERCISES where you will save the results of your analyses from your R CONSOLE window as you work through this chapter. To do this using RGUI, click on the FILE menu and select SAVE WORKSPACE. To do this in RStudio, click on SESSION and select SAVE WORKSPACE AS. In both cases, save it as a WORKSPACE file with the name CHAPTER_SEVEN_EXERCISES.RDATA in your WORKING DIRECTORY folder (this will be the one called STATS_FOR_ BIOLOGISTS_TWO that you have just created). If you are using RStudio, you will also want to save the contents of your SCRIPT EDITOR window (where you will enter and edit the R code you will use to carry out specific commands). To do this, click on the FILE menu and select SAVE AS. Save your file as an R SCRIPT file with the name CHAPTER_SEVEN_EXERCISES.R in your WORKING DIRECTORY folder. As you work through the exercises in this chapter, remember to regularly save the contents of your R CONSOLE window (which will contain the R objects you have created up to that point) to your WORKSPACE file and, if you are using RStudio, the contents of your SCRIPT EDITOR window to your R SCRIPT file.

Finally, you need to remove any data that are currently held in R's temporary memory. To do this, enter the following command into R:

```
rm(list=ls())
```

If you are using RGUI, you can simply type this code after the command prompt at the bottom of the R CONSOLE window (it looks like this: >) and then press the ENTER key on your keyboard to run it. If you are using RStudio, you can type this command into the SCRIPT EDITOR window (the upper left hand window). To run this command, select it and then click on the RUN button at the top of this window. This will run it in the R CONSOLE window (the lower left hand one in the main RStudio user interface). You are now ready to start the exercises in this chapter.

EXERCISE 5.1: HOW TO MAKE A MAP SHOWING THE DISTRIBUTION OF POINT LOCATIONS IN RELATION TO OTHER FEATURES :

The first type of map that you are likely to want to be able to make in R is a map which shows the spatial distribution of point data in a data set in relation to other features in the local environment. The most difficult part of creating such maps is usually getting your data into R in the first place and converting them into a format that can be plotted using the various graphing tools available in R. When these data are feature data (also known as vector data) which have been saved in the shapefile format, there are two ways they can be imported into in R. These are the 'fortify' approach (based on the `fortify` command from the `ggplot2` package) and the 'sf' approach (based on the `read_sf` command from the `sf` package). Both approaches will allow you to produce maps from the data you import using similar commands. If you find that the lines or polygons in a shapefile do not plot properly on your graph when you import them using one approach, the best option is usually to try the other one to see whether this solves any issues you are having with it.

In this exercise, you will create a map that shows the distribution of nest boxes in an area of native oak woodland on the shores of Loch Lomond in central Scotland. These nest boxes are part of an on-going project being conducted by researchers from the University of Glasgow SCENE field station to study factors which affect the breeding success of hole-nesting birds, such as the blue tit. This map will show not only the location of each nest box, but also the outline of the oak woodland study area, the location of a small body of freshwater bordering it (called the Dubh Loch), the shoreline of Loch Lomond (a much larger body of freshwater which is nearby), the routes taken by roads which pass near the study area and the location of a forestry track that runs through the middle of it. The spatial data for these local environmental features are all provided as shapefiles, and in this exercise, you will use the 'fortify' approach to import them into R before making a map from them. For the data set showing the spatial distribution of the nest boxes themselves, you will start by importing these data from a shapefile, but in later parts of this exercise, you will also learn how to import them from a list of positions held in a spreadsheet or table. If you wish to learn how you can make the types of shapefiles you will use in this exercise, you can find out how to do this in Exercise Two of a book called *GIS For Biologists: A Practical Introduction for Undergraduates*. Similarly, if you wish to learn how to make this type of map in GIS software, you can find instructions on how to make this same map in Exercise Two of a workbook called *An Introduction to Integrating QGIS and R for Spatial Analysis*.

To create a map showing the location of nest boxes in a patch of native oak woodland in central Scotland from a series of existing spatial data sets stored in the shapefile format, work through the flow diagram that starts on the next page.

NOTE: When importing a shapefile into R using the 'fortify' approach, use the suffix `_shp` in the name of the object created by the `readORG` command (see step 3) to allow you to easily identify its data format, and separate it from other data formats used when making maps. Similarly, when a shapefile is converted into a table or data frame using the `fortify` command as part of this approach (this is also done in step 3), use the suffix `_df` to allow you to easily identify it as a data frame R object.

Data sets held in shapefiles or other spatial data formats

For this example, you will use six data sets that are stored in the shapefile spatial data format. These are called `scene_oak_woodland.shp`, `dubh_loch.shp`, `loch_lomond_shoreline.shp`, `scene_roads.shp`, `forestry_track.shp` and `scene_nestbox_locations.shp`. These shapefiles are located in the WORKING DIRECTORY folder you created during the introduction to this chapter.

Before you start any analysis in R, you first need to set the WORKING DIRECTORY. To do this, enter the text `setwd("` and then type the address of your WORKING DIRECTORY, using slashes (/) as the folder separators, before entering a second quotation mark followed by a closing bracket, like this `")`. For example, if your WORKING DIRECTORY has the address C:\STATS_FOR_BIOLOGISTS_TWO, your `setwd` command should look like this:

```
setwd("C:/STATS_FOR_BIOLOGISTS_TWO")
```

If you are using RGUI, enter your `setwd` command in the R CONSOLE window (remembering to use the address of your own WORKING DIRECTORY folder in it) and then press the ENTER key on your keyboard. If you are using RStudio, enter your `setwd` command into the SCRIPT EDITOR window. To run it, select it and then click on the RUN button at the top of this window. You will enter all the remaining commands for this exercise in a similar manner, depending on the user interface you are using.

1. Set the WORKING DIRECTORY for your analysis project

To check that your WORKING DIRECTORY has been set properly, enter the command `getwd()` and carefully check that the address it returns is the same as the one for the STATS_FOR_BIOLOGISTS_TWO folder you created at the start of this chapter.

Before you move on to step 2, make sure that all the data you wish to use in your analysis project are located in this WORKING DIRECTORY folder. In this case, these are the files called `scene_oak_woodland.shp`, `dubh_loch.shp`, `loch_lomond_shoreline.shp`, `scene_roads.shp`, `forestry_track.shp` and `scene_nestbox_locations.shp`. **NOTE:** If the data you are going to import into R in steps 3, 4 and 5 of this flow diagram are not located in the WORKING DIRECTORY you set in this step, the code provided for importing them will not work.

In order to be able to import shapefiles into R and use them to make a map, you will need to use a variety of different packages and their associated command libraries. As a result, the first step in this process is to ensure that you have the required packages installed in your version of R. To do this, enter the command `library()` into R. In the R PACKAGES AVAILABLE window which opens, look for the following packages: `ggplot2`, `Rcpp`, `rgdal`, `ggmap`, `data.table`, `proj4` and `ggspatial`. If any of these packages are not listed in this window, you can install them by entering the following command into R (**NOTE**: If you already have any of these packages installed in your version of R, remove their name(s) from this command before you run it to ensure you do not re-install them as this can cause problems with R):

```
install.packages(c("ggplot2",
"Rcpp","rgdal","ggmap","data.table",
"proj4","ggspatial"))
```

This CODE BLOCK 132 in the document R_CODE_DATA _VISUALISATION_WORKBOOK.DOC. **NOTE**: If you already have all these packages installed in your version of R, you do not need to run this `install.packages` command and you can move on to the next part of this step.

Once you have ensured that all the required packages are installed in your version of R, you will need to load their associated command libraries (as well as the `ggplot2` command library) into your analysis project. To do this, enter the following block of code into R (**NOTE**: Each library command needs to be entered on a separate line):

```
library(ggplot2)
library(Rcpp)
library(rgdal)
library(ggmap)
library(data.table)
library(proj4)
library(ggspatial)
library(ggplot2)
```

This code has to be entered exactly as it is written here or it will not work. If you wish to use the copy-and-paste approach for entering this command, copy the text directly below CODE BLOCK 133 in the document R_CODE_DATA _VISUALISATION_WORKBOOK.DOC and paste it into R.

2. Ensure that the packages required to import shapefiles and make maps are installed in your version of R and load their associated command libraries into your analysis project

Once you have all the required command libraries loaded into your analysis project, you are ready to import your first shapefile using the 'fortify' approach. This first shapefile should contain data which are either polygons or lines, and not points. This is because the process for importing a point shapefile into R is slightly different (see step 5). In this case, you will start by importing a shapefile containing a polygon which represents the native oak woodland study area called `scene_oak_woodland.shp`. To import this shapefile into your analysis project, enter the following command into R:

```
scene_oak_woodland_shp <- readOGR(dsn=".",
        layer="SCENE_OAK_WOODLAND")
```

This is CODE BLOCK 134 in the document R_CODE_DATA_VISUALISATION_WORKBOOK.DOC. This command will create a new object in R called `scene_oak_woodland_shp` containing the data from the specified shapefile. Once you have imported a shapefile into R, you need to check what map projection is associated with it, and that this is the same as the one associated with all the other data sets you wish to use to build your map. To do this, enter the following command into R:

```
summary(scene_oak_woodland_shp)
```

This is CODE BLOCK 135 in the document R_CODE_DATA_VISUALISATION_WORKBOOK.DOC. The proj4 string for the map projection for your shapefile is listed in the information that this command returns in the R CONSOLE window. Next, you need to convert your shapefile into a type of R object called a data frame. This is done using the `fortify` command by entering the following code into R:

```
scene_oak_woodland_df <-
    fortify(scene_oak_woodland_shp)
```

This is CODE BLOCK 136 in the document R_CODE_DATA_VISUALISATION_WORKBOOK.DOC. Once the shapefile has been converted into a data frame object (in this case, one called `scene_oak_woodland_df`), you need to rename the columns called `lat` and `long` that have been added to this data frame as `x` and `y`. This is because, depending on the map projection of the original shapefile, the coordinates contained in these columns may be eastings and northings rather than latitude and longitude. To do this, enter the following command into R:

```
setnames(scene_oak_woodland_df,old=
    c('lat','long'),new=c('y','x'))
```

This is CODE BLOCK 137 in the document R_CODE_DATA_VISUALISATION_WORKBOOK.DOC. Finally, you need to view the R object containing the data frame version of your shapefile in order to check that it has been processed correctly. To do this, enter the following command into R:

```
View(scene_oak_woodland_df)
```

This is CODE BLOCK 138 in the document R_CODE_DATA_VISUALISATION_WORKBOOK.DOC.

3. Import your first shapefile into R using the 'fortify' approach

4. Repeat step 3 for all other
shapefiles containing polygon
or line features

To import additional spatial data stored as shapefiles into R,
you can repeat step 3, replacing the names of the shapefile
and the R objects to be created with the appropriate new
names. Do this to import the data in the shapefile containing
polygon data called `dubh_loch.shp` and the shapefiles
containing line data called `loch_lomond_`
`shoreline.shp`, `scene_roads.shp`, `foresty_`
`track.shp`. The final R objects that you create from these
shapefiles should be called `dubh_loch_df`, `loch_`
`lomond_shoreline_df`, `scene_roads_df` and
`forestry_track_df`. If you use different names for these
R objects, the blocks of code provided for the remaining
steps in this exercise will not work properly. The blocks of
code required to import each of these shapefiles using the
correct names are provided in CODE BLOCK 139 to
CODE_BLOCK 142 in the document R_CODE_DATA
_VISUALISATION_WORKBOOK.DOC.

Even though you are still using the 'fortify' approach, shapefiles containing point data need to be imported in a slightly different way to those containing line and polygon data. Specifically, the `as.data.frame` command is used instead of the `fortify` command. The shapefile containing the point data which you wish to import in this example is called `scene_nestbox_locations.shp`. To import this point shapefile into your analysis project, enter the following block of code into R:

```
scene_nestbox_locations_shp <-
readOGR(dsn=".",layer="SCENE_NESTBOX_
                  LOCATIONS")
```

This is CODE BLOCK 143 in the document R_CODE_DATA _VISUALISATION_WORKBOOK.DOC. This command will create a new object in R called `scene_nestbox_ locations_shp` containing the data from the specified shapefile. Once you have imported a shapefile into R, you need to check what map projection is associated with it, and whether this is the same as the one associated with all the other data sets you have already imported. To do this, enter the following command into R:

```
summary(scene_nestbox_locations_shp)
```

This is CODE BLOCK 144 in the document R_CODE_DATA _VISUALISATION_WORKBOOK.DOC. Next, you need to convert your shapefile into an R object called a data frame. This is done using the `as.data.frame` command by entering the following code into R:

```
scene_nestbox_locations_df <-
as.data.frame(scene_nestbox_locations_shp)
```

This is CODE BLOCK 145 in the document R_CODE_DATA _VISUALISATION_WORKBOOK.DOC. Once the shapefile has been converted into a data frame object (in this case, one called `scene_nestbox_locations_df`), you need to rename the columns called `coords.x1` and `coords.x2` that have been added to this data frame as `x` and `y`. To do this, enter the following command into R:

```
setnames(scene_nestbox_locations_df,old=
c('coords.x1','coords.x2'),new=c('x','y'))
```

This is CODE BLOCK 146 in the document R_CODE_DATA _VISUALISATION_WORKBOOK.DOC. Finally, you need to view the R object containing the data frame version of your shapefile in order to check that it has been processed correctly. To do this, enter the following command into R:

```
View(scene_nestbox_locations_df)
```

This is CODE BLOCK 147 in the document R_CODE_DATA _VISUALISATION_WORKBOOK.DOC.

5. **Import any shapefiles containing point data into R using the 'fortify' approach**

263

Creating maps in R requires a lot of code, as a result, rather than building up a single block of code that will create a map all in one go, you will create an initial map (in this case, called `map_1`) which will be saved as an R object that you will then modify using additional commands and blocks of code. To do this for the data being used in this example, enter the following code into R:

```
map_1 <- ggplot() + geom_polygon(data=
    scene_oak_woodland_df,aes(x=x,y=y),
      colour="black",fill="green") +
geom_polygon(data= dubh_loch_df,aes(x=x,
    y=y),colour="black",fill="blue") +
    geom_path(data=loch_lomond_shoreline_df,
aes(x=x, y=y),colour="blue",size=0.75) +
    geom_path(data=scene_roads_df,aes(x=x,
        y=y),colour="black",size=1) +
geom_path(data=forestry_track_df,aes(x=x,
      y=y),colour="brown",size=0.5) +
geom_point(data=scene_nestbox_locations_df,
      aes(x=x, y=y),shape=21,size=1.5,
          colour="black",fill="red")
```

6. Create an initial map from the data sets you have just imported

This is CODE BLOCK 148 in the document R_CODE_DATA _VISUALISATION_WORKBOOK.DOC, and it contains seven commands separated by + symbols. These are the `ggplot` command, and then a separate graphing command for each of the data sets you wish to plot on your map. The `ggplot` command is used to create a blank map on to which the other commands will add various features. The `geom_polygon` command is used to plot data sets which consist of polygons (in this case, the data sets called `scene_oak_woodland_df` and `dubh_loch_df`), while the `geom_path` command is used to plot data sets which consist of lines (in this case, `loch_lomond_shoreline_ df`, `scene_roads_df` and `forestry_ track_df`) and the `geom_point` command is used to plot data sets which consist of point data (in this case, `scene_nestbox_ locations_df`). Within each of these commands, the `data` argument is used to specify the name of the data set to be plotted, while the `x` and `y` arguments in the `aes` element specify the columns containing the X and Y coordinates, respectively. As with other types of data visualisations created using these commands, the `colour`, `fill`, `size` and/or `shape` augments are used to define exactly how each data set will be displayed.

To view the initial map created by this block of code, enter the name of the object created by it into R. In this case, enter the name `map_1`.

7.	Set the extent of your map

When you first create an initial map (in this case, called map_1), you will no doubt be rather underwhelmed by its appearance. This is because it is unlikely to look right until you define the exact area you want it to cover. This is done by using the coord_cartesian command to set the extent of the final map. To do this for the map being created in this example, enter the following command into R:

```
map_1 <- map_1 +
coord_cartesian(xlim=c(236900,239000),
       ylim=c(694900,696600))
```

This is CODE BLOCK 149 in the document R_CODE_DATA _VISUALISATION_WORKBOOK.DOC. It contains the name of the map you wish to modify the extent of (in this case, it is the one called map_1 that you created in step 6) followed by a + symbol and the coord_cartesian command. This means that the output of this new command will modify the extent of the map created in step 6 (rather than generating a new map). In the coord_cartesian command, the xlim and ylim arguments are used to specify the east-west and north-south limits of your map, respectively. The values included in these arguments will be in the map units of the map projection that the original shapefiles were in (see the outputs of the summary commands in steps 3, 4 and 5). You can work out the best values to include in these arguments either by examining the spatial limits of the data sets you are plotting (again, this information can be found in the outputs of the summary commands from steps 3 to 5), or through a process of trial-and-error.

To view the updated map created by this block of code, enter the name of the object modified by it into R. In this case, enter the name map_1.

In order to show which part of the world a map represents, you need to add what is known as a graticule around its edge. This is a scale that can be used to read off the coordinates for different locations on the map. For this example, you will do this by adding axes to the left, right, top and bottom of your map and formatting them. This will create a graticule based on the map units of the map projection of your original shapefiles (see the outputs of the `summary` commands in steps 3, 4 and 5). If you wish to create a graticule that uses latitude and longitude instead, you will need to import your shapefiles using the 'sf' approach and add it using the `coord_sf` command instead (see Exercise 5.2 for more details). To add a graticule to the map you are creating in this example using the map units of the projection/coordinate system of the original shapefiles, enter the following block of code into R:

```
map_1 <- map_1 + scale_y_continuous(name=
    "",sec.axis=sec_axis(~./1,name="")) +
   scale_x_continuous(name="",sec.axis=
sec_axis(~./1,name="")) + theme(axis.text.y=
(element_text(angle=90,vjust=0,hjust=0.5)),
  axis.text.y.right=(element_text(angle=-90,
       vjust=0,hjust=0.5)),axis.line=
     element_line(colour="black",size=1))
```

8. Add a graticule around the edge of your map to show which part of the world it represents

This is CODE BLOCK 150 in the document R_CODE_DATA _VISUALISATION_WORKBOOK.DOC. It contains the name of the map you wish to add a graticule to (in this case, `map_1`) followed by a + symbol and then three different commands. This means that the output of these new commands will be added to the map updated in step 7 (rather than generating a new map). The commands included in this block of code are `scale_y_continuous`, `scale_x_continuous` and a `theme` style command. The `scale_y_continuous` command contains two arguments. The first of these is `name`, and this has no value, meaning that no name is added to the Y axes of the map. The second is `sec.axis`, and this is used to add a Y axis to both the left and right hand sides of your map.

The `scale_x_continuous` command contains two arguments. The first of these is `name`, and this has no value, meaning that no name is added to the X axes of the map. The second is `sec.axis`, and this is used to add a X axis to both the top and bottom of your map.

Finally, the `theme` style command contains arguments which set the colour and thickness of the lines representing each axis, and modify the position of the text on the Y axes.

To view the updated map created by this block of code, enter the name of the object modified by it into R. In this case, enter the name `map_1`.

266

```
9.    Format
      the background of
      your map
```

After you have added a graticule around the edge of your map, you need to format its background. In general, you will want to change the default values so that the background area of the map is blank and there are no grid lines on it. To do this for the map you are creating in this example, enter the following block of code into R:

```
map_1 <-map_1 + theme(panel.background=
    element_blank(),panel.grid.major=
    element_line(colour=NA,size=0.5),
panel.grid.minor=element_line(colour=NA,
        size=0.5),panel.ontop=TRUE)
```

This is CODE BLOCK 151 in the document R_CODE_DATA _VISUALISATION_WORKBOOK.DOC. It contains the name of the map you wish to format the background of (in this case, `map_1`) followed by a + symbol and a single `theme` command which contains three different style arguments. The `panel.background=element_blank()` argument removes the existing background from the map, while the two `panel.grid` arguments include the term `colour=NA` to remove the existing gridlines from the map. **NOTE:** If you wish to display gridlines on your map, change the `colour` term in one or both of these arguments to "`black`" or the name of any other colour you wish them to be drawn in.

To view the updated map created by this block of code, enter the name of the object modified by it into R. In this case, enter the name `map_1`.

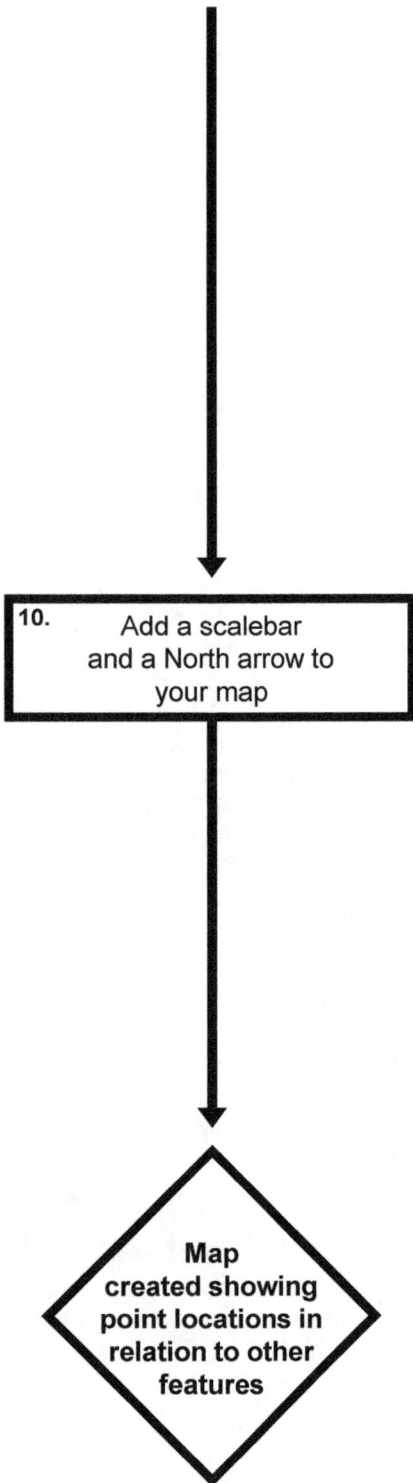

The final elements you need to add to your map are a scalebar and a North arrow. These will help anyone looking at it to understand its contents better. To do this for the map you are creating in this example, enter the following block of code into R:

```
map_1 <- map_1 + annotation_scale(
plot_unit="m",bar_cols=c("black","white"),
 location="bl") + annotation_north_arrow(
style=north_arrow_orienteering,which_north=
"grid",height=unit(1,"cm"),width=unit(0.75,
             "cm"),location="tr")
```

This is CODE BLOCK 152 in the document R_CODE_DATA _VISUALISATION_WORKBOOK.DOC. It contains the name of the map you wish to add a scalebar and a North arrow to (in this case, `map_1`) followed by a + symbol and then two `annotation` commands from the `ggspatial` package installed into R in step 2 of this flow diagram. These are the `annotation_scale` command, which adds a scalebar, and `annotation_north_arrow` command which adds a North arrow. For the `annotation_scale` command, the `plot_unit` argument is use to set the units for the scale bar (in this case, `"m"` for metres), while the `bar_cols` argument sets to colours that will be used for it. The `location` argument is then used to determine the position of the scalebar on your map. The options for this argument are `tl` (top left), `tr` (top right), `bl` (bottom left) and `br` (bottom right). In this case, the option used is `bl` meaning the scalebar is positions in the bottom left corner of the map.

For the `annotation_north_arrow` command, the `style` argument sets the style of the North arrow which will be added to the map, while the `which_north` argument lets you select whether the arrow points at true north or grid north. The `height` and `width` arguments determine the size of the North arrow and, as with the `annotation_scalebar` command, the `location` argument is used to determine the position of the North arrow on your map.

To view the updated map created by this block of code, enter the name of the object modified by it into R. In this case, enter the name `map_1`.

10. Add a scalebar and a North arrow to your map

Map created showing point locations in relation to other features

The final map created by working through this flow diagram should look like the image at the top of the next page.

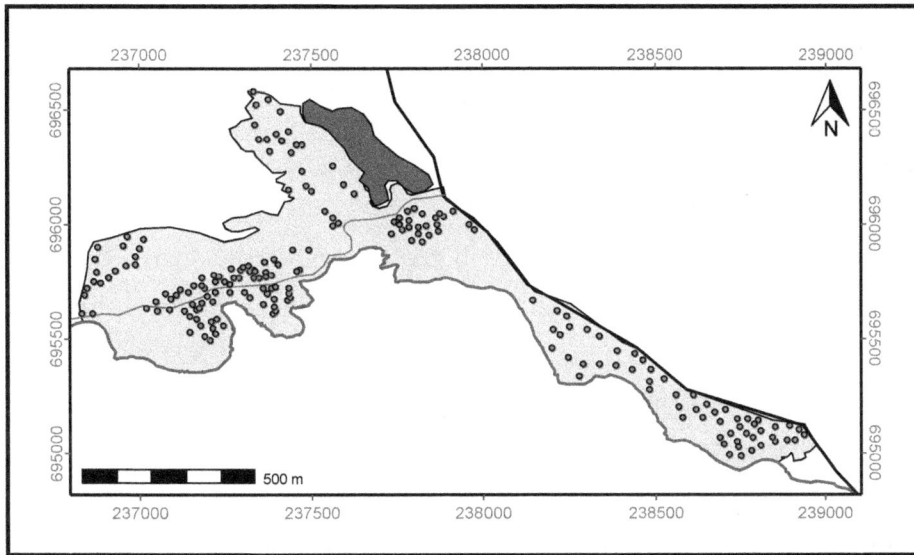

Once you have created a map to show the distribution of point locations in relation to other features, you can export it from R so that you can include it in a manuscript or presentation. If you are using RGUI, you can do this by clicking on the R GRAPHICS window containing your map to select it, before clicking on FILE on the main menu bar and selecting SAVE AS. This will allow you to save it in a variety of different formats. If you are using RStudio, you can export your map by clicking on the EXPORT button at the top of the window displaying it and selecting SAVE AS IMAGE. Alternatively, if you wish to save a high resolution version of your map, you can use the `ggsave` command from the `ggplot2` package. This command will let you save your map using a specific resolution, size and format. Information on how to use this command can be found at *www.rdocumentation.org/packages/ggplot2/version3.3.3/topics/ggsave.*

When including a map in a manuscript, it is important that you provide an appropriate figure legend for it. This legend should provide all the information required for the reader to interpret its contents. For the above map, an appropriate legend would be:

***Figure 1:** The location of nest boxes used to study the breeding success of hole-nesting birds in an area of native oak woodland in Central Scotland. Red points: nest box locations; Green polygon: the area of woodland; Blue polygon: the Dubh Loch; Blue line: The shoreline of Loch Lomond; Black line: Local roads; Brown line: A local forestry track. The coordinates are in the British National Grid map projection.*

All the spatial data used to create the above map were stored as shapefiles. However, there will be times when you wish to plot point locations which are held as a list of coordinates in a spreadsheet or table rather than in a specific spatial data format. For example, it may be that you have locational data which are not in a pre-existing shapefile, but instead have been recorded with a GPS receiver and entered into a spreadsheet. Such point data can easily be added to a map, but the exact way that you will do this will depend on whether your data are provided in the coordinates of a specific map projection (such as the British National Grid map projection used for the above example) or in latitude and longitude. This is done by varying the code used to import the data for point locations in step 5 of the above flow diagram. Details of the code required to import point locations from a shapefile as well as from a list of coordinates in the same map projection as your map and in latitude and longitude are provided in the table below.

Data storage format	Code required to import point locations into R so they can be added to a map
| Shapefile in same map projection as the rest of your spatial data | In order to import point data saved in a shapefile with the same map projection as the rest of your spatial data, you can use the 'fortify' approach. In this approach, the `readORG` command is used to import the data into R as a shapefile. The `summary` command is then used to check the map projection of the data, before the `as.data.frame` command is used to convert the shapefile into a data frame containing a list of coordinates that can be plotted with the `geom_point` graphing command. Finally, the `setnames` command is use to change the names of the columns used to store the coordinates, before the `View` command is use to view the final data set to allow you to check it has been imported correctly. For example, to import a shapefile called `scene_nestbox_locations.shp` into R using the 'fortify' approach, the code block would look like this (**NOTE:** This code block assumes that the shapefile you are importing is stored in the WORKING DIRECTORY of your analysis project):<br><br>```R
scene_nestbox_locations_shp <- readOGR(dsn=".",
        layer="scene_nestbox_locations")
summary(scene_nestbox_locations_shp)
       scene_nestbox_locations_df <-
  as.data.frame(scene_nestbox_locations_shp)
setnames(scene_nestbox_locations_df,old=
   c('coords.x1','coords.x2'),new=c('x','y'))
       View(scene_nestbox_locations_df)
``` |
```

Data storage format	Code required to import point locations into R so they can be added to a map
Shapefile in same map projection as the rest of your spatial data	Point locations stored in a shapefile with the same map projection as the rest of your spatial data can also be imported using the 'sf' approach. In this approach, the `read_sf` command from the `sf` package is used to import the data into R as a simple feature data set. The resulting data set can then be plotted on a map using the `geom_sf` command. For example, to import a shapefile called `scene_nestbox_locations.shp` into R using the 'sf' approach, the code would look like this: `scene_nestbox_locations_sf <- read_sf(dsn=".",layer= "scene_nestbox_locations")`

List of coordinates in latitude and longitude	Point locations stored as a list of latitude and longitude coordinates can be imported by first using a `read.table` command before using the `st_to_sf` command from the `sf` package to convert the resulting table object to a simple feature object. This object is then transformed into the required map projection using the `st_transform` command, before the `st_coordinates` command is used to add coordinates in the new map projection. Next, the `cbind` command is used to join the data from the original table created by the `read.table` command to this list of coordinates before the `setnames` command is used to rename the columns that store the new X and Y coordinates. Finally, the `View` command allows you to examine the final data set to ensure this process has been carried out correctly. These data can then be plotted using the `geom_point` graphing command. For example, to import a data set containing latitude and longitude coordinates called `nestbox_locations_lat_long.csv` and transform these coordinates into the British National Grid map projection, the code would look like this:
	
	`nestbox_locations_lat_long <- read.table(file="nestbox_ locations_lat_long.csv",sep=",",header=TRUE)`
	`nestbox_locations_lat_long_sf <- st_as_sf(nestbox_locations_lat_long,coords=c("Longitude", "Latitude"),crs=4326)`
	`nestbox_locations_osgb36_sf <- st_transform( nestbox_locations_lat_long_sf,27700)`
	`nestbox_locations_osgb36_coords <- st_coordinates( nestbox_locations_ogsb36_sf)`
	`nestbox_locations_osgb36_coords <- cbind(nestbox_ locations_osgb36_coords,nestbox_locations_lat_long)`
	`setnames(nestbox_locations_osgb36_coords,old=c('X','Y'), new=c('x','y'))`
	`View(nestbox_locations_osgb36_coords)`
	
	NOTE: In order to be able to transform latitude and longitude coordinates into the coordinates of a different map projection, you will need to know the EPSG code for this projection and include it at the end of the `st_transform` command. For the British National Grid map projection, this code is 27700. You will also need to include the EPSG code for the original map projection your data were recorded in the `crs` argument of the `st_as_sf` command. For the WGS 84 datum, this code is 4326. The EPSG code for different map projections and datums can be found on the *epsg.io* website.

Data storage format	Code required to import point locations into R so they can be added to a map

List of coordinates in same map projection as the rest of your spatial data	Point locations stored as a list of coordinates in the same map projection as the rest of your spatial data can be imported using the standard `read.table` command and then plotted with the `geom_point` command. For example, to import a table called `nestbox_locations.csv` with a list of coordinates in the British National Grid map projection into your analysis project, the code would look like this:
	
	`nestbox_locations <- read.table(file=`
	`"nestbox_locations.csv",sep=",",header=TRUE)`
	
	NOTE: In order to be able to plot these data using the `geom_point` command, the X and the Y coordinates will need to be stored in different columns in the original data table.

For the next part of this exercise, you will customise the workflow from the above flow diagram to explore how to import point locations stored in different formats and with coordinates from different map projections into R so that they can be plotted on a map. This will be done by modifying how the point data are imported in step 5. Firstly, you will explore how to import a point data set stored in a spreadsheet or table where the X and Y coordinates are in the same map projection as the other spatial data you wish to plot on your map. In this case, this is the British National Grid map projection. The data set that you will import is held in a comma separated values file called `nestbox_locations.csv`. To import this data set into R and use it to create a map similar to the one created in the original workflow, you first need to enter the following `read.table` command into R:

```
nestbox_locations <-
read.table(file="nestbox_locations.csv",sep=",",
header=TRUE)
```

Once you have run this `read.table` command, you need to check that the data have been imported correctly. This can be done by entering the following command into R:

```
View(nestbox_locations)
```

You can also use this command to allow you to identify the names of the columns which contain the X and Y coordinates that you will use to plot these data. In this case, they are columns called `x` and `y`, respectively. Once you have these point location data imported, you can then make your map using them and the existing shapefiles you imported in earlier steps of the above work flow. As before, you start by creating an initial map which displays

272

the polygon, line data and points you wish to display on it. To do this, edit the command from step 6 of the above flow diagram (this is CODE BLOCK 148 from the document R_CODE_DATA_VISUALISATION_WORKBOOK.DOC) so that it looks like this (the required modifications are highlighted in **bold**):

```
map_2 <- ggplot() + geom_polygon(data=
scene_oak_woodland_df,aes(x=x, y=y),colour="black",
fill="green") + geom_polygon(data=dubh_loch_df,aes(x=x,
y=y),colour="black",fill="blue") + geom_path(data=
loch_lomond_shoreline_df,aes(x=x,y=y),colour="blue",
size=0.75) + geom_path(data=scene_roads_df,aes(x=x,y=y),
colour="black",size=1) + geom_path(data=forestry_track_df,
aes(x=x,y=y),colour="brown",size=0.5) + geom_point(
data=nestbox_locations,aes(x=x,y=y),shape=21,size=1.5,
colour="black",fill="red")
```

The code from the remaining steps (steps 7 to 10) can then be run as before, with the only the name of the map needing to be updated from map_1 to map_2. Thus, these blocks of code needs to be edited to look like this (the required modifications are highlighted in **bold**):

```
map_2 <- map_2 + coord_cartesian(xlim=c(236900,239000),
ylim=c(694900,696600))

map_2 <- map_2 + scale_y_continuous(name="",sec.axis=
sec_axis(~./1,name="")) + scale_x_continuous(name="",
sec.axis=sec_axis(~./1,name="")) + theme(axis.text.y=
(element_text(angle=90,vjust=0,hjust=0.5)),
axis.text.y.right=(element_text(angle=-90,vjust=0,
hjust=0.5)),axis.line=element_line(colour="black",size=1))

map_2 <- map_2 + theme(panel.background=
element_blank(),panel.grid.major=element_line(colour=NA,
size=0.5),panel.grid.minor=element_line(colour=NA,
size=0.5),panel.ontop=TRUE)

map_2 <- map_2 + annotation_scale(plot_unit="m",
bar_cols=c("black","white"),location="bl") +
annotation_north_arrow(style=north_arrow_orienteering,
which_north="grid",height=unit(1,"cm"),width=unit(0.75,
"cm"),location="tr")
```

Once you have run each of these edited blocks of code, you can enter the name `map_2` into R in order to display your new map. It should look identical to the map shown above which was produced by working through the main flow diagram.

Next, you will create a third version of this map, this time using data stored in a table where the coordinates are in latitude and longitude rather than in the British National Grid map projection being used by the other data sets. This means that these coordinates need to be transformed into the British National Grid map projection before they can be used to make the map. These data are held in a comma separated values file called `nestbox_locations_lat_long.csv`. To import this data set into R and use it to create a map similar to the one created in the original workflow, you first need to enter the following `read.table` command into R:

```
nestbox_locations_lat_long <- read.table(file=
"nestbox_locations_lat_long.csv",sep=",",header=TRUE)
```

After you have run this `read.table` command, you need ensure that you have the `sf` package installed in your version of R. To do this, enter the command `library()` into R. If the `sf` package is not listed in the R PACKAGES AVAILABLE window that opens, you can download and install it by entering the following command into R:

```
install.packages("sf")
```

Once you have ensured that the `sf` package is installed in your version of R, you can load its command library into your analysis project by entering the following command into R:

```
library(sf)
```

You can then use the `st_to_sf` command from this package to convert the R object created by the above `read.table` command into a simple feature R object by entering the following code into R:

```
nestbox_locations_lat_long_sf <- st_as_sf(
nestbox_locations_lat_long,coords=c("Longitude",
"Latitude"),crs=4326)
```

The simple feature object created by this command can then be transformed into British National Grid map projection (which as the EPSG code 27700) using the st_transform command by entering the following code into R:

```
nestbox_locations_osgb36_sf <- st_transform(
    nestbox_locations_lat_long_sf,27700)
```

Next, you need to extract the coordinates for each point in the map projection that you have just transformed your data into. To do this, enter the following command into R:

```
nestbox_locations_osgb36_coords <- st_coordinates(
    nestbox_locations_osgb36_sf)
```

The object created by this st_coordinates command will only have the X and Y coordinates of the new map projection in it, and not any of the other data from your original data set. To join the original data to this new table of coordinates, you need to use the cbind command. To do this, enter the following command into R:

```
nestbox_locations_osgb36_coords <-cbind(
nestbox_locations_osgb36_coords,nestbox_locations_lat_long)
```

Finally, you need to update the names of the columns containing the X and Y coordinates of the projected coordinate system so that they are called x and y. To do this, enter the following command into R:

```
setnames(nestbox_locations_osgb36_coords,old=c('X','Y'),
    new=c('x','y'))
```

At this point, you should view your data to ensure that they have been imported correctly. This can be done by entering the following View command into R:

```
View(nestbox_locations_osgb36_coords)
```

Once you have these data imported, you can then make your map using both the existing shapefiles you imported in earlier steps of the above work flow and the point data layer that you just imported and transformed from latitude and longitude coordinates to British National Grid coordinates. As before, you start by creating an initial map which displays the polygon, line and points data sets you wish to display on it. To do this, edit the block of code from step 6 of the above flow diagram (this is CODE BLOCK 148 from the document R_CODE_DATA_VISUALISATION_WORKBOOK.DOC) so that it looks like this (the required modifications are highlighted in **bold**):

```
map_3 <- ggplot() + geom_polygon(data=
scene_oak_woodland_df,aes(x=x,y=y),colour="black",
fill="green") + geom_polygon(data=dubh_loch_df,aes(x=x,
y=y),colour="black",fill="blue") + geom_path(data=
loch_lomond_shoreline_df,aes(x=x, y=y),colour="blue",
size=0.75) + geom_path(data=scene_roads_df,aes(x=x,
y=y),colour="black",size=1) + geom_path(data=
forestry_track_df,aes(x=x,y=y),colour="brown",size=0.5) +
geom_point(data=nestbox_locations_osgb36_coords,aes(x=x,
y=y),shape=21,size=1.5,colour="black",fill="red")
```

The code from the remaining steps (steps 7 to 10) can then be run as before, with the only the name of the map needing to be updated from map_1 to map_3. This means that the code for these steps needs to be edited so that they look like this (the required modifications are highlighted in **bold** – **NOTE**: If you wish, these can be run as a single block of code):

```
map_3 <- map_3 + coord_cartesian(xlim=c(236900,239000),
ylim=c(694900,696600))

map_3 <- map_3 + scale_y_continuous(name="",sec.axis=
sec_axis(~./1, name="")) + scale_x_continuous(name="",
sec.axis=sec_axis(~./1,name=""))

map_3 <- map_3 + theme(axis.text.y=(element_text(angle=90,
vjust=0,hjust=0.5)),axis.text.y.right=(element_text(
angle=-90,vjust=0,hjust=0.5)),axis.line=
element_line(colour="black",size=1))

map_3 <- map_3 + theme(panel.background=element_blank(),
panel.grid.major=element_line(colour=NA,size=0.5),
panel.grid.minor=element_line(colour=NA,size=0.5),
panel.ontop=TRUE)

map_3 <- map_3 + annotation_scale(plot_unit="m",bar_cols=
c("black","white"),location="bl") + annotation_north_arrow(
style=north_arrow_orienteering,which_north="grid",height=
unit(1,"cm"),width=unit(0.75,"cm"),location="tr")
```

Once you have run this code, you can enter the name map_3 into R in order to display your new map. It should look identical to the map shown above that was produced by working through the main flow diagram.

If you wish to use different symbols on your map for different groups in a point data set, you can do this by using the subset command to create separate objects for each group of data which can then be plotted separately using different geom_point commands. For example, to make a map that shows which nest boxes were occupied by a particular species of bird, the blue tit, and which were not, you can use the information in a column called BT_OCC in the nestbox_locations_osgb36_coords data set. To do this, first enter the following two subset commands into R:

```
occupied_boxes <- subset(nestbox_locations_osgb36_coords,
                BT_OCC=="1")
unoccupied_boxes <- subset(nestbox_locations_osgb36_coords,
                BT_OCC=="0")
```

These commands will create two new data sets, one called occupied_boxes which contain the nest boxes occupied by blue tits (which have a value of 1 in the column called BT_OCC), and one called unoccupied_boxes which contains the unoccupied nest boxes (these have a value of 0 in the BT_OCC column). Next, you can then remake your initial map by editing the block of code from step 6 in the above flow diagram so that it includes two geom_point commands, one for each subset. The edited block of code should look like this (the required modifications are highlighted in **bold**).

```
Map_4 <- ggplot() + geom_polygon(data=
scene_oak_woodland_df,aes(x=x,y=y),colour="black",
fill="green") + geom_polygon(data=dubh_loch_df,aes(x=x,
y=y),colour="black",fill="blue") + geom_path(data=
loch_lomond_shoreline_df,aes(x=x,y=y),colour="blue",
size=0.75) + geom_path(data=scene_roads_df,aes(x=x,
y=y),colour="black",size=1) + geom_path(data=
forestry_track_df,aes(x=x,y=y),colour="brown",
size=0.5) + geom_point(data=unoccupied_boxes,aes(x=x,y=y),
shape=21,size=1.5,colour="black",fill="red") +
geom_point(data=occupied_boxes,aes(x=x,y=y),shape=21,
size=3,colour="black",fill="blue")
```

The code from the remaining steps (steps 7 to 10) can then be run as before, with the only the name of the map needing to be updated from `map_1` to `map_4`. This means that the code for these steps needs to be edited so that they look like this (the required modifications are highlighted in **bold** – **<u>NOTE</u>**: If you wish, these can be run as a single block of code):

```
map_4 <- map_4 + coord_cartesian(xlim=c(236900,239000),
          ylim=c(694900,696600))

map_4 <- map_4 + scale_y_continuous(name="",sec.axis=
sec_axis(~./1,name="")) + scale_x_continuous(name="",
sec.axis=sec_axis(~./1,name="")) + theme(axis.text.y=
      (element_text(angle=90,vjust=0,hjust=0.5)),
   axis.text.y.right=(element_text(angle=-90,vjust=0,
hjust=0.5)),axis.line=element_line(colour="black",size=1))

map_4 <- map_4 + theme(panel.background=
element_blank(),panel.grid.major=element_line(colour=NA,
   size=0.5),panel.grid.minor=element_line(colour=NA,
           size=0.5),panel.ontop=TRUE)

map_4 <- map_4 + annotation_scale(plot_unit="m",
   bar_cols=c("black","white"),location="bl") +
annotation_north_arrow(style=north_arrow_orienteering,
which_north="grid",height=unit(1, "cm"),width=unit(0.75,
              "cm"),location="tr")
```

Once you have run each of these blocks of code, you can enter the name `map_4` into R in order to display your new map. It should look like this:

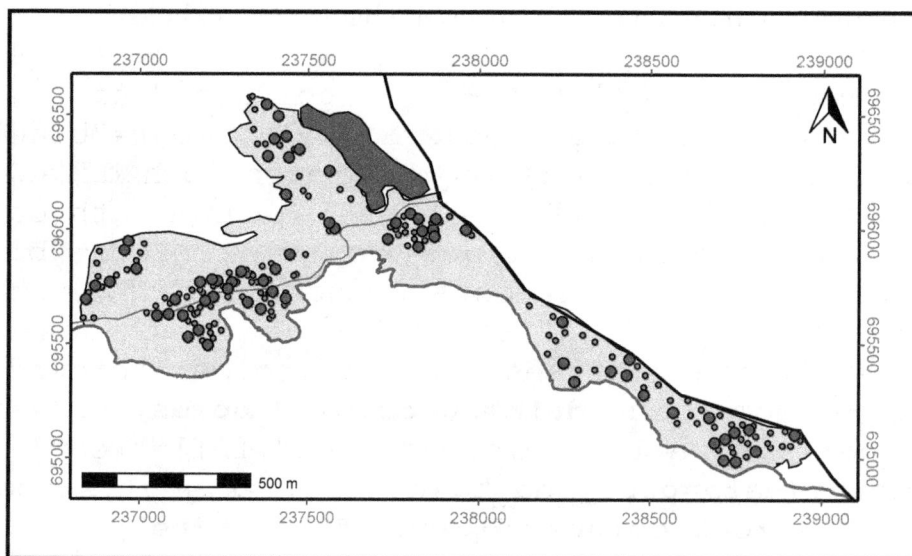

EXERCISE 5.2: HOW TO MAKE A MAP WHICH USES PIE CHARTS AS MARKERS FOR A POINT LOCATION DATA SET:

When you are plotting point locations, you will usually use points to display them. However, you can also create maps which use pie charts as the markers for point spatial data. This allows you to display additional data on your map for each point. For example, you can use the pie chart markers to display the biodiversity of species recorded at different points in space, the prevalence of different diseases, the demographics of different populations or any other similar variables. This is done using a graphing command called `geom_scatterpie`, which is part of the `scatterpie` package. In this exercise, you will use this command to create a map showing the diversity of cetacean species recorded during surveys in five oceanographically distinct areas in the northern North Sea using pie charts as markers for the centroids (the central geographic point) of each of these areas. In this exercise, you will use the 'sf' approach for importing spatial data held in shapefiles into R. This will give you more experience with using this approach. To create this map, work through the flow diagram that starts on the next page.

NOTE: When importing a shapefile into R using the 'sf' approach (which stands for Simple Feature), use the suffix `_sf` in the name of the object created by the simple feature import command (see step 3) to allow you to easily identify its data format, and separate it from other data formats used when making maps, such as shapefiles (which should have the suffix `_shp`) and tables or data frames (which should have the suffix `_df`).

NOTE: This work flow assumes that you have the `ggplot2`, `Rcpp`, `rgdal`, `ggmap`, `data.table`, `proj4`, `ggspatial` and `sf` packages installed in your version of R and that their command libraries has been loaded into your analysis project. If you have not already done this, you will need to do it before working through this flow diagram. Instructions for how to do this for the first seven of these packages can be found in the flow diagram for Exercise 5.1, while the instructions for the `sf` package can be found on page 274.

Data sets held in shapefiles or other spatial data formats

For this example, you will use three data sets that are stored in the shapefile spatial data format. These are called `north_sea_study_area.shp`, `north_sea_land.shp` and `study_area_centroids.shp`. These shapefiles are located in the WORKING DIRECTORY folder you created during the introduction to this chapter.

1. Set the WORKING DIRECTORY for your analysis project

Before you start any analysis in R, you first need to set the WORKING DIRECTORY. To do this, enter the text `setwd("` and then type the address of your WORKING DIRECTORY, using slashes (/) as the folder separators, before entering a second quotation mark followed by a closing bracket, like this `")`. For example, if your WORKING DIRECTORY has the address C:\STATS_FOR_BIOLOGISTS_TWO, your `setwd` command should look like this:

```
setwd("C:/STATS_FOR_BIOLOGISTS_TWO")
```

If you are using RGUI, enter your `setwd` command in the R CONSOLE window (remembering to use the address of your own WORKING DIRECTORY folder in it) and then press the ENTER key on your keyboard. If you are using RStudio, enter your `setwd` command into the SCRIPT EDITOR window. To run it, select it and then click on the RUN button at the top of this window. You will enter all the remaining commands for this exercise in a similar manner, depending on the user interface you are using.

To check that your WORKING DIRECTORY has been set properly, enter the command `getwd()` and carefully check that the address it returns is the same as the one for the STATS_FOR_BIOLOGISTS_TWO folder you created at the start of this chapter.

Before you move on to step 2, make sure that all the data you wish to use in your analysis project are located in this WORKING DIRECTORY folder. In this case, these are the files called `north_sea_study_area.shp`, `north_sea_land.shp` and `study_area_centroids.shp`. **NOTE:** If the data you are going to import into R in steps 2 and 3 of this flow diagram are not located in the WORKING DIRECTORY you set in this step, the code provided for importing them will not work.

2. Import any polygon and line shapefiles you wish to have on your map into R using the 'sf' approach

Once you have set your WORKING DIRECTORY, you are ready to import any polygon and line shapefiles you wish to display on your map into R. For this exercise, you will do this using the 'sf' approach. To import a polygon shapefile with the outlines of the five oceanographically distinct areas in the northern North Sea, enter the following command into R:

```
north_sea_study_area_sf <- read_sf(dsn=".",
      layer="north_sea_study_area")
```

This is CODE BLOCK 153 in the document R_CODE_ DATA_VISUALISATION_WORKBOOK.DOC. This command will create a simple feature object in R called `north_sea_study_area_sf` containing the data from the shapefile specified in the `layer` argument.

You can now repeat this process for a polygon shapefile which shows the areas of land in the study area. To do this, enter the following command into R:

```
north_sea_land_sf <- read_sf(dsn=".",
      layer="north_sea_land")
```

This is CODE BLOCK 154 in the document R_CODE_ DATA_VISUALISATION_WORKBOOK.DOC. This command will create a simple feature object in R called `north_sea_land_sf` containing the data from the specified shapefile.

To import the shapefile containing the point locations marking the centroid of each part of the study area, you will need to use the 'fortify' approach for point data (see step 5 in Exercise 5.1). This uses the `as.data.frame` command instead of the fortify command. The shapefile containing the point data you wish to import in this example is called `study_area_centroids.shp`. To import this point shapefile into your analysis project, enter the following block of code, consisting of four separate commands, into R:

```
study_area_centroids_shp <- readOGR(dsn=".",
        layer="study_area_centroids")

    summary(study_area_centroids_shp)

        study_area_centroids_df <-
    as.data.frame(study_area_centroids_shp)

    setnames(study_area_centroids_df,old=
  c('coords.x1','coords.x2'),new=c('x','y'))
```

This is CODE BLOCK 155 in the document R_CODE_DATA_VISUALISATION_WORKBOOK.DOC. Once your data have been imported, you need to check that all the columns which contain numeric data have been classified as such. To do this, enter the following command into R:

```
str(study_area_centroids_df)
```

This is CODE BLOCK 156 in the document R_CODE_DATA_VISUALISATION_WORKBOOK.DOC. You now need to re-classify any mis-classified columns containing numeric data using the `as.numeric` command. In this case, you need to re-classify the columns called `bnd`, `cd`, `hp`, `mw` and `wbd`. To do this, enter the following block of code, containing five separate `as.numeric` commands, into R:

```
        study_area_centroids_df$bnd <-
    as.numeric(study_area_centroids_df$bnd)
        study_area_centroids_df$cd <-
    as.numeric(study_area_centroids_df$cd)
        study_area_centroids_df$hp <-
    as.numeric(study_area_centroids_df$hp)
        study_area_centroids_df$mw <-
    as.numeric(study_area_centroids_df$mw)
        study_area_centroids_df$wbd <-
    as.numeric(study_area_centroids_df$wbd)
```

This is CODE BLOCK 157 in the document R_CODE_DATA_VISUALISATION_WORKBOOK.DOC. Finally, you need to view the R object containing the data frame version of your shapefile in order to check that it has been processed correctly. To do this, enter the following command into R:

```
View(study_area_centroids_df)
```

This is CODE BLOCK 158 in the document R_CODE_DATA_VISUALISATION_WORKBOOK.DOC.

3. Import the shapefile containing point data you wish to display using pie charts

4. **Ensure that the `scatterpie` package is installed in your version of R and load its command library into your analysis project**

In order to be able to create a map of point locations which use pie charts as markers, you will need to ensure you have the `scatterpie` package installed in your version of R. To do this, enter the command `library()` into R. If the `scatterpie` package is not listed in the R PACKAGES AVAILABLE window which opens, you can install it by entering the following command into R:

```
install.packages("scatterpie")
```

This is CODE BLOCK 159 in the document R_CODE_DATA _VISUALISATION_ WORKBOOK.DOC.

Once you have ensured that the `scatterpie` packages is installed in your version of R, you will need to load its associated command library into your analysis project. To do this, enter the following block of code into R:

```
library(scatterpie)
```

This is CODE BLOCK 160 in the document R_CODE_DATA _VISUALISATION _WORKBOOK.DOC.

283

```
┌─────────────────────────────┐
│ 5.    Create an initial map │
│  displaying the spatial data you │
│       have just imported    │
└─────────────────────────────┘
```

Creating maps in R requires a lot of code, as a result, rather than building up a single block of code that will create a map all in one go, you will create an initial map (in this case, called `scatterpie_1`) that will be saved as an R object which you will then modify using additional commands and blocks of code. To do this for the data being used in this example, enter the following code into R:

```
scatterpie_map_1 <- ggplot() +
geom_sf(data=north_sea_study_area_sf,colour=
    "black",fill=NA) + geom_sf(data=
    north_sea_land_sf,colour="black",
              fill="green")
```

This is CODE BLOCK 161 in the document R_CODE_ DATA_VISUALISATION_WORKBOOK.DOC, and it contains three commands separated by + symbols. These are the `ggplot` command, and then a separate graphing command for each of the data sets you wish to plot on your map. In this case, as these are simple feature data sets, these are two `geom_sf` commands, one for the data set called `north_sea_study_area` and another for the data set called `north_sea_land`. Within each of these commands, the `data` argument is used to specify the name of the data set to be plotted. **NOTE:** When plotting simple feature data sets using the `geom_sf` command, there is no need to include arguments to specify the columns containing the X and Y coordinates. As with other graphing commands, the `colour` and `fill` augments are used to define exactly how each data set will be displayed.

To view the initial map created by this block of code, enter the name of the object created by it into R. In this case, enter the name `scatterpie_map_1`.

284

When you create your initial map, you will no doubt be rather underwhelmed by its appearance. This is because it is unlikely to look right until you define the exact area you want it to cover. As you have imported your polygon shapefiles into R using the 'sf' approach, this is done by using the `coord_sf` command from the `sf` package. To do this for the map being created in this example, enter the following block of code into R:

```
scatterpie_map_1 <- scatterpie_map_1 +
   coord_sf(xlim=c(-150000,70000),ylim=
c(50000,400000),expand=TRUE,crs=NULL,datum=
sf::st_crs(4326),label_axes="ENEN",ndiscr=
                100,clip="on")
```

This is CODE BLOCK 162 in the document R_CODE_DATA_VISUALISATION_WORKBOOK.DOC. It contains the name of the map you wish to change the extent of (in this case, it is the one called `scatterpie_map_1` which you created in step 5) followed by a + symbol and the `coord_sf` command. This means that this new command will be applied to the map created in step 5 (rather than generating a new map). In the `coord_sf` command, the `xlim` and `ylim` arguments are used to specify the east-west and north-south limits of your map, respectively. The values included in these arguments will be in the units of the map projection that the original shapefiles were in (this information can be found by running a `summary` command using the name of the R object containing the simple feature spatial data set). You can work out the best values to include in these arguments either by examining the spatial limits of the data sets you are plotting (again, this information can be found by using a `summary` command), or through a process of trial-and-error.

The `coord_sf` command can also be used to set the map projection which will be used for the graticule which will be added around the edge of your final map. In this case, by including the `sf::st_crs(4236)` argument in this command, the graticule will display latitude and longitude values rather than coordinates of the map projection used by the original polygon and line shapefiles.

To view the updated map created by this block of code, enter the name of the object created by it into R. In this case, enter the name `scatterpie_map_1`.

6. Set the extent and coordinate system for the graticule of your final map

In order to display point data on a map with pie chart markers, you need to use the `geom_scatterpie` command from the `scatterpie` package. In this example, you wish to plot the point data called `study_area_centroids_df` with pie chart markers which show the proportion of all cetaceans recorded in each part of the study area that belong to different species. To do this, enter the following block of code into R:

```
scatterpie_map_1 <- scatterpie_map_1 +
      geom_scatterpie(data=
   study_area_centroids_df,aes(x=x,
      y=y,group=region,r=20000),
        cols=c("bnd","cd","hp",
     "mw","wbd"),colour="black",
              alpha=0.8)
```

This is CODE BLOCK 163 in the document R_CODE_DATA_VISUALISATION_WORKBOOK.DOC. It contains the name of the map you wish to add the new data to (in this case, `scatterpie_map_1`) followed by a + symbol and then the `geom_scatterpie` command. This means that the output of this command will be added to this existing map (rather than generating a new map). In the `scatterpie` command, the `data` argument is used to identify the data which will be added to the map as points with pie chart markers. In this case, this is the data set called `study_area_centroids_df`.

In the `aes` element of this command, the `x` and `y` arguments are used to identify the columns containing the X and Y coordinates for the data, while the `group` argument is used to identify the column which contains the value that identifies which group each row of data in the data set belongs to. In this case, it is a column called `region`. The `r` argument sets the size of the pie charts which will be displayed on the map. This is in the units of the map projection that the map is being made in (i.e. that of the spatial data sets which are being plotted on it). In this case, this is metres, and a value of `20000` (or 20,000 metres) is used in order to ensure the pie charts are large enough to be read without being so large that they will cover each other up. **NOTE:** If you find that your pie chart markers are too big or too small for your map, you can change the value in this `r` argument to make them a more suitable size.

The `cols` argument is used to identify the columns which contain the values that will be used to determine the width of each pie slice, while the `colour` argument determines the outline colour for the pies and the `alpha` argument determines their level of transparency. In this case, the `alpha` argument has a value of `0.8`, meaning that they are slightly transparent, so you can see any other data they spatially overlap with.

To view the updated map created by this block of code, enter the name of the object created by it into R. In this case, enter the name `scatterpie_map_1`.

7. Add the data you wish to display with pie chart markers to your map

286

In order to show which part of the world a map represents, you need to add what is known as a graticule around the edge. This is a scale that can be used to read off the coordinates for different locations on the map. For this example, you will do this by adding axes the left, right, top and bottom of your map and formatting them. To add a graticule to the map you are creating in this example, enter the following block of code into R (**NOTE:** Due to the use of the argument `sf::st_crs(4236)` in the `coord_sf` command in step 6, the graticules will display latitude and longitude rather than the eastings and northings of the map projection of the data being displayed on the map):

```
scatterpie_map_1 <- scatterpie_map_1 +
  scale_y_continuous(name="",sec.axis=
      sec_axis(~./1,name="")) +
  scale_x_continuous(name="",sec.axis=
    sec_axis(~./1,name="")) + theme(
axis.text.y=(element_text(angle=90,vjust=0,
      hjust=0.5)),axis.text.y.right=
      (element_text(angle=-90,vjust=0,
  hjust=0.5)),axis.line=element_line(
          colour="black",size=1))
```

8. Add a graticule around the edge of your map to show which part of the world it represents

This is CODE BLOCK 164 in the document R_CODE_ DATA_VISUALISATION_WORKBOOK.DOC. It contains the name of the map you wish to add a graticule to (in this case, `scatterpie_map_1`) followed by a + symbol and then three different commands. This means that the output of these new commands will be added to this existing map (rather than generating a new map). The commands included in this block of code are `scale_y_continuous`, `scale_x_continuous` and a `theme` style command. The `scale_y_command` contains two arguments. The first of these is `name`, and this has no value, meaning that no name is added to the Y axes of the map. The second is `sec.axis`, and this is used to add a Y axis to both the left and right hand sides of your map.

The `scale_x_continuous` command contains two arguments. The first of these is `name`, and this has no value, meaning that no name is added to the X axes of the map. The second is `sec.axis`, and this is used to add a X axis to both the top and bottom of your map.

Finally, the `theme` style command contains arguments which set the colour and thickness of the lines representing each axis, and modify the position of the text on the Y axes.

To view the updated map created by this block of code, enter the name of the object created by it into R. In this case, enter the name `scatterpie_map_1`.

287

After you have added the graticule around the edge of your map, you need to format its background. In general, you will want to change the default values so that the background area of the map is blank and there are no grid lines on it. To do this for the map you are creating in this example, enter the following block of code into R:

```
scatterpie_map_1 <- scatterpie_map_1 +
theme(panel.background=element_blank(),
panel.grid.major=element_line(colour=NA,
        size=0.5),panel.grid.minor=
    element_line(colour=NA,size=0.5),
            panel.ontop=TRUE)
```

This is CODE BLOCK 165 in the document R_CODE_ DATA_VISUALISATION_WORKBOOK.DOC. It contains the name of the map you wish to format the background of (in this case, `scatterpie_map_1`) followed by a + symbol and then a single theme command which contains three different style arguments. The `panel.background= element_blank()` argument removes the existing background from the map, while the two `panel.grid` arguments include the term `colour=NA` to remove the existing gridlines from the map. **NOTE**: If you wish to display gridlines on your map, change the `colour` term in one or both of these arguments to `"black"` or the name of any other colour you wish them to be drawn in.

To view the updated map created by this block of code, enter the name of the object created by it into R. In this case, enter the name `scatterpie_map_1`.

9.	Format the background of your map

288

10. Change the title of the legend which shows what the colours on the pie charts represent

Finally, you need to change the title of the legend which shows what the colours on the pie charts represent. This is done using a `theme` command with the `legend.title` argument in it. In this example, you wish to remove the current title. To do this, enter the following block of code into R:

```
scatterpie_map_1 <- scatterpie_map_1 +
  theme(legend.title=element_blank())
```

This is CODE BLOCK 166 in the document R_CODE_ DATA_VISUALISATION_WORKBOOK.DOC. To view the updated map created by this block of code, enter the name of the object created by it into R. In this case, enter the name `scatterpie_map_1`.

Map created which uses pie charts as markers for point location data

The final map created by working through this flow diagram should look like this:

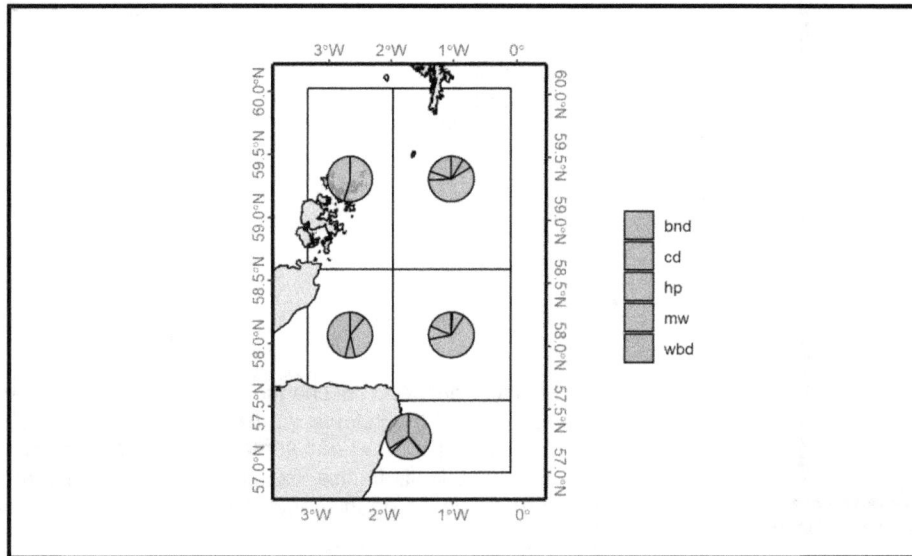

Once you have created a map with pie charts markers for a point data set, you can export it from R so that you can include it in a manuscript or presentation. If you are using RGUI, you can do this by clicking on the R GRAPHICS window containing your map to select it, before clicking on FILE on the main menu bar and selecting SAVE AS. This will allow you to save it in a variety of different formats. If you are using RStudio, you can export your map by clicking on the EXPORT button at the top of the window displaying it and selecting SAVE AS IMAGE. Alternatively, if you wish to save a high resolution version of your map, you can use the `ggsave` command from the `ggplot2` package. This command will let you save your map using a specific resolution, size and format. Information on how to use this command can be found at *www.rdocumentation.org/packages/ggplot2/version3.3.3/topics/ggsave.*

When including a 'scatterpie' map in a manuscript, it is important that you provide an appropriate figure legend for it. This legend should provide all the information required for the reader to interpret its contents. For the above map, an appropriate legend would be:

Figure 1: *The relative diversity of cetacean species in five oceanographically distinct regions in the northern North Sea based on the number of individuals recorded during a series of surveys. BND: Bottlenose dolphins; CD: Common dolphin; HP: Harbour porpoise; MW: Minke whale; WBD: White-beaked dolphin.*

There are a number ways that you can customise a 'scatterpie' map. Firstly, you can customise the colours used for the wedges on the pie chart markers for the different data points. This can be done with the `scale_fill_manual` command. In this command, you can specify the exact colours you wish to use for the wedges representing each column of data. To do this for the above example, you can enter a block of code containing this command that will update your existing map. The code you would need to add would look like this:

```
scatterpie_map_1 <- scatterpie_map_1 +
scale_fill_manual(values=c("blue","red","yellow",
                "purple","brown"))
```

When you run this new code block, it will modify the existing map called `scatterpie_map_1` created by working through the above flow diagram. In order to see the updated map, you need to enter the name `scatterpie_map_1` into R. When you do, you should get an updated map that looks like this:

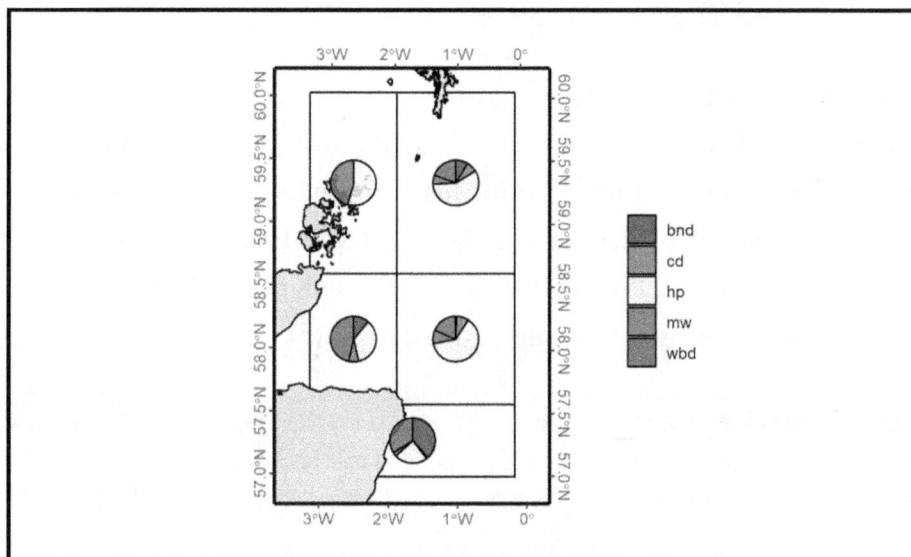

If you examine this new map, you will see that the colours used to represent the wedge for each species on the pie charts has been updated using the list of colours provided in the above block of code.

As well as changing the colours used for the pie charts, you can also change their size. In particular, you can use the size of the pies to represent an additional variable, such as sample size, area or population size. This is done by specifying a column of data in the r argument of the aes element in the geom_scatterpie command. In the example being used in this exercise, you will explore how to do this by scaling the size of the pies so that they represent the total number of cetaceans recorded in each region of the study area. By doing this, you can get a greater understanding of which region is most important for which species in terms of the absolute numbers and as well as the numbers relative to other species within each region. To do this, you need to first create a new column in the study_area_centroids_df data set which contains the total number of cetaceans recorded in each region by adding together the number recorded for each species in each one. To do this, enter the following pair of commands into R:

```
study_area_centroids_df$total_ind=(study_area_centroids_df$
bnd+study_area_centroids_df$cd+study_area_centroids_df$hp+s
  tudy_area_centroids_df$mw+study_area_centroids_df$wbd)
              View(study_area_centroids_df)
```

Once you have checked that this calculation has been done correctly by examining the updated version of the study_area_centroids_df data set, you are ready to create a new version of your scatterpie map. To do this, you need to edit the blocks of code from steps 5 to 10 the above flow diagram (these are CODE BLOCKS 161 to 166 in the document R_CODE_DATA_VISUALISATION_WORKBOOK.DOC) so that they look like this (the required modifications are highlighted in **bold**):

```
scatterpie_map_2 <- ggplot() + geom_sf(data=
north_sea_study_area_sf,colour="black",fill=NA) +
   geom_sf(data=north_sea_land_sf,colour="black",
                  fill="green")
```

```
scatterpie_map_2 <- scatterpie_map_2 + coord_sf(xlim=
c(-150000,70000),ylim=c(50000,400000),expand=TRUE,crs=
   NULL,datum=sf::st_crs(4326),label_axes="ENEN",
                ndiscr=100,clip="on")
```

```
scatterpie_map_2 <- scatterpie_map_2 +
geom_scatterpie(data=study_area_centroids_df,aes(x=x,y=y,
      group=region,r=total_ind),cols=c("bnd","cd",
      "hp","mw","wbd"),colour="black",alpha=0.8)
```

```
scatterpie_map_2 <- scatterpie_map_2 +
scale_y_continuous(name="",sec.axis=sec_axis(~./1,name=
"")) + scale_x_continuous(name="",sec.axis=sec_axis(~./1,
   name="")) + theme(axis.text.y=(element_text(angle=90,
   vjust=0,hjust=0.5)),axis.text.y.right=(element_text(
      angle=-90,vjust=0,hjust=0.5)),axis.line=
         element_line(colour="black",size=1))
```

```
scatterpie_map_2 <- scatterpie_map_2 +
theme(panel.background=element_blank(),panel.grid.major=
   element_line(colour=NA,size=0.5),panel.grid.minor=
   element_line(colour=NA,size=0.5),panel.ontop=TRUE)
```

```
scatterpie_map_2 <- scatterpie_map_2 + theme(legend.title=
                  element_blank())
```

You will now need to add an edited version of the `scale_fill_manual` command from page 291 to specify the colours that will be used to represent each species on the pie charts that mark each point on the map. This command should look like this (the required modifications are highlighted in **bold**):

```
scatterpie_map_2 <- scatterpie_map_2 +
scale_fill_manual(values=c("blue","red","yellow","purple",
                  "brown"))
```

Finally, you can enter the name of the map created by these blocks of code to allow you to view your final map. This name is `scatterpie_map_2`. Once all these blocks of code have been run in the required order, you should have a map that looks like the image at the top of the next page.

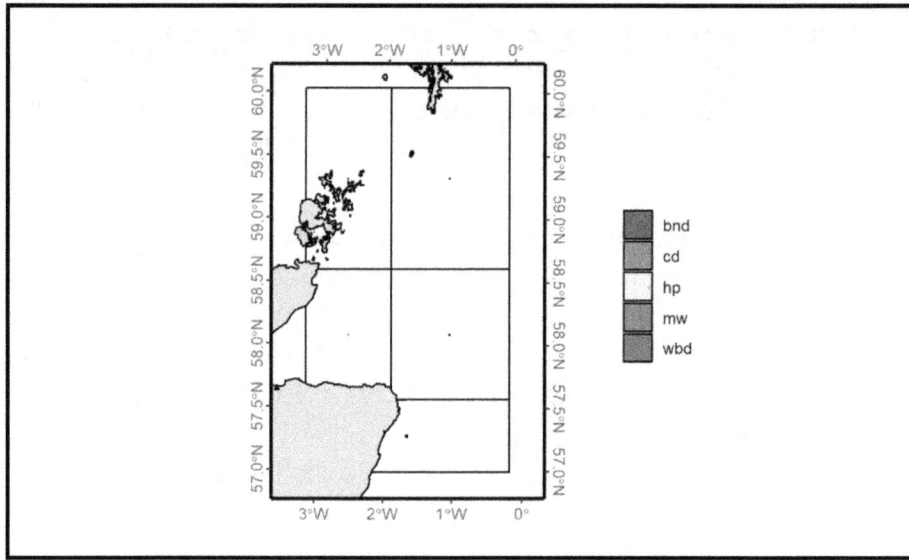

When you first look at this map, it may appear that the pie chart markers have all disappeared. However, if you examine it more closely, you will see this is not the case. Instead, they have just been drawn very small. This is because the maximum number of individuals recorded in any of the five regions is only 939 individuals, so this is the maximum radius (in metres) used for any of the pie charts. In order to make the pie charts a more appropriate size, you can add a 'scaling factor' to the r argument in the geom_scatterpie command. In this case, you can add a scaling factor of 100, so that the r argument is r=total_ind*100. Edit the r argument in the geom_scatterpie command in the above block of code so that it looks like this (the required modifications are highlighted in **bold**):

```
scatterpie_map_2 <- scatterpie_map_2 +
geom_scatterpie(data=study_area_centroids_df,aes(x=x,y=y,
    group=region,r=total_ind*100),cols=c("bnd","cd",
      "hp","mw","wbd"),colour="black",alpha=0.8)
```

You now need to re-run all the other code blocks used to make your map (all the ones that start with scatterpie_map_2 from page 292 and 293), excluding the original geom_scatterpie command. If you do not do this, then rather replacing the existing pie chart markers on the map with new, larger ones, they will be added to it as a new set of points. Once you have done this, run the new version of the geom_scatterpie command. This should result in an updated map that looks like the image at the top of the next page.

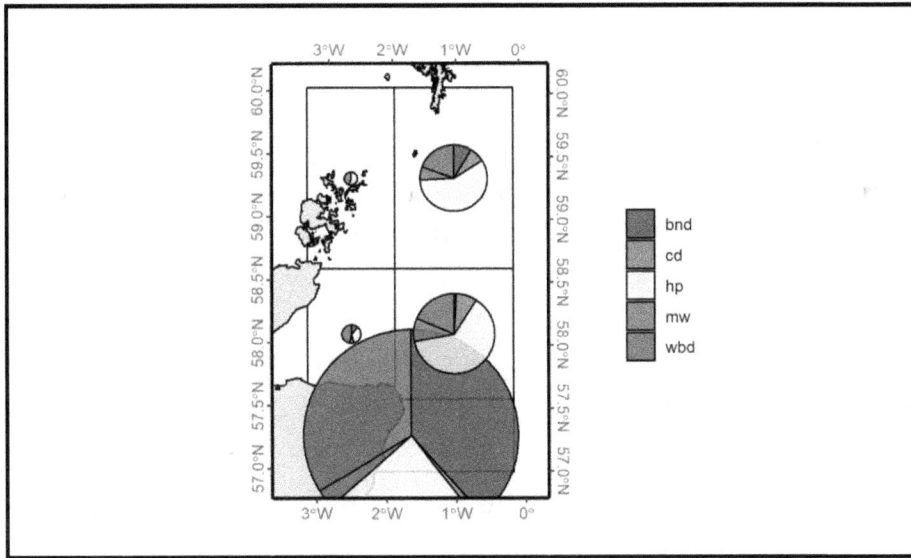

While adding a scaling factor of 100 now makes all the pie markers more visible, it has revealed a different problem. This is that the range of values for the total number of individuals in each region is too great to allow you to use these raw values to determine the relative size of each pie. In such circumstances, you need to transform your variable in some way so that the difference between the maximum and minimum values is reduced. This can be done by applying the same types of transformations you would apply to normalise any right-skewed data set (see page 122 in *An Introduction to Basic Statistics for Biologists using R*). For this particular data set, the best transformation is a square root transformation. To apply this to the data in the `total_ind` column of the `study_area_ centroids_df` data set, enter the following pair of commands into R:

```
study_area_centroids_df$sqrt_total_ind=
sqrt(study_area_centroids_df$total_ind)
        View(study_area_centroids_df)
```

Once you have checked that this calculation has been done correctly by examining the updated version of the `study_area_centroids_df` data set, you are ready to create a new version of your scatterpie map. In order to make pie charts with a radius based on this new column, you need to edit the `geom_scatterpie` command from page 294. The required editing involves changing the `r` argument so that it refers to the new column created by the above command (called `sqrt_total_ind`), and to add an appropriate scaling factor (in this case, use a scaling factor of `1500`). This will give you a new version of the `geom_scatterpie` command which should it looks like this (the required modifications are highlighted in **bold**):

```
scatterpie_map_2 <- scatterpie_map_2 +
geom_scatterpie(data=study_area_centroids_df,aes(x=x,y=y,
group=region,r=sqrt_total_ind*1500),cols=c("bnd","cd",
"hp","mw","wbd"),colour="black",alpha=0.8)
```

As before, you now need to re-run all the other code blocks used to make your map (all the ones that start with `scatterpie_map_2` from page 292 and 293), excluding the original `geom_scatterpie` command), and then run this new version of the `geom_scatterpie` command. When you do this, you should end up with an updated map that looks like this:

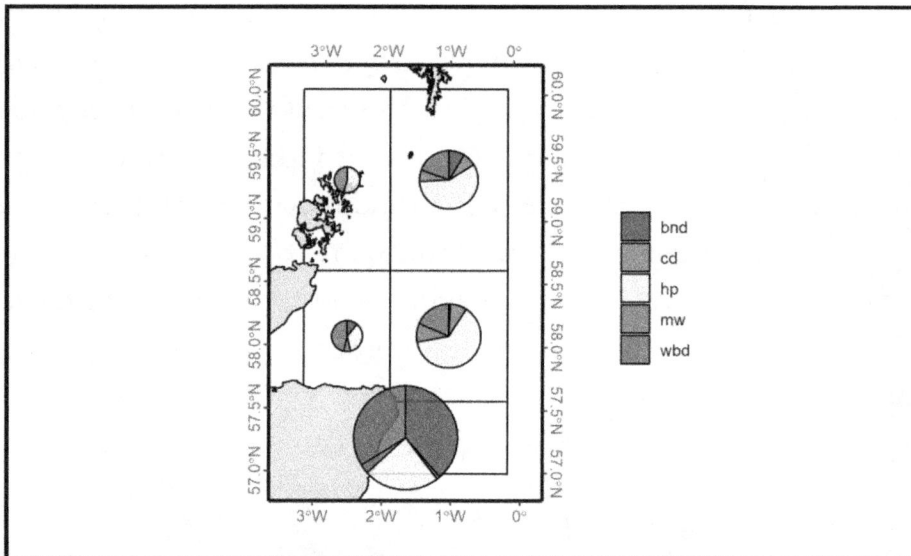

When you examine this new version of your map, you will see that the pie chart markers are still scaled to represent where the most and the least number of cetaceans were recorded, but the difference between the smallest and the largest symbols is now greatly reduced, giving the map a much improved appearance and making the symbols easier to compare.

EXERCISE 5.3: HOW TO MAKE A MAP DISPLAYING A RASTER DATA SET:

So far, when making maps in this chapter, you have been working with data which can be represented as features, such polygons, lines and points. However, there is another format of spatial data that is commonly used in biological research. This is the raster data format. Rather than containing a set of distinct features, a raster data set consists of a grid of data made of individual cells or pixels. This means that they need to be handled in a different way to data sets containing features. In addition, rather than being stored as shapefiles, they are typically saved in other formats, such as geotiffs (these are tiff files that have been georeferenced). Raster data sets are commonly used in biological research to represent continuous surfaces of data. This can include environmental variables, such as land elevation or water depth, or summaries of data from other sources, such as the density of records for a particular individual, species, or cases of a disease.

In this exercise, you will learn how to import raster data sets into R and plot them on a map. This will be done using data sets which track the spread of a fictional zoonotic disease across the continental United States over a six month period. You will be provided with one raster data set per month which consists of a kernel density estimate (or KDE) of the number of records of the disease in 100 by 100 kilometre grid cells (if you wish to find out how these raster data sets were created, you can find out in Exercise 5 of *GIS For Biologists: A Practical Introduction for Undergraduates*). You will start by creating a map showing the distribution of the disease in a single month based on the appropriate raster data set. To do this, work through the flow diagram that starts on the next page.

NOTE: When importing a raster data set into R, use the suffix _r in the name of the object created by the `raster` command (see step 3) to allow you to easily identify its data format, and separate it from other spatial data formats used when making maps, such as shapefiles (which should have the suffix _shp), simple features (which should have the suffix _sf) and tables or data frames (which should have the suffix _df).

NOTE: This workflow assumes that you have the `ggplot2`, `Rcpp`, `rgdal`, `ggmap`, `data.table`, `proj4`, `ggspatial` and `sf` packages installed in your version of R and that their command libraries has been loaded into your analysis project. If you have not already done this, you will need to do it before working through this flow diagram.

Instructions for how to do this for the first seven of these packages can be found in the flow diagram for Exercise 5.1, while the instructions for the `sf` package can be found on page 274.

Spatial data sets, including at least one raster data set, you wish to make a map from

For this example, you will use one raster data set that is stored as a geotiff, and one feature data set that is stored in the shapefile spatial data format. These are called `month_4_mask.tiff` and `united_states_outline_albers.shp` respectively. These data sets are located in the WORKING DIRECTORY folder you created during the introduction to this chapter.

1. Set the WORKING DIRECTORY for your analysis project

Before you start any analysis in R, you first need to set the WORKING DIRECTORY. To do this, enter the text `setwd("` and then type the address of your WORKING DIRECTORY, using slashes (/) as the folder separators, before entering a second quotation mark followed by a closing bracket, like this `")`. For example, if your WORKING DIRECTORY has the address C:\STATS_FOR_BIOLOGISTS_TWO, your `setwd` command should look like this:

```
setwd("C:/STATS_FOR_BIOLOGISTS_TWO")
```

If you are using RGUI, enter your `setwd` command in the R CONSOLE window (remembering to use the address of your own WORKING DIRECTORY folder in it) and then press the ENTER key on your keyboard. If you are using RStudio, enter your `setwd` command into the SCRIPT EDITOR window. To run it, select it and then click on the RUN button at the top of this window. You will enter all the remaining commands for this exercise in a similar manner, depending on the user interface you are using.

To check that your WORKING DIRECTORY has been set properly, enter the command `getwd()` and carefully check that the address it returns is the same as the one for the STATS_FOR_BIOLOGISTS_TWO folder you created at the start of this chapter.

Before you move on to step 2, make sure that all the data you wish to use in your analysis project are located in this WORKING DIRECTORY folder. In this case, these are the files called `month_4_mask.tiff` and `united_states_outline_albers.shp`. **NOTE:** If the data you are going to import into R in steps 3 and 6 of this flow diagram are not located in the WORKING DIRECTORY you set in this step, the code provided for importing them will not work.

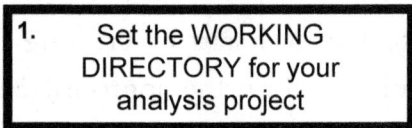

2. Ensure that the `raster` package is installed in your version of R and load its command library into your analysis project

In order to be able to import raster data sets into R, you will need to ensure you have the `raster` package installed in your version of R. To do this, enter the command `library()` into R. If the `raster` package is not listed in the R PACKAGES AVAILABLE window that opens, you can install it by entering the following command into R

```
install.packages("raster")
```

This is CODE BLOCK 167 in the document R_CODE_DATA _VISUALISATION_ WORKBOOK.DOC and paste it into R.

Once you have ensured that the `raster` package is installed in your version of R, you will need to load its associated command library into your analysis project. To do this, enter the following block of code into R:

```
library(raster)
```

This is CODE BLOCK 168 in the document R_CODE_ DATA_VISUALISATION_WORKBOOK.DOC .

3. **Import your raster data set into R**

Once you have the `raster` command library loaded into your analysis project, you are ready to import your raster data set into it. The first step in this process is to check what map projection it is in. This is done using the `GDALinfo` command. To do this for the raster data set being used in this example, enter the following command into R:

```
GDALinfo("month_4_mask.tif")
```

This is CODE BLOCK 169 in the document R_CODE_ DATA_VISUALISATION_WORKBOOK.DOC. This command will return the metadata associated with the raster file named in it, including its map projection as a proj4 string. In this case, the projection is the USA Albers Equal Area projection which has a proj4 string that says `proj=aea +lat_0=37.5+lon_0=-96+lat_1=29.5+lat_2=45.5+ x_0=0+y_0=0+datum=NAD83+units=m+no_defs`.

After you have checked the projection information, you can import your raster data set by entering the following command into R (remembering to use the suffix `_r` on the name of the object it will create so you know it contains a raster data set) :

```
month_4_r <- raster("MONTH_4_MASK.tif")
```

This is CODE BLOCK 170 in the document R_CODE_ DATA_VISUALISATION_WORKBOOK.DOC. This command will create a raster object in R called `month_4_r`. containing the data from the raster data set called `MONTH_ 4_MASK.tif`. This contains a kernel density estimate (KDE) for the distribution of the novel zoonotic disease in the fourth month of the six month period where its distribution was being tracked.

Next, you need to create an initial plot of your raster data set to ensure it has been imported correctly. To do this, enter the following command into R:

```
plot(month_4_r)
```

This is CODE BLOCK 171 in the document R_CODE_ DATA_VISUALISATION_WORKBOOK.DOC. This command will create and display a preliminary plot of the contents of your raster data set which you can examine to ensure it has been imported correctly.

Once you have imported a raster data set into R, you need to convert it into a data frame and remove any missing values. To do this for the raster data set being used in this example, enter the following block of code containing three separate commands into R (**NOTE:** Each of these commands needs to be entered on a separate line):

```
month_4_df <- as.data.frame(month_4_r,
               xy=TRUE)
month_4_df <- subset(month_4_df,
        MONTH_4_MASK!="NA")
          View(month_4_df)
```

This is CODE BLOCK 172 in the document R_CODE_ DATA_VISUALISATION_WORKBOOK.DOC. The first of these commands, `as.data.frame`, will convert the named raster data set (in this case, `month_4_r`) into a table that will be stored as an R data frame object. The `xy=TRUE` argument is included in this command so that a list of X and Y coordinates for the centre of each cell in the raster data set is added to the R table object created by this command (in this case, `month_4_df`). Remember to use the `_df` suffix to denote that the R object created by this command is a data frame.

The second command is a `subset` command which is used to remove any rows from this new R object that have a value of `NA` (meaning they contain missing values). This is set by the argument `MONTH_4_MASK!="NA"`. **NOTE:** If missing values in a raster data set are denoted by a term other than `NA`, you would need to include this alternative term in this argument.

The final command in this code block is a `View` command. This allows you to view the contents of the final version of the table you generated from your raster data set so you can examine it to make sure that all the steps in the raster data conversion process have worked correctly.

4. Convert your raster data set into a table and remove any missing values

5. **Create an initial map displaying your raster data set**

Now that your raster data set has been converted into an R data frame object (in this case, called `month_4_df`), you can create a map which displays the contents of this table. Creating maps in R requires a lot of code, as a result, rather than building up a single block of code that will create a map all in one go, you will create an initial map (in this case, called `raster_map_1`) that will be saved as an R object which you will then modify using additional commands and blocks of code. To do this, enter the following block of code into R:

```
raster_map_1 <- ggplot() +
geom_tile(data=month_4_df,aes(x=x,y=y,
fill=MONTH_4_MASK)) + scale_fill_gradient(
low="gray",high="red")
```

This is CODE BLOCK 173 in the document R_CODE_DATA_VISUALISATION_WORKBOOK.DOC. It contains three commands separated by + symbols. These commands are the `ggplot` command, the `geom_tile` command and the `scale_fill_gradient` command. The `ggplot` command is used to create a blank map onto which the other commands will add various features. The `geom_tile` command plots a data set as a grid of data, with each row of data in it being plotted as a grid cell. In this command, the `data` argument is used to identify the raster data set which will be plotted. In this case, it is the raster data set called `month_4_df` created in step 4. The `x` and `y` arguments in the `aes` element of the `geom_tile` command sets the columns which contain the X and Y coordinates for the raster data set, while the `fill` argument identifies the column which contains the values that will be represented by the grid cells plotted on the map.

The `scale_fill_graident` command is used to determine how the raster data set will be displayed. In this command, the `low` argument determines the colour for the lowest values in the data set (in this case, `grey`), while the `high` argument determines the colour to be used for the highest values (in this case, `red`). Grid cells with values in between these two extremes will be shaded using an appropriate gradient between these two colours.

The map created by this block of code will be saved in a new object called `raster_map_1`. This means that once you have run it, you will need to enter the name of this object into R in order to view it.

6. Import any polygon and line shapefiles you wish to have on your map into R using the 'sf' approach and add them to your initial map

Once you have plotted your raster data set on your initial map, you are ready to import any polygon and line shapefiles you wish to display on it. For this exercise, you will do this using the 'sf' approach. To import a polygon shapefile with the outlines of the United States of America, enter the following command into R:

```
united_states_sf <- read_sf(dsn=".",
layer="UNITED_STATES_OUTLINE_ALBERS")
```

This is CODE BLOCK 174 in the document R_CODE_ DATA_VISUALISATION_WORKBOOK.DOC. This command will create a simple feature object in R called `united_ states_sf` containing the data from the specified shapefile.

Once you have imported your shapefile as a simple feature object, you can then add it to your initial map using the `geom_sf` command. To do this for the example being used in this exercise, enter the following command into R:

```
raster_map_1 <- raster_map_1 +
geom_sf(data=united_states_sf,colour=
"black",fill=NA)
```

This is CODE BLOCK 175 in the document R_CODE_ DATA_VISUALISATION_WORKBOOK.DOC. To view the updated map created by this command, enter the name of the object created by it into R. In this case, enter the name `raster_map_1`.

```
7.    Set the extent and
coordinate system for the
graticule of your final map
```

After you have added all the required data sets to your map, you need to set its extent and the coordinate system that will be used for the graticules around its edge. In this case, you wish to have a graticule which displays latitude and longitude rather than the coordinate system of the map projection of the raster data set. As a result, the easiest way to do this is to use the `coord_sf` command. To do this for the map being created in this example, enter the following command into R:

```
raster_map_1 <- raster_map_1 +
coord_sf(xlim=NULL,ylim=NULL,expand=TRUE,
crs=NULL,datum=sf::st_crs(4326),label_axes=
"ENEN",ndiscr=100,clip="on")
```

This is CODE BLOCK 176 in the document R_CODE_ DATA_VISUALISATION_WORKBOOK.DOC. It contains the name of the map you wish to change the extent of (in this case, it is the one called `raster_map_1` which you created in steps 5 and 6) followed by a + symbol and the `coord_sf` command. This means that the output of this new command will be added to this existing map (rather than generating a one). In the `coord_sf` command, the `xlim` and `ylim` arguments are used to specify the east-west and north-south limits of your map, respectively. In this case, you will set these to NULL meaning that the extent of the map will be set by the extent of the raster and polygon data sets which have already been plotted on it.

The `coord_sf` command can also be used to determine the coordinate system which will be used for the graticule that will be added around the edge of your map in step 8. In this case, by including the `sf::st_crs(4236)` argument in this command, the graticule will display latitude and longitude values rather than the coordinates of the map projection of the spatial data sets being plotted on it.

To view the updated map created by this block of code, enter the name of the object created by it into R. In this case, enter the name `raster_map_1`.

In order to show which part of the world a map represents, you need to add what is known as a graticule around its edge. This is a scale that can be used to read off the coordinates for different locations on the map. For this example, you will do this by adding left, right, top and bottom axes to your map and formatting them. To add a graticule to the map you are creating in this example using latitude and longitude, enter the following block of code into R:

```
raster_map_1 <-raster_map_1 +
scale_y_continuous(name="",sec.axis=
    sec_axis(~./1,name="")) +
scale_x_continuous(name="",sec.axis=
    sec_axis(~./1,name="")) +
theme(axis.text.y=(element_text(angle=90,
 vjust=0,hjust=0.5)),axis.text.y.right=
    (element_text(angle=-90,vjust=0,
hjust=0.5)),axis.line=element_line(colour=
                "black",size=1))
```

This is CODE BLOCK 177 in the document R_CODE_ BASIC_STATS_WORKBOOK.DOC. It contains the name of the map you wish to add a graticule to (in this case, `raster_ map_1`) followed by a + symbol and then three different command. This means that the output of this new command will be added to an existing map (rather than generating a new one). The commands included in this block of code are `scale_y_continuous`, `scale_x_ continuous` and a `theme` style command. The `scale_y_ continuous` command contains two arguments. The first of these is `name`, and this has no value, meaning that no name is added to the Y axes of the map. The second is `sec.axis`, and this is used to add a Y axis to both the left and right hand sides of your map.

The `scale_x_continuous` command contains two arguments. The first of these is `name`, and this has no value, meaning that no name is added to the X axes of the map. The second is `sec.axis`, and this is used to add a X axis to both the top and bottom of your map.

Finally, the `theme` style command contains arguments which set the colour and thickness of the lines representing each axis, and the position of the text on each axis.

To view the updated map created by this block of code, enter the name of the object created by it into R. In this case, enter the name `raster_map_1`.

---

8. Add a graticule around the edge of your map to show which part of the world it represents

---

After you have added the graticule around the edge of your map, you need to format the background. In general, you will want to change the default values so that the background area of the map is blank and there are no grid lines on it. To do this for the map you are creating in this example, enter the following block of code into R:

```
raster_map_1 <- raster_map_1 +
theme(panel.background=element_blank(),
panel.grid.major=element_line(colour=NA,
       size=0.5), panel.grid.minor=
    element_line(colour=NA,size=0.5),
           panel.ontop=TRUE)
```

9. Format the background of your map

This is CODE BLOCK 178 in the document R_CODE_ DATA_VISUALISATION_WORKBOOK.DOC. It contains the name of the map you wish to format the background of (in this case, `raster_map_1`) followed by a + symbol and then a single theme command which contains three different style arguments. The `panel.background=element_blank ()` argument removes the existing background from the map, while the two `panel.grid` arguments include the term `colour=NA` to remove the existing gridlines from the map. **Note:** If you wish to display gridlines on your map, change the `colour` term in one or both of these arguments to `"black"` or the name of any other colour you wish them to be drawn in.

To view the updated map created by this block of code, enter the name of the object created by it into R. In this case, enter the name `raster_map_1`.

```
10.    Change the title of
the legend which shows what
the colours used to display the
raster data set represent
```

Finally, you need to change the title of the legend which shows what the colours used to display your raster data set represent. This is done using a `theme` command with the `legend.title` argument. In this example, you wish to remove the current title. To do this, enter the following block of code into R:

```
raster_map_1 <- raster_map_1 +
theme(legend.title=element_blank())
```

This is CODE BLOCK 179 in the document R_CODE_ DATA_VISUALISATION_ WORKBOOK.DOC. To view the updated map created by this block of code, enter the name of the object created by it into R. In this case, enter the name `raster_map_1`.

```
Map
created which
displays data from
a raster data
set
```

The final map created by working through the above flow diagram should look like this:

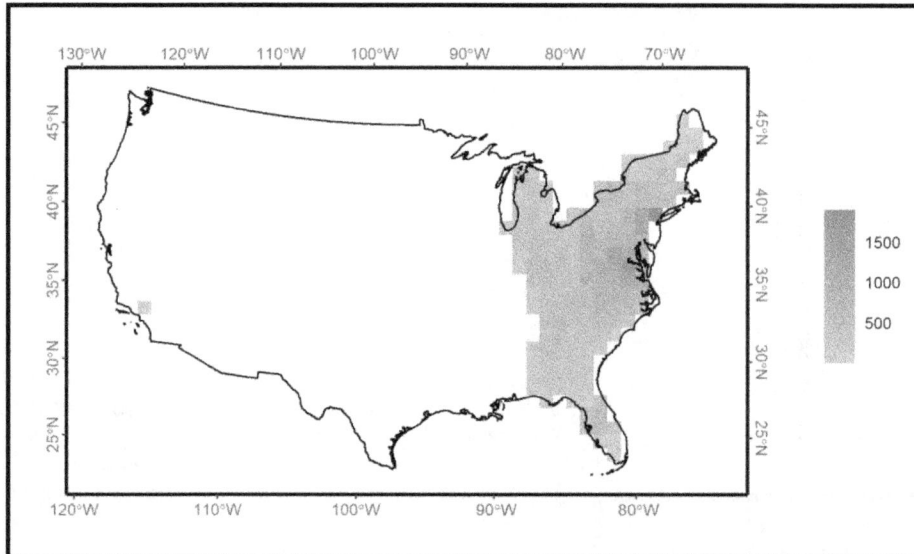

Once you have created a map from a raster data set, you can export it from R so that you can include it in a manuscript or presentation. If you are using RGUI, you can do this by clicking on the R GRAPHICS window containing your map to select it, before clicking on FILE on the main menu bar and selecting SAVE AS. This will allow you to save it in a variety of different formats. If you are using RStudio, you can export your map by clicking on the EXPORT button at the top of the window displaying it and selecting SAVE AS IMAGE. Alternatively, if you wish to save a high resolution version of your map, you can use the `ggsave` command from the `ggplot2` package. This command will let you save your map using a specific resolution, size and format. You will use this command as part of the next step you will complete in this exercise.

When including a map made from a raster data set in a manuscript, it is important that you provide an appropriate figure legend for it. This legend should provide all the information required for the reader to interpret its contents. For the above map, an appropriate legend would be:

Figure 1: *The distribution of cases of a novel zoonotic disease in month 4 of an epidemic in the continental United States of America. Shading represents the density of records per 100 by 100 km grid cells., ranging from 1 case (grey) to 1889 cases (red) per grid cell.*

NOTE: The maximum value for a raster data layer can be obtained either from the original raster file saved on your computer using the `GDALinfo` command from the `rgdal` package (see step 3 of the above flow diagram), or from a raster data set that you have imported into R using the `summary` command. For example, for the `month_4_r` dataset you imported into R as part of this exercise, the `summary` command would be `summary(month_4_r)`.

When working with raster data sets, it is common to have a number of different ones which represent data from different periods of time. Together, they can be used to create a time series of distributional data. While these can be displayed individually, or as part of a multi-panel figure (for example, using the `ggarrange` command from the `ggpubr` packages – see Exercise 1.4), you can also use them to create an animated sequence to show changes over time. This can be done in R using tools from a number of different packages. However, regardless of the package you use, you will first have to create a set of standardised maps, with each one displaying the data from a different time period. To allow you to explore how this can be done, you will now create an animation showing the spread of the novel zoonotic disease which is being used as the basis for the example in this exercise over a six month period. This will be done with the tools in the `magick` package as these tools provide the simplest way to create an animation in R from a series of images, such as maps, representing data from different periods of time.

In order to create an animation of the spread of this disease across the continental United States, you will first need to import the raster data sets for months one to six of this epidemic (remembering that you have already imported the raster data set for month four). This is done by editing the commands from steps 3 and 4 from the above flow diagram (these are CODE BLOCKS 169 to 172 in the document R_CODE_DATA_ VISUALISATION_WORKBOOK.DOC) to replace the names of the raster data set to be imported and the R objects to be created by the commands with those for a different month. For example, to import the data for month one, the code from these steps would need to be edited so that it looks like the block of code at the top of the next page (the required modifications are highlighted in **bold**).

```
GDALinfo("month_1_mask.tif")
month_1_r <- raster("MONTH_1_MASK.tif")
plot(month_1_r)
month_1_df <- as.data.frame(month_1_r, xy=TRUE)
month_1_df <- subset(month_1_df, MONTH_1_MASK>0)
View(month_1_df)
```

Run this modified block of code to import the data from the raster data set for month one, and then repeated the process to import the data from the raster data sets for month two (called month_2_mask.tif), month three (called month_3_mask.tif), month five (called month_5_mask.tif) and month six (called month_6_mask.tif). The data sets created by this import process should be called month_1_df, month_2_df, month_3_df, month_5_df and month_6_df respectively.

Once the raster data sets for the additional months have been imported into R and converted into data frame objects with the suffix _df, you need to work out the maximum density of cases per grid cell contained in the different raster data sets. This is so that you can set all the raster data sets to be displayed using the same scale (in order to make them directly comparable). To find out the maximum density values for each individual raster data set, you can use the summary command. For example, to find out the maximum value in the raster data set for month one, enter the following command into R:

```
summary(month_1_r)
```

Repeat this for the remaining five monthly raster data sets (including the one from month four created when working through the main flow diagram). When you do this, you should find that the maximum value for any raster data set is 22,867.00 per grid cell. As a result, when creating your maps, you will use a scale of 0 to 25,000 for all your raster data sets. This is set by adding the argument limits=c(0,25000) to the scale_fill_gradient command in the code used to create your initial map in step 5 of the above flow diagram.

You are now ready to create a map displaying the data from each monthly data set using the same scale to display all the raster data sets. This is done by editing the commands in steps 5 to 10 of the above flow diagram (these are CODE BLOCKS 173 to 179 in the document

R_CODE_DATA_VISUALISATION_WORKBOOK.DOC). For example, the edited code required to create a map displaying the data from month one would look like this (the required modifications are highlighted in **bold**):

```
month_1_map <- ggplot() + geom_tile(data=month_1_df,
aes(x=x,y=y,fill=MONTH_1_MASK)) + scale_fill_gradient(low=
"grey90",high="red",limits=c(0,25000),guide="colourbar")

month_1_map <- month_1_map + geom_sf(data=united_states_sf,
                    colour="black",fill=NA)

month_1_map <- month_1_map + coord_sf(xlim=NULL,ylim=NULL,
    expand=TRUE,crs=NULL,datum=sf::st_crs(4326),label_axes=
                "ENEN",ndiscr=100,clip="on")

month_1_map <- month_1_map + scale_y_continuous(name="",
    sec.axis=sec_axis(~./1,name="")) + scale_x_continuous(
        name="",sec.axis=sec_axis(~./1,name="")) +
    theme(axis.text.y=(element_text(angle=90,vjust=0,
  hjust=0.5)),axis.text.y.right=(element_text(angle=-90,
vjust=0,hjust=0.5)),axis.line=element_line(colour="black",
                    size=1))

month_1_map <- month_1_map +
theme(panel.background=element_blank(),panel.grid.major=
    element_line(colour=NA,size=0.5),panel.grid.minor=
    element_line(colour=NA,size=0.5),panel.ontop=TRUE)

month_1_map <- month_1_map + theme(legend.title=
                element_blank())
```

By running this modified code, you will created a new map which displays the data from month one rather than month four. However, as you will be creating maps for a series of months, you will need to add a label to allow the viewer to identify which month each map represents. This is done by adding a new element to your map using an `annotate` command. This can be done for the map called `month_1_map` be entering the block of code at the top of the next page.

```
month_1_map <- month_1_map + annotate("text",x=-1800000,
          y=-1150000,label="Month 1",size=8)
```

Once you have run this command, you can view your final map displaying the data from month one. To do this, enter the following code into R:

```
month_1_map
```

The map displayed by this code should look like this:

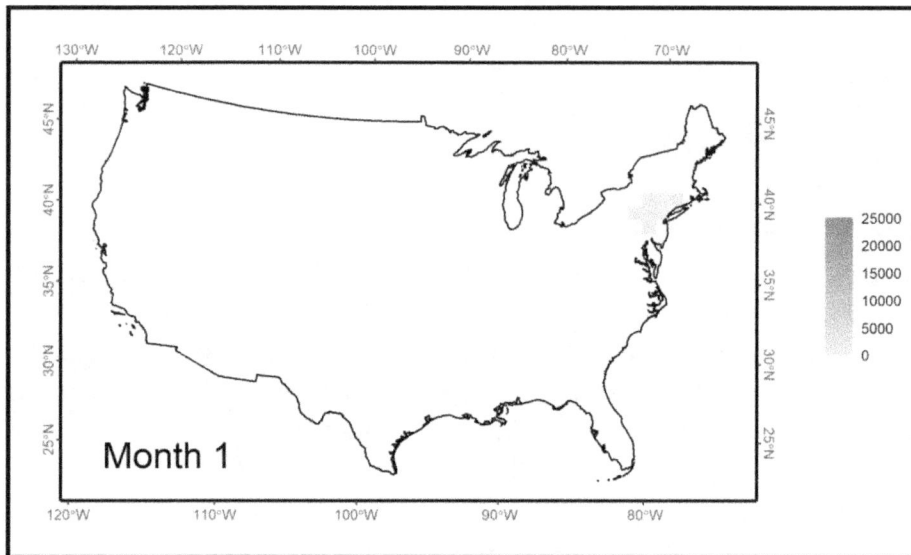

Repeat this process to create maps for the remaining five months (including remaking the map for month 4 as you are using a new scale and you are adding a new label to it) to create maps called `month_2_map`, `month_3_map`, `month_4_map`, `month_5_map` and `month_6_map`.

The map called `month_2_map` should look like this:

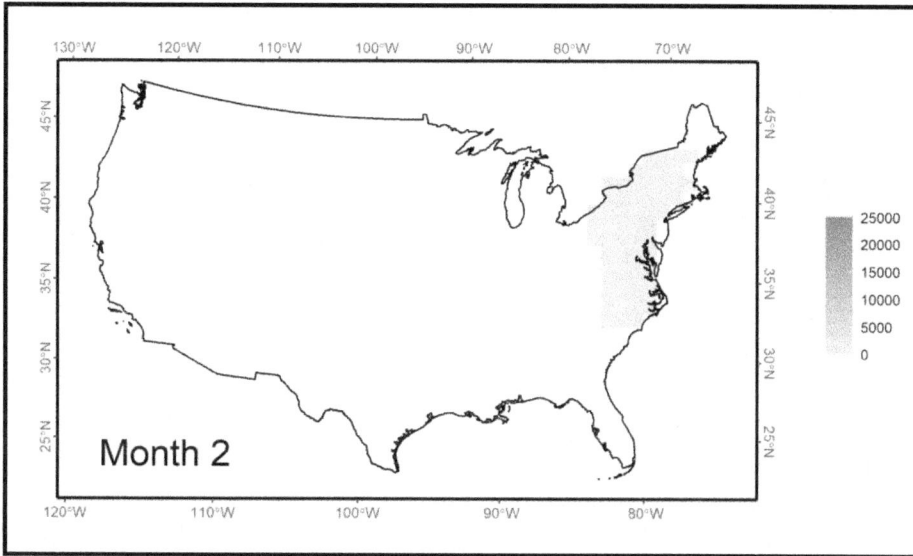

The map called `month_3_map` should look like this:

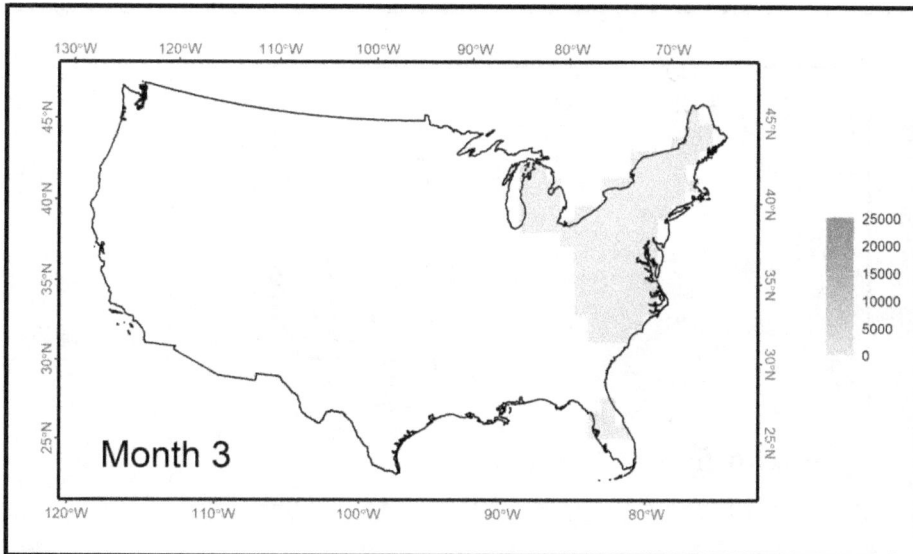

The map called `month_4_map` should look like this:

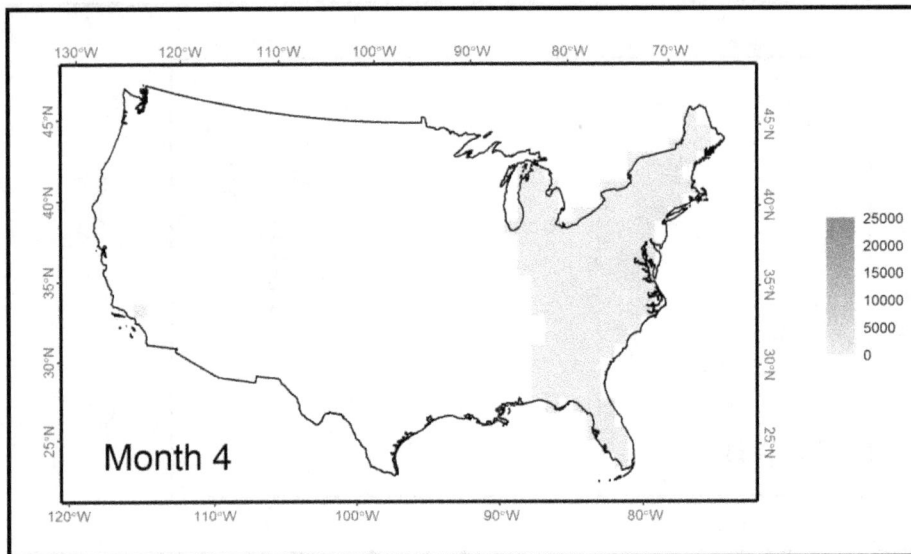

The map called `month_5_map` should look like this:

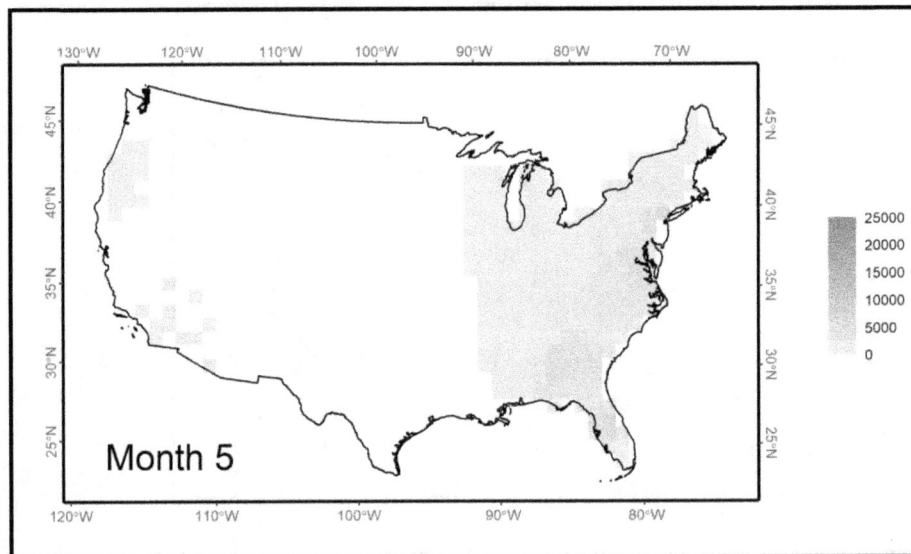

The map called `month_6_map` should look like this:

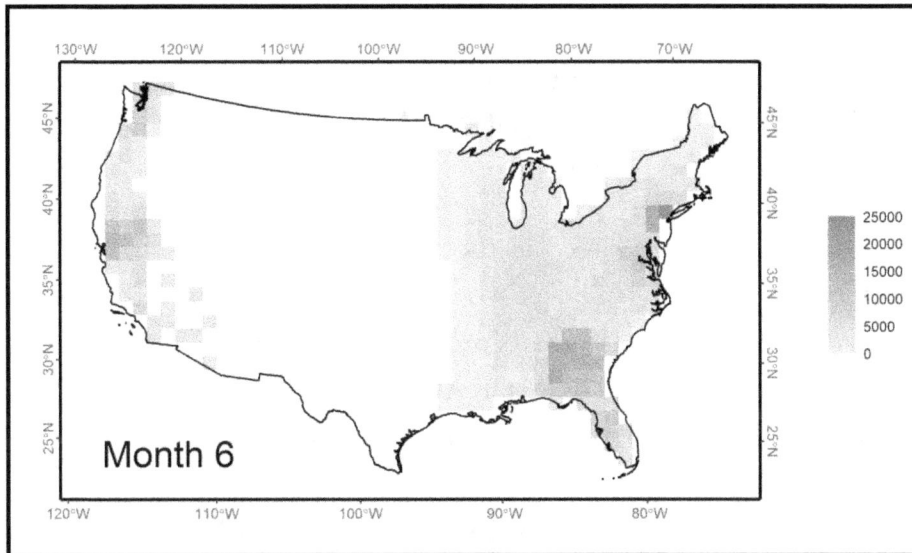

Once you have successfully created a map for each month, you can use them to create an animation. The first step in this process is to create a new sub-folder in your WORKING DIRECTORY called `ex_5_3_animation` where you will save copies of these maps. To create this new sub-folder, enter the following command into R:

```
dir.create("ex_5_3_animation",recursive=TRUE)
```

Once this new sub-folder has been created, you can save each map into it using the `ggsave` command. To do this for the map called `month_1_map`, enter the following command into R:

```
ggsave(plot=month_1_map,filename=
"ex_5_3_animation/month_1_map.png",units=c("cm"),dpi=300,
device="png")
```

This will save the map contained in the R object called `month_1_map` as a .png file with the name `month_1_map.png`. Repeat this process by editing this `ggsave` command to save the maps for the other month as .png files called `month_2_map.png`, `month_3_map.png`, `month_4_map.png`, `month_5_map.png`, `month_6_map.png` in the sub-folder in your WORKING DIRECTORY called `ex_5_3_animation`. For example, the edited version of this code which will create the file called `month_2_map.png` should look like this (the required modifications are highlighted in **bold**):

```
ggsave(plot=month_2_map,filename=
"ex_5_3_animation/month_2_map.png",units=c("cm"),dpi=300,
device="png")
```

Once all of your maps have been saved in the appropriate format with the required names and in the specified sub-folder, you need to ensure that you have the `magick` package installed in your version of R. To do this, enter the command `library()` into R. If the `magick` package is not listed in the R PACKAGES AVAILABLE window that opens, you can install it by entering the following block of code into R:

```
install.packages("magick")
```

After you have ensured that the `magick` package is installed in your version of R, you can load its command library into your analysis project by entering the following code into R:

```
library(magick)
```

Next, you need to read all the map files you wish to include in your animation into R. In this case, these are the .png files you created of the maps for each month using the above `ggsave` commands. To do this, enter the following block of code, containing two separate commands which need to be entered on separate lines, into R:

```
monthly_maps <- list.files("ex_5_3_animation",
                full.names=TRUE)
img_list <- lapply(monthly_maps,image_read)
```

Finally, you need to join the image files you have just imported together and save them as an animation. To do this, enter the following block of code, which contains two commands that need to be entered on separate lines, into R:

```
img_joined <- image_join(img_list)
img_animated <- image_animate(img_joined,fps=0.5,loop=0)
```

This will create a new R object called `img_animated` which will contain the animation. In the `image_animate` command, the `fps` argument is used to set the number of frames per second in your animation. In this case, it is set to `0.5`, meaning that each map will be displayed for 2 seconds. The `loop` argument is to determine how many times the sequences of images will be run through. In this case, a value of `0` is used for this argument to create a loop that will continue until the animation is closed. The resulting animation can be viewed by entering the following code into R:

```
img_animated
```

However, when you run this code, you will most likely find that the size of the animation is too large to fit in the VIEWER panel which will open. As a result, it is better to save the animation to a file in your WORKING DIRECTORY and then open it from there. To do this, enter the following command into R:

```
image_write(image=img_animated,path="monthly_maps.gif")
```

This will save the contents of the R object called `img_animated` as a .gif file called `monthly_maps.gif` in your WORKING DIRECTORY folder. If you navigate to this folder, you can open this file in a web browser or other suitable gif viewing program and see the results of your animation. It should move sequentially through the maps for each month every two seconds before looping back to the map for month one and starting the process again. You can find a copy of what it should look like at *www.gisinecology/stats-for-biologists-2/exercise-5-3-animation.*

--- Appendix I ---

A List Of All The R Packages Used For The Exercises In This Workbook

Below, you will find a list of all the R packages used in this workbook. This will allow you to easily identify which packages you may need to install to complete a specific task when working with your own data to create graphs and maps using R. In addition, it provides you with a resource for finding out what other tools these packages contain that you might find useful for making high quality and informative data visualisations from biological data.

corrgram: This package provides a number of tools which can be used when creating pair plots or correlograms. This includes the `corrgram` tool used in Exercise 3.5 to create a pair plot with correlation coefficients on it. You can find out more about the contents of this package and how to use it at *www.rdocumentation.org/packages/corrgram/versions/1.14*.

dplyr: This package contains tools which can be used to process and summarise your data prior to making graphs. This includes the `ifelse` and `case_when` commands used in Exercise 2.1 to classify a continuous variables into groups to allow you to make bar graphs from them. You can find out more about the contents of this package and how to use it at *www.rdocumentation.org/packages/dplyr/versions/0.7.8*.

extrafonts: This package adds additional fonts to the basic fonts which come packaged with the core R code. Commands from this package are used in Exercise 1.2 in this book when creating a publication quality graph. You can find out more about the contents of this package and how to use it at *www.rdocumentation.org/packages/extrafont/versions/0.17*.

ggplot2: This package contains tools which allow you to create more advanced graphs than the basic graphing commands that are part of the core R code. Commands from this package are used in every exercise in this workbook. These include the `ggplot`, `geom_histogram`, `geom_density`, `geom_line`, `geom_point`, `geom_path`, `geom_bar`, `geom_polygon`, `geom_freqpolygon`, and `geom_boxplot` commands. You

318

can find out more about the commands available in this package at *ggplot2.tidyverse.org/ reference.*

ggpubr: This package contains a number of tools which can be used to help produce publication quality graphs using the tools from the `ggplot2` package. The `ggarrange` command from this package is used in this book to make multi-panel figures containing a number of different graphs (e.g. Exercise 1.4). You can find out more about the contents of this package and how to use it at *www.rdocumentation.org/packages/ggpubr/versions/ 0.4.0* and *rpkgs.datanovia.com/ggpubr/.*

ggrepel: This package contains two tools which allow you to add non-overlapping labels to graphs created using the `ggplot2` package. This includes the `geom_text_repel` command used in Exercise 4.4. You can find out more about the contents of this package and how to use it at *www.rdocumentation.org/packages/ggrepel/versions/0.9.1.*

ggspatial: This package contains a variety of tools for processing and handling spatial data, and making maps from them. This includes the `annotation_scale` and `annotation_north_arrow` commands used in Exercise 5.1 to add a scale bar and a North arrow to a map. You can find out more about the contents of this package and how to use it at *www.rdocumentation.org/packages/ggspatial/versions/1.1.1.*

magick: This is an advanced image processing package. It is used in this workbook to create an animation based on a series of maps in Exercise 5.3. You can find out more about this package and how to use it at *cran.r-project.org/web/packages/magick/vignettes/ intro.html.*

plotrix: This package is another data processing and graphing package. It contains a wide variety of useful tools, some of which do not have equivalents in other similar packages. This includes `std.error` command used to calculate standard errors in summary tables so they be used to add error bars showing this summary statistic to bar graphs, as is done in Exercise 2.3 of this workbook. You can find out more about the contents of this package and how to use the commands it contains at *www.rdocumentation.org/packages/plotrix/versions/3.8-1* and *rdrr.io/cran/plotrix/man/.*

plyr: This package contains another set of tools for processing and summarising your data prior to making graphs from them. This includes the `ddply` command used to create a summary table in Exercise 2.3 which can then be used to make a bar graph of summary statistics with error bars. You can find out more about the contents of this package and how to use it at *www.rdocumentation.org/packages/plyr/versions/1.8.6.*

raster: This package contains a range of tools to help you import and work with spatial data that are in the raster format. This includes the `raster` command used to import raster data sets into R in Exercise 5.3. You can find out more about this package and how to use it at *www.rdocumentation.org/packages/raster/versions/3.4-10.*

Rcpp: This package provides a range of tools which are required for a number of the other packages used in the workbook to function properly. As a result, while you will not use tools from it directly, it is still important to have it installed in order for these other packages to work as they should.

rgdal: This packages allows you to access GDAL tools for processing spatial data in R. This includes the `readORG` command used to import shapefiles into R in Exercises 5.1 to 5.3. You can find out more about the contents of this package and how to use it at *www.rdocumentation.org/packages/rgdal/versions/1.5-23.*

Rmisc: This package contains another set of tools for processing and summarising your data prior to making graphs from them. This includes the `group.ci` command which can be used to calculate confidence intervals for different groups of data that can then be used to plot error bars showing the confidence intervals on various types of graph. You can find out more about the contents of this package and how to use it at *www.rdocumentation.org/packages/Rmisc/versions/1.5.*

scatterpie: This packages contains tools which allow you to use pie charts as markers for point data displayed on a map or a graph. This includes the `geom_scatterpie` command used to do this in Exercise 5.2. You can find out more about this package and how to use it at *www.rdocumentation.org/packages/scatterpie/versions/0.1.6.*

sf: This package contains a variety of tools that work with the simple feature spatial data format. This includes the `read_sf` command used in Exercise 5.2 to import spatial data in the shapefile format into R using the 'sf' approach. You can find out more about this package and how to use it at *www.rdocumentation.org/packages/sf/versions/1.0-0*.

tidyquant: This is a package which provides tools which are typically used for analysing financial data. However, a number of these tools are also useful for analysing biological data. This includes the `roll_mean` command used in Exercise 3.4 to calculate a running average when creating graphs of time series data. You can find out more about the contents of this package and how to use it at *www.rdocumentation.org/packages/tidyquant/versions/1.0.3*.

Data Sources for Beaked Whale Strandings Data from the United Kingdom and the Republic of Ireland used in Exercise 3.4

The data on beaked whale strandings from the United Kingdom and the Republic of Ireland between 1913 and 2002 used in Exercise 3.4 were collated by Colin D MacLeod from a range of sources. This includes the Irish Whale and Dolphin Group website (*www.iwdg.ie*) and the UK strandings reporting scheme. Data from 1913 to 1989 for the UK were provided courtesy of the Trustees of the Natural History Museum, London. UK strandings data from 1990 to 2002 were collated by the Cetacean Strandings Investigation Programme, which was co-funded by DEFRA and the Devolved Governments of Scotland and Wales.

These data were originally collated for MacLeod *et al.* 2004. Geographic and temporal variations in strandings of beaked whales (Ziphiidae) on the coasts of the UK and the Republic of Ireland from 1800-2002. *J. Cetacean Res. Manage.* **6**: 79-86.

www.ingramcontent.com/pod-product-compliance
Lightning Source LLC
Chambersburg PA
CBHW081803200326
41597CB00023B/4125